MBL Lectures in Biology
Volume 7

THE ORIGIN AND EVOLUTION OF SEX

MBL LECTURES IN BIOLOGY

Volume 1
The Origins of Life and Evolution
Harlyn O. Halvorson and K.E. Van Holde, *Editors*

Volume 2
Time, Space, and Pattern in Embryonic Development
William R. Jeffery and Rudolf A. Raff, *Editors*

Volume 3
Microbial Mats: Stromatolites
Yehuda Cohen, Richard W. Castenholz, and Harlyn O. Halvorson, *Editors*

Volume 4
Energetics and Transport in Aquatic Plants
John A. Raven

Volume 5
The Visual System
Alan Fein and Joseph S. Levine, *Editors*

Volume 6
Blood Cells of Marine Invertebrates: Experimental Systems in Cell Biology and Comparative Physiology
William D. Cohen, *Editor*

Volume 7
The Origin and Evolution of Sex
Harlyn O. Halvorson and Alberto Monroy, *Editors*

THE ORIGIN AND EVOLUTION OF SEX

Proceedings of a meeting held July 30—August 4, 1984
at the Marine Biological Laboratory
Woods Hole, Massachusetts

Editors

Harlyn O. Halvorson

Department of Biochemistry
Rosensteil Basic Medical Sciences Research Center
Brandeis University
Waltham, Massachusetts

Alberto Monroy

Stazione Zoologica
Villa Communale
Naples, Italy

Alan R. Liss, Inc. • New York

Address All Inquiries to the Publisher
Alan R. Liss, Inc., 41 East 11th Street, New York, NY 10003

Printed in the United States of America.

Library of Congress Cataloging in Publication Data

Main Entry under title:

Origin and evolution of sex.

Includes bibliographies and index.
1. Sex (Biology) — Congresses. I. Halvorson, Harlyn O. II. Monroy, Alberto.
QH481.075 1985 574.1′66 85-10167
ISBN 0-8451-2206-1

Contents

Contributors

Edward A. Adelberg, Department of Human Genetics, School of Medicine, Yale University, New Haven, CT 06510 **[87, 89]**

Graham Bell, Biology Department, McGill University, Montreal, Quebec, Canada H3A 181 **[221, 257]**

Harris Bernstein, Department of Microbiology and Immunology, University of Arizona, Tucson, AZ 85721 **[29]**

Peter J. Bruns, Section of Genetics and Development, Cornell University, Ithaca, NY 14853 **[141, 153]**

Henry Byerly, Department of Philosophy, University of Arizona, Tucson, AZ 85721 **[29]**

Alvin J. Clark, Department of Molecular Biology, University of California, Berkeley, CA 94720 **[47]**

Don B. Clewell, Departments of Oral Biology and Microbiology/Immunology and The Dental Research Institute, Schools of Dentistry and Medicine, University of Michigan, Ann Arbor, MI 48109 **[13]**

Thomas W. Cline, Department of Biology, Princeton University, Princeton, NJ 08544 **[301]**

Ursula W. Goodenough, Department of Biology, Washington University, St. Louis, MO 63130 **[123]**

Constance Grabowy, Division of Cancer Genetics, Dana-Farber Cancer Institute, and Department of Microbiology and Molecular Genetics, Harvard Medical School, Boston, MA 02115 **[113]**

Harlyn O. Halvorson, Department of Biology and the Rosenstiel Basic Medical Sciences Research Center, Brandeis University, Waltham, MA 02254 **[ix, 3]**

James B. Hicks, Cold Spring Harbor Laboratory, Cold Spring Harbor, NY 11724 **[97]**

Frederick Hopf, Optical Sciences Center, University of Arizona, Tucson, AZ 85721 **[29]**

John M. Ivy, Cold Spring Harbor Laboratory, Cold Spring Harbor, NY 11724 **[97]**

Kenneth W. Jones, Department of Genetics, University of Edinburgh, Edinburgh, Scotland **[199]**

Amar J.S. Klar, Cold Spring Harbor Laboratory, Cold Spring Harbor, NY 11724 **[97]**

S.S. Koide, Center for Biomedical Research, The Population Council, New York, NY 10021 **[271]**

Lynn Margulis, Department of Biology, Boston University, Boston, MA 02215 **[69]**

Duane W. Martindale, Section of Genetics and Development, Cornell University, Ithaca, NY 14853 **[141]**

Helen Martindale, Section of Genetics and Development, Cornell University, Ithaca, NY 14853 **[141]**

Anne McLaren, MRC Mammalian Development Unit, Wolfson House, London NW1 2HE, England **[289]**

The number in brackets is the opening page number of the contributor's article.

Horacio Merchant-Larios, Instituto de Investigaciones Biomedicas, Universidad Nacional Autonoma de Mexico, Apartado Postal 70228K D.F. 04510, Mexico **[271, 329]**

Richard E. Michod, Department of Ecology and Evolutionary Biology, University of Arizona, Tucson, AZ 85721 **[29]**

Alberto Monroy, Stazione Zoologica, Villa Communale, Naples, Italy **[ix, 157]**

Lorraine Olendzenski, Department of Biology, Boston University, Boston, MA 02215 **[69]**

Dorion Sagan, Department of Biology, Boston University, Boston, MA 02215 **[69]**

Ruth Sager, Division of Cancer Genetics, Dana-Farber Cancer Institute, Department of Microbiology and Molecular Genetics, Harvard Medical School, Boston, MA 02115 **[93, 113]**

Virginia L. Scofield, Department of Microbiology and Immunology, School of Medicine, University of California, Los Angeles, CA 90024 **[213]**

Sheldon J. Segal, Population Sciences, Rockefeller Foundation, New York, NY 10036 **[263]**

Jeffrey N. Strathern, Cold Spring Harbor Laboratory, Cold Spring Harbor, NY 11724 **[97]**

Teruko Taketo, Center for Biomedical Research, The Population Council, New York, NY 10021 **[271]**

Pierre Tardent, Zoological Institute, University of Zurich-Irchel, Winterthurerstrasse 190, CH 8057 Zurich, Switzerland **[163]**

Norton D. Zinder, The Rockefeller University, New York, NY 10021 **[7]**

Preface

This is the third symposium in the Evolution series conducted as part of the microbiology course at the Marine Biological Laboratory. The first symposium, "The Origin of Life and Evolution," was jointly sponsored by microbiology and physiology courses in 1979. Two years later, the Microbiology and Marine Ecology and the Ecosystems Center held a symposium that compared the first evidences of life on earth with the modern cyanobacter in microbial mats. This workshop had its origins in 1983 when we had agreed that the critical factor in evolution was the development of a sexual mechanism to stabilize the species. Sex is distributed throughout the biological kingdoms, and when it has been analyzed in any detail, is elaborate and involves complex, finely tuned mechanisms. There is no disagreement that from bacteria to man, sex involves DNA replication—and recombination, presumably by the algal mechanisms that have been developed in bacteria.

What is new in the origin of sex? With molecular genetics and gene cloning it is now possible for the first time to analyze in detail chromosome structure and homology. Probes, at the level of genes and chromosome segments, allow us to explore similarities and mechanisms at the genetic level that reveal the developmental programs in sexuality. Understanding the signals—temporal expression or selection pressures—provides a rational basis for posing experimental and meaningful questions about the evolution of sex. We now have a great deal of knowledge based largely on our investigations at the genetic and molecular level. The elegance of understanding of sex in bacteria, yeast and Drosophila are not accidental—but reflect our efforts in genetic studies on these systems. It is clear that sex in *Streptococcus, Tetrahymena, Chlamydomonas,* etc., will be equally complex. With the advent of improved cell manipulations, genetic analysis, and refined molecular probes this complexity can be further revealed.

The evidence is very strong that basic mechanisms, such as protein synthesis and nucleic acid synthesis are highly conserved. One would predict that essential mechanisms for DNA replication, repair, and excision were also used for sex in higher organisms. Many direct tests of these mechanisms—with molecular probes—are now possible. Further, chromosome mobilization is a part of sex in bacteria—and in another form—the basis of the cassette model for mating type in yeast. One is led to ask, does mobilization of DNA, known to occur in simple eukaryotic cells, play a role in sex expression in higher cells?

Which of the regulatory mechanisms have evolved to control sex? Do they vary more widely than the elements themselves? How widespread is DNA modulation? It is now clear that in bacteria alternate regulatory mechanisms have developed in different species to achieve exactly the same result.

As our understanding of development of sex increases, we have a greater hope for finding or evolving intermediates between eukaryotes and procaryotes in a field that was explored by the late Roger Stanier and summarized in the Symposium on Microbial Mats: Stromatolites.

Progression from unicellular to multicellular eukaryotes is marked by such a diversification of strategies of sex differentiation that it becomes sometimes difficult not to lose the Arianna's thread. One of the most challenging, intractable problems is that of the origin of sex chromosomes. However, it can now be attacked, and is in fact under scrutiny, through the analysis of the conservation and distribution of DNA sequences specific for the sex chromosomes. It is also likely that the molecular approach to sex determination will have important consequences for the understanding of sex abnormalities in humans that appear to be linked to chromosomal abnormalities.

Harlyn O. Halvorson
Alberto Monroy

Acknowledgments

This book arose as a result of a workshop held as part of the Microbiology: Molecular Aspects of Cellular Diversity course at the Marine Biological Laboratory, Woods Hole, Massachusetts, in July, 1984. This workshop was supported by grants from the Rosenstiel Foundation and the Rockefeller Foundation. We would like to express our appreciation to all the contributors for the help that they have given in the preparation of this book, and especially to Lorraine Olendzenski and Michael Sheldon for their assistance during the workshop.

BACTERIAL CONJUGATION: BEGINNINGS OF SEXUALITY IN PROKARYOTES

The Origin and Evolution of Sex, pages 3–6

Beginnings of Sexuality in Prokaryotes

Harlyn O. Halvorson

Department of Biology and the Rosenstiel Basic Medical Sciences Research Center, Brandeis University, Waltham, Massachusetts 02254

In the last few decades, Schopf [1972], MacGregor [1940], Knoll and Barghoorn [1975], and others found from examination of the oldest known unmetamorphosed sedimentary rocks that there is not only a discontinuity in the carbon-isotope ratios, which is consistent with autotrophic organisms, but also that these rocks contain microscopic organic spheroids, some of which are strikingly reminiscent to dividing cells [for review see Awramik, 1984]. The ancient structures with which some microfossils are associated have been interpreted as stromatolite rocks that are true fossils of microbial mat communities. From these we are left with the conclusion that cellular evolution began as early as 3.5×10^9 years ago. These early undifferentiated fossils are remarkably similar to modern prokaryotes (cyanobacteria). Some are filamentous. Although less clear, the fossil record suggests that it was not until about 10^9 billion years ago, during the late Proterozoic Aeon, that larger flora emerged. Eukaryotic forms appeared in late Proterozoic shales (Knoll and Vidal, 1983; Vidal, 1984) and from the older Austrian dolomites (Schopf, 1978). Presumably, this is in the time during which mitosis and meiosis first arose. Evolutionary biologists have stressed the development of recombination of genetically varied progeny. This could have arisen through fusion not unlike phage infection or the fusion of two haploid cells. Alternatively, algae could have developed a nucleus, chromosomes, and mitotic apparatus. A defect in cytokinesis during mitosis would have led to a homozygous diploid. Variant chromosomes could have arisen leading, without cell division, to chromosomal separation and ultimately to chromosomal reunion. If two or more chromosomes were involved, and cell division occurred after the separation of homologous chromosomes, gametes would be generated. As pointed out by Bell [1982], this scheme proceeds by known protista steps and puts primary emphasis on division rather than fusion.

In a broader sense, sex is the process whereby DNA is transmitted from one cell to another by mechanisms other than cell division. Because extant bacteria possess recombination mechanisms, this process may have existed as early as 3.5 billion years ago. Genetic exchange can be accomplished by processes such as bacterial conjugation, transformation, and viral recombination. Bacterial matings may be uncommon in nature and restricted to only a few strains as far as we know at the present time. Levin [1981] has argued that even in enteric bacteria, recombination occurs at negligible rates. According to his view, transformation is the exception and sex in bacteria is a laboratory curiosity. Unfortunately, it is unlikely that we will find molecular details of the early stages of the sexual cycle preserved intact in the fossil record. Therefore, an understanding of the origin of sex will have to come from study of contemporary organisms.

Nevertheless, one cannot help but be impressed with the biochemical and molecular similarities between bacteria and eukaryotic cells. Mechanisms exist in bacterial and mammalian cells to carry out related chemical conversions. The key mechanisms of the synthesis of proteins and nucleic acids are remarkably similar. Within bacteria, because of their ability to exchange DNA by transformation, we can readily recognize close relatives in, for example, the *Acinetobacter* family (Juni, 1978). The evolution of pathways, such as the β-keto adipic pathway (Yeh and Ornston, 1981), is readily identified in bacteria. Even genes with similar functions appear to share evolutionary history. In highly conserved genes, such as ribosomal RNA, members of the Archaebacteria can be identified and close relatives readily detected (Woese and Fox, 1977). DNA sequences for essential function, such as cytochrome C (Fitch and Margoliash, 1967) and hemogloblins (Kimura, 1979) have been analyzed to measure the rate of evolution. As our body of knowledge on gene sequences has rapidly expanded, we have gained confidence in concluding that what exists has built upon what went before. The age in which bacteria were the sole replicative life forms on earth were periods of experimentation, in part in unique inhospitable environments. For these bacteria, change was followed by selection in which efficiency was monitored carefully.

It has been assumed frequently [e.g., Haldane, 1954] that sex began very early following the origin of life. Dougherty [1955] proposed, later supported by others [e.g., Smith, 1976], that sex arose as a mechanism to overcome genetic damage. Bernstein [1981; Bernstein et al., 1984a, 1984b] suggested that sexual reproduction originated as a recombinational repair process, first in RNA protocells and later in duplex DNA microorganisms. It is likely that early in evolution there were strong selective pressures to develop mechanisms for protecting DNA. For this process to evolve, two homologous DNA helices must be present in the same cell. Damage in one could be repaired by information in the other. The process of recombination repair and excision repair have persisted in present-day organisms. During the same evolutionary period, structures and pathways for cell recognition were developed. What is not as yet clear is whether the genes that evolved for these processes were the same ones that evolved for sexuality in eukaryotic cells. Are the genes that evolved for sex in bacteria modified and used for sex in protiston eukaryotes—or in animals or plants? Are there conserved sequences in bacterial sex genes that are recognized in genes regulating sex in higher forms? Alternatively, are separate independent mechanisms evolved for nucleated forms of life? Are the basic mechanisms for DNA replication, repair, and recombination highly conserved genes throughout evolution? With the advent of modern techniques in molecular biology, these questions are answerable.

On the other hand, one could imagine that animals or plants might utilize earlier evolved mechanisms for the exchange of genetic information, for example, cytoskeletal proteins, as the microtubules of the mitotic spindle are used for chromosomal separation during meiosis and mitosis. The elements involved have strong resemblances to structural components in bacteria for motion [Margulis, 1981], cell recognition and response to environmental signals. One could well imagine that structures evolved for another purpose (motility, environmental recognition, etc.) have been incorporated into sexual mechanisms in more complex forms.

Sex represents the most important challenge to the modern theory of evolution. This dilemma was expressed elegantly by Graham Bell in his recent book "The Masterpiece of Nature—the Evolution and Genetics of Sexuality" [1982]. "In the first instance it was long assumed that it (sex) evolved, not only as the result of the normal Darwinian process of natural selection, but through competition between populations or species, an hypothesis elsewhere almost universally discredited. Secondly, attempts to develop a Darwinian theory of sex were hampered by the realization that sexual reproduction usually implies an enormous reduction in fitness because sexual females transmit genetic material only half as fast as asexual females."

A review of the spectrum of mechanism available in bacteria includes the variety of ways in which DNA is organized and transmitted through plasmids and viruses to another bacteria, how such transmitted DNA participates in recombination, and how recipient cells recognize chemical signals that determine mating opportunities. Finally, the nature of genes in bacteria that control the actual mating response will be reviewed.

Our goals are the following

1) To analyze the molecular basis of sex and sexuality in bacteria and

2) To set the foundation for comparison of sex and sexuality in more complex biological forms.

REFERENCES

Awramik SM (1984): Ancient stromatolites and microbial mats. In Cohen Y, Castenholz RW, Halvorson HO (eds): "Microbiol Mats: Stromatolites." New York: Alan R. Liss, Inc., pp 1–22.

Bell F (1982): "The Masterpiece of Nature: The Evolution and Genetics of Sexuality." Berkeley: University of California Press.

Bernstein H (1981): Deoxyribonucleic acid repair in bacteriophage. Microbiol Rev 45:72–98.

Bernstein H, Byerly HC, Hopf FA, Michod RE (1984a): Origin of sex. J. Theor Biol (in press).

Bernstein H, Byerly HC, Hopf FA, Michod RE (1984b): The evolutionary role of recombinations, repair and sex. Int Rev Cytol (in press).

Dougherty EC (1955): Comparative evolution and the origin of sexuality. Syst Zool 4:145–169.

Fitch WM, Margoliash E (1967): Construction of phylogenetic trees. Science 155:279–284.

Haldane JBS (1954): The origins of life. New Biol 16:12.

Juni E (1978): Ann Rev Microbiol 32:349–371.

Kimura M (1979): A neutral theory of molecular evolution. Sci Am 241:98–126.

Knoll AH, Barghoorn ES (1975): Precambrian eucaryotic organisms: a reassessment of the evidence. Science 190:52–54.

Knoll AH, Vidal G (1983): Microbiota of the late precambrian Hunnberg Formation, Nordaustlandot, Svalbard, Norway. Proterozoic Plankton Geol Soc of Am 161:265–277.

Levin B (1981): Periodic selection, infectious gene exchange and the genetic structure of *E. coli* populations. Genetics 99:1–23.

MacGregor AM (1940): A precambrian algal limestone in Southern Rhodesia. Trans Geol Soc S Africa 43:9.

Margulis L (1982): "Early Life." Science Books International, Inc.

Schopf JW (1972): Evolutionary significance of the Bitter Springs (Late precambrian) microflora. Proc. XXIV Int. Geol. Congr. Sect. 1, Precambrian Geol., Montreal.

Schopf JW (1978): The evolution of the earliest cells. Sci Am 239:110–139.

Smith JM (1976): "The Evolution of Sex." London: Cambridge University Press.

Veh WK, Ornston LN (1981): Evolutionary homologous $\alpha_2\beta_2$ oligomeric structures in β-ketoadipate succinyl-CoA transferase from *Acinetobacter calcoacoticus* and *Pseudomonas putida*. J Biol Chem 256:1565–1569.

Vidal G (1984): The oldest eukaryotic cells. Sci Am 250:48–57.

Woese CR, Fox GE (1977): Phylogenetic structure of the prokaryotic domain: The primary kingdoms. Proc Natl Acad Sci USA 74:5088–5090.

The Origin and Evolution of Sex, pages 7–12

The Origin of Sex: An Argument

Norton D. Zinder

The Rockefeller University, New York, New York 10021

Sexual reproduction pervades the world of living organisms, yet its origin and evolution remain obscure. The major problem lies in this question: What is the advantage of transmitting only half of one's genes to one's offspring? Various authors have dubbed this the cost of meiosis [Williams, 1975], the cost of sex [Bell, 1982], and the cost of producing males [Maynard-Smith, 1978]. This cost approaches twofold as a maximum, as it only includes the genes that are different in females and males.

Still, it is a paradox. Classifical genetic scholars, such as R. A. Fisher (unpublished), H. J. Muller [1964], Crow and Kimura [1965], Cavalli-Sforza and Bodmer [1971], have postulated that the advantage of sexual reproduction is that it increases the rate of evolution by combining together, far more quickly than could asexual reproduction, new, useful gene mutations. Fisher just stated this point; the others derived equations demonstrating this gene flow. They concluded that the advantage there was a twofold increase in the rate of gene mixing per segregating useful gene with sexual reproduction. One of the parameters in the equations was population size, and population size in a binary event like sex is of considerable importance. Although Crow and Kimura [1965] deemed moderate populations adequate, Cavalli-Sforza and Bodmer [1971] concluded that small populations were best; however, recently, Maynard-Smith [1978] has concluded that populations must be very large i.e., outside the range of reality, for sex to promote the admixture of useful genes. Evidently, we are at a mathematical impasse.

Recent scholarship has used comparative biology as a guide to the value of sex [Williams, 1975; Bell, 1982]. Studies have been made of the distribution of sexual and asexual organisms and the relative kinds of environment they inhabit. Sex is viewed from this perspective as being of immediate value when it provides genetic diversity in a changing environment. Evolutionary considerations are secondary. Molecular biological considerations have led some to postulate that sex arose because it provides an efficient mechanism for the repair of damage to DNA [Bernstein et al., 1981]. DNA damage could provide a "strong" selective force for chromosome mixing; however, it is necessary that both of the parental chromosomes be damaged. With early sex probably not very efficient, such continuous heavy damage could well drive the population to extinction. Given a similar rate of replication, chromosome damage to haploids would tend to select diploids.

Most of the hypotheses noted above tacitly assume already existing full-blown sexual species. I now present an argument as to how sexuality might have arisen in prokaryotic organisms. A somewhat similar argument was developed by Rose [1983]. I will define sex as any horizontal transmission of DNA. The scenario goes as follows. Consider a protobacterium not too unlike those of the present. It is haploid, has a complement of DNA of about a thousand genes and can double in a reasonably short time. Physical or chemical factors cause the cell to dissolve, releasing and breaking its DNA into pieces (Fig. 1). These pieces can be taken up with some efficiency by nearby organisms. Among the fragments taken up is the one containing the replicator. In the right system, it will be replicated and lead to large numbers, perhaps in turn causing its host to lyse and release these replicons. For them, clearly, there is a strong drive to continue to pass themselves on in a horizontal manner. At that time, perhaps it was easier than currently for macromolecules

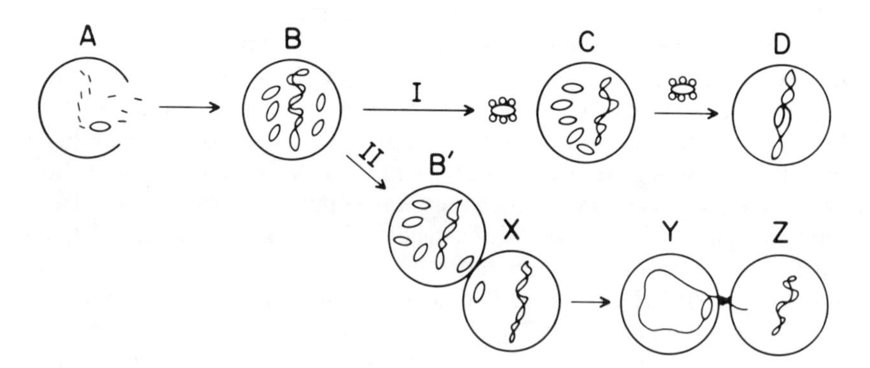

Fig. 1. Scheme for evolution of gene transfer in bacteria. I, Pathway to bacteriophage. II, Pathway to plasmid: A, bacterium lysing; O is replicator segment, --, other genomic fragments; B, transformation of new host and replication of replicator; C, released replicators have coopted bacterial proteins to form protophage; continued cycling leads to phage; D, lysogenization; B′, cycling of plasmid; X, transfer of plasmid, passive at first then involves mechanism; Y, plasmid with transfer genes enters host chromosome mobilizing it for transfer to Z.

to enter and leave cells. We note that for the simplest of the bacteriophages, the RNA phages [Atkins et al., 1979] and the icosahedral single-stranded DNA phages [Barrell et al., 1976], the gene for host cell lysis is internal to other genes. It is as if these phages early had had little need of lysis genes, and when later these genes became necessary, the size the phage genomes could attain was limited by the size of the protein shells.

By passing between host bacteria over time, these replication fragments could follow two different evolutionary pathways to assure their survival by replication. First, they might coopt cell proteins that were able to bind to, thereby protect their DNA, and thus become a protophage. Binding, though probably weak at first, still would be protective when these fragments released into the environment. Following genetic recombination with the host genome, those DNA fragments that picked up the genes specifying the DNA binding proteins would assure their further survival on transmission, for the protein genes would be carried along and could now evolve according to the requirements of the protophage. What drives the evolution of a phage is the phage's purposes of replication, but because the phage can also transfer host genes, it is also a sexual process.

The second means by which replication fragments might preserve themselves is to become plasmids. These obligate intracellular parasites can increase their number and distribution by transferring themselves from cell to cell during cell contact. Moreover, as for the protophage, they can coopt preadaptive membrane proteins that happen to aid this process. In time, by recombination and mutation, a contact-transfer apparatus is born, the genes for which are on the plasmid, thereby setting up an escape hatch for the plasmid as soon as it enters a new host cell.

The argument presented here is that phages and plasmids evolve to maintain their own transfer. In doing so, they acquire the genes necessary for their individual lifestyles. Thus, the coat proteins of a phage and the pili and accessory transfer proteins of a plasmid play an analogous role. They expedite the transfer of these replicons from cell to cell.

What is in the fragment transfers for the host bacteria? The majority of phages are temperate and have learned to enter the bacterial chromosome and stay in a quiescent state. Some of their genes function and induce new bacterial properties. These include antigenic changes which can modify sensitivity to other phages. Perhaps more important are those controlling bacterial toxins associated with such diseases as diphtheria and scarlet fever, not too long ago scourges of children. Currently, these toxins seem to play no role in the life cycle of these phages; perhaps in the past their ability to enhance bacterial virulence had a selective value for the bacteria.

When leaving their hosts, some phages pick up nearby genes, whereas others replace their entire genome with bacterial genes (about 1% of the bacterial genome per transducing particle). Phage particles, lacking metabolic systems, are generally quite stable to wide-ranging environmental changes, while living bacteria are not. These transducing particles can conserve these units of bacterial genome and transfer them to other hosts. Whether the availability of these infective genes floating in the environment is significant in evolution is not an answerable question, but the fact is, by existing, they could be.

Plasmids not only have acquired the genes to promote their transfer but, since they are more dependent on their hosts than phage, seem far more solicitous of their host's well being [Bukhari et al., 1977]. Variabily they carry genes to detoxify heavy metals, destroy antibiotics, digest hydrocarbons, and even occasionally provide the requisite elements for invading host tissues. Other virulence-associated traits are also plasmid controlled. All of these phenotypes are, in a sense, accessory "aggressive" factors providing strong selective advantages to the host as well as themselves in certain special, stringent environmental conditions. These past years, with increasing amounts of antibiotics being placed in the environment, an ever-increasing fraction of antibiotic-resistant bacteria have been isolated.

The plasmids also promote transfer of the bacterial chromosomes they are associated with, but at low rates. When they enter a bacterial chromosome, it becomes essentially one large plasmid which, when signalled by contact with another bacterium, transfers the bacterial as well as the plasmid genes. In the laboratory, the frequency of occurrence of such plasmid (conjugal) integrates seems too low ($\sim 10^{-6}$) to be of great significance. However, evolutionary time is long, and selective pressures can be very large.

The kind of sexuality described above would be most effective with a mixture of sexual and asexual bacterial growth. Moreover, it would occur with a generally haploid population. It is, in fact, sexuality which might initiate diploidy. What diploidy does best is cover partially defective genes, simple recessives. As a consequence of this, diploidy allows the genome to grow, for it can now withstand the existence of much larger numbers of hypomorphic genes. Thus, a cell produced by mating might suddenly find itself with a large selective advantage if it failed to reduce to haploidy. All of its many minor genetic lesions would be complemented so that it could grow faster and better. From there, such a cell would be set to enlarge its genome and evolve away from its parent. Should diploidy arise from misreplication of a haploid cell, not only would there be no *immediate* advantage (over time there is), but there might be the disadvantage of having two chromosomes to replicate instead of one before cellular metabolism had had a chance to adjust.

There is no way to put bacterial evolution and the role of sex into a simple environmental context. To this day, little is known of the ecology and population biology of these organisms. At the extremes, their growth can be characterized in three different ways. There is unrestricted exponential growth when nutrients are plentiful and physical parameters are right. Probably most usual is a kind of steady-state growth due to the fact that some component of the environment is limiting. Last, we have catastrophic destruction due to some physical or chemical activity. During active growth, as shown by Atwood et al. [1951], bacteria undergo periodic selection. That is, within the population there arises some mutant organism with a small selective advantage and which therefore takes over the population in time. Such selection has the interesting consequence of cleansing the population of the majority of its accumulated neutral mutations since they existed in the unselected portion of population. Atwood et al. were able to show that in some instances the organisms selected actually grew faster whereas others were probably just better competitors. Because genes producing such selective effects arise at low frequency, the low frequency sexual processes, transductions, and conjugations could readily supply the gene conferring the selective advantage. Moreover, in contrast to mutation, sexually derived genes have already been tested for value in real bacterial populations. Similar advantages apply to a sexually derived source of genes for the static bacterial populations. Such genes might even be able to partially relieve the factor limiting the population's growth. Near destruction of a bacterial population gives little opportunity for an effect of rare, sexually derived genes. However, the small numbers of organisms that are the founders of a new population might not be affected by previous competitive situations and thereby might provide an opportunity for specific surviving elements to grow up—genetic drift.

There is no way to connect prokaryotic sex with eukaryotic sex, but then again almost none of the processes that differentiate these two groups can be readily connected. Our argument here is that perhaps, as for prokaryotes, sex in higher forms may be no more than an elaboration of a set of processes which gained adaptive value only after existing.

ACKNOWLEDGMENTS

Thanks are due to my colleague, Michael Young, for encouraging me all during the endeavor to think this notion through. Supported in part by grants from the National Science Foundation and the National Institutes of Health.

REFERENCES

Atkins, JF, Steitz JA, Anderson CW, Model P (1979): Binding of mammalian ribosomes to MS2 phage RNA reveals an overlapping gene encoding a lysis function. Cell 18:247–256.

Atwood KC, Schneider LK, Ryan FJ (1951): Selective mechanisms in bacteria. Cold Spring Harbor Symp Quant Biol 16:345–355.

Barrell BG, Air GM, Hutchison CA (1976): Overlapping genes in bacteriophage φX174. Nature 264:34–37.
Bell G (1982): "Masterpiece of Nature: The Evolution and Genetics of Sexuality." Berkeley: University of California Press.
Bernstein H, Byers G, Michod R (1981): Evolution of sexual reproduction: Importance of DNA repair, complementation and variation. Am Naturalist 117:537–549.
Bukhari AI, Shapiro JA, Adhya SL (1977): "DNA Insertion Elements, Plasmids and Episomes." New York: Cold Spring Harbor Laboratory.
Cavalli-Sforza L, Bodmer W (1971): "The Genetics of Human Populations." San Francisco: WH Freeman & Co.
Crow J, Kimura M (1965): Evolution in sexual and asexual populations. Am Naturalist 99:439–450.
Maynard-Smith J (1978): "The Evolution of Sex." London: Cambridge University Press.
Muller HJ (1964): The relation of recombination to mutational advance. Mutat Res 1:2–9.
Rose M (1983): The contagious mechanism for the origin of sex. J Theor Biol 101:137–146.
Williams GC (1975): "Sex and Evolution." Princeton: Princeton University Press.

The Origin and Evolution of Sex, pages 13–28

Sex Pheromones, Plasmids, and Conjugation in *Streptococcus faecalis*

Don B. Clewell

Departments of Oral Biology and Microbiology/Immunology and The Dental Research Institute, Schools of Dentistry and Medicine, University of Michigan, Ann Arbor, Michigan 48109

INTRODUCTION

In highly evolved multicellular organisms, cellular communication via hormonal substances is intimately involved in fertility and development. Chemical communication also can occur between individual organisms, and the molecules representing these signals are referred to as pheromones [Karlson and Luscher, 1959]. In the case of unicellular organisms, communication is, by definition, via pheromones; and the involvement of such substances in certain stages of development has been implicated in a number of microbial species [Ensign, 1978; Hirosawa and Wolk, 1979; Kaiser et al., 1979; Biro et al., 1980; Aaronson, 1981; Stephens et al., 1982]. When the substances facilitate mating between two organisms, they are called sex pheromones; these compounds and their behavior have been studied in fungi, algae, and ciliates [Yanagishima, 1978; Kochert, 1978; Nanney, 1977]. Much attention has been focused on the yeast α and *a* factors and the mating phenomena associated with their pheromonal activities [Manney et al., 1981; Kurjan and Herskowitz, 1982; Sprague et al., 1983].

Whereas extracellular compounds that affect competence for transformation are well known in *Streptococcus pneumoniae* and *Streptococcus sanguis* [Lacks, 1977], substantive evidence for bacterial sex pheromones that promote conjugative transfer of genetic material has been reported only in *Streptococcus faecalis* [Dunny et al., 1978; Dunny et al., 1979]; however, sex-related chemotactic factors in *Escherichia coli* [Collins and Broda, 1975] and

Salmonella typhimurium [Bezdek and Soska, 1972] have been suggested. Certain species of *Agrobacterium* that cause plant hyperlasias are characterized by their ability to insert segments of plasmid DNA into the plant genome, and "opines" specifically excreted by resulting plant tumors can induce transfer of plasmid DNA between two bacterial strains [Nester and Kosuge, 1981]. In this communication, the *S. faecalis* sex pheromones and their relationship to the transfer of plasmid DNA will be reviewed and discussed.

During the past 10 years, conjugative plasmids have been identified in several different species of streptococci; the properties of many of these elements have been summarized in a comprehensive review [Clewell, 1981]. Two general catagories of plasmids have been distinguished. One type, with a size in the range of 22–30 kb, does not transfer well between cells in liquid broth; donors and recipients must be mixed on solid surfaces (e.g., on a filter membrane placed on a semisolid medium). These plasmids generally encode MLS resistance (i.e., resistance to macrolides, lincosamides, and streptogramin B) and usually are able to transfer between a variety of streptococcal species. Generally, transfer frequencies are in the range of 10^{-5} to 10^{-3} (intraspecies). Some of these plasmids can transfer to other genera, including *Staphylococcus* and *Bacillus* [Engel et al., 1980; Schaberg et al., 1982; Landman et al., 1981; Clewell et al., 1985]; interspecies transfer usually occurs at much lower frequency than intraspecies transfer.

The other type of conjugative plasmid thus far has been observed only in *S. faecalis;* these elements are generally of a size greater than 45 kb; they transfer in broth matings at frequencies as high as 10^{-3} to 10^{-1} within a few hours and confer a mating response (Fig. 1) to small heat-stable compounds (sex pheromones) excreted by potential recipient cells [Dunny et al., 1978, 1979]. The pheromone induces the synthesis of a proteinaceous adhesin that coats the donor cell surface [Yagi et al., 1983]. The adhesin is referred to as "aggregation substance" (AS) and is believed to facilitate the formation of mating aggregates upon random collision between the nonmotile donors and recipients. Once the recipient acquires a copy of the plasmid, the production of the related pheromone ceases, but different pheromones specific for donors harboring different classes of plasmids continue to be excreted [Dunny et al., 1979; Clewell et al., 1982b]. The different pheromones are designated by their relationship to the plasmid system originally used to resolve them. For example, cAD1, cOB1, cPD1, and cAMγ2 refer to the pheromones to which specific responses occur by strains respectively harboring the plasmids pAD1, pOB1, pPD1, and pAMγ2.

When pheromone-responding donor strains are exposed to culture filtrates of recipients, the resulting induction leads to self-aggregation or "clumping"; for this reason, the sex pheromones also are referred to as clumping inducing agents (CIAs). The CIA response provides the basis of a convenient assay

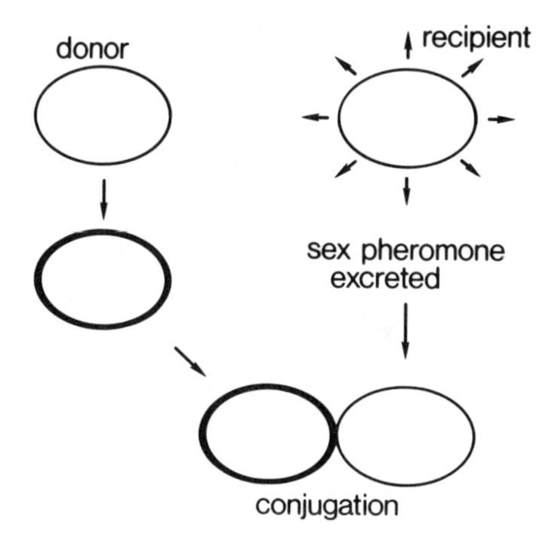

Fig. 1. Expression of sex pheromones by a recipient *S. faecalis* strain and response by a donor containing a pheromone-sensitive conjugative plasmid.

for pheromones, and a microtiter dilution method is used for quantitation [Dunny et al., 1979]. The pheromone titer is taken as the reciprocal of the highest dilution still giving rise to a visible clumping response. The induced adhesin (AS) is presumed to adhere to a substance, referred to as "binding substance" (BS), which is located on the surface of both donor and recipient cells.

Figure 2 illustrates a previously proposed [Dunny et al., 1979] hypothetical model suggesting how pheromones, aggregation, and mating may interrelate. Applied here toward the 56.7 kb hemolysin-bacteriocin plasmid pAD1 [Tomich et al., 1979; Clewell et al., 1982a], a determinant designated *IcAD1* has been proposed to be involved in the "shutoff" of the chromosome-determined pheromone cAD1. The pheromone cPD1 (which induces donors harboring the 54.6 kb bacteriocin plasmid pPD1) and other pheromones (not indicated) continue to be excreted. A determinant involved in controlling expression of AS also was proposed; designated *RcAD1*, it is most simply conceived as encoding a substance that represses the synthesis of AS. Whereas the *AS* determinant is likely to be plasmid-borne, as shown in the model, the possibility of a chromosomal locus has not been ruled out entirely.

THE AGGREGATION EVENT

Aggregates of *S. faecalis* 39-5 (contains pPD1) resulting from pheromone induction become dissociated by exposure to EDTA, and cells so treated can be pelleted and resuspended in buffer (0.03 M Tris, pH 7.0) without reaggregating [Yagi et al., 1983]. Referred to here as "dissociated cells," aggre-

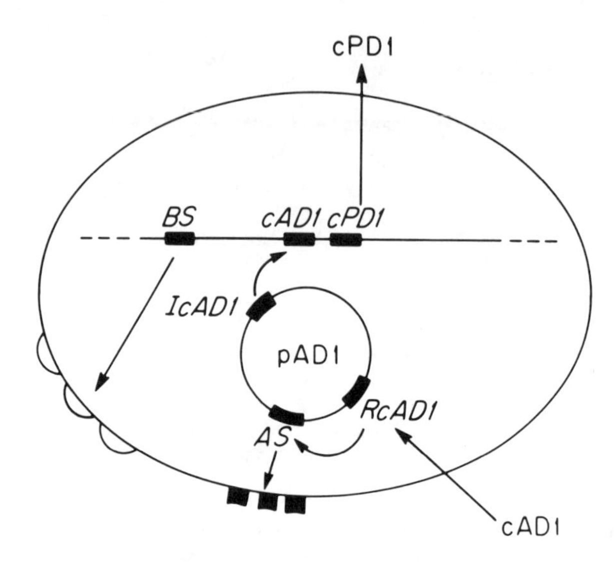

Fig. 2. A model illustrating certain aspects of the pheromone response of pAD1 and the "shutting-off" of endogenous cAD1. *IcAD1* is a determinant for the modification (inactivation) of endogenous cAD1. The cell continues to excrete different pheromones such as cPD1. *RcAD1* determines a regulatory protein involved in controlling the synthesis of "aggregation substance" (AS), via a direct or indirect interaction with exogenous cAD1. The dark surface structures represent AS and are believed to represent an adhesin that uniformly coats the cell as a result of induction by cAD1. The clear surface structures represent binding substance (BS); it is encoded by the chromosome and is always present on the surface of both recipients and donors. BS interacts with AS in the generation of mating aggregates. (Reproduced from Ike and Clewell, [1984], with permission of the publisher.)

gation does not recur unless the cells are provided with a minimum of 5 mM phosphate and a divalent cation such as Mg^{++}, Mn^{++}, Ca^{++}, or Co^{++} at a concentration of at least 0.1 mM. Phosphate and divalent cations may act to stabilize the conformation of AS or its receptor (BS) and/or modulate the repulsive effects of neighboring cell surface molecules.

Exposure of dissociated cells to proteases (trypsin or pronase), detergent (0.05% sodium dodecyl sulfate), or heat (98°C, 5 min) destroyed the ability of the cells to reaggregate when provided with phosphate and divalent cation, whereas lysozyme or lipase had no effect [Yagi et al., 1983]. In addition, a variety of different sugars had no effect on the ability to reaggregate.

An antiserum raised against induced 39-5 cells, and absorbed with uninduced cells, was used in immunoelectron microscopy studies to reveal a "fuzzy" surface material (possibly a microfibrillar substance) uniformly coating the cells [Yagi et al., 1983]. This substance was absent from uninduced cells and is likely to correspond to AS. The observation that Fab fragments of the immunoglobulin prevented aggregation adds support for this view

[Kessler and Yagi, 1983]. An immunodominant protein with a molecular wt of 78,000 could be extracted from induced cells; and the timing of appearance of this antigen coincided with the timing of aggregation after exposure to pheromone at 45–60 min. post induction [Kessler and Yagi, 1983]. It is noted that both rifampin and chloramphenicol block induction of aggregation if present at the time of exposure to pheromone [Dunny et al., 1978].

Surface material induced in strains harboring several different conjugative plasmids (pAMγ1, pAMγ2, pAMγ3, pAD1, and pOB1), conferring responses to pheromones differing from cPD1, reacted to some extent with the antiserum raised against induced pPD1-containing cells [Yagi et al., 1983]. A significant degree of structural similarity thus is implied.

THE CHEMICAL NATURE OF THE SEX PHEROMONES

On the basis of the microtiter dilution assay, pheromone titers of 64 to 256 generally are observed in filtrates of late exponential cultures. In all cases tested, the activity is not affected by 100°C (10 min). The two pheromones cPD1 and cAD1 were found to be sensitive to chymotrypsin, pepsin, pronase, leucine amino peptidase, and carboxypeptidases A and B; however, they were both resistant to trypsin [Ike et al., 1983; R. Craig, personal communication]. On the basis of gel filtration chromatography, cPD1 and cAD1 were sized at approximately 800 and 1050 daltons, respectively [Ike et al., 1983; Suzuki et al., 1984] recently have purified and characterized cPD1. Its amino acid sequence was found to be H-Phe-Leu-Val-Met-Phe-Leu-Ser-Gly-OH (molecular wt 912). A synthetic octapeptide showed the same biological activity and chromatographic behavior as the isolated cPD1, and activity was detectable at a concentration of 4×10^{-11} M. The very hydrophobic nature of the peptide is particularly noteworthy.

Although the pheromones represent small peptides, their synthesis via ribosomal assembly, rather than by enzymatic condensation of amino acids, seems a reasonable possibility. Support for this view is the fact that pheromone activity in culture media closely parallels the growth of the cells and stops increasing if transcription or translation is blocked with rifampin or tetracycline, respectively (B. White and D. Clewell, unpublished data). The small size of these substances, however, suggests the possible involvement of larger precursor polypeptides which are subsequently processed. In this regard, there could be some resemblance to the yeast α factor, a peptide sex pheromone synthesized as a precursor containing four tandemly arranged copies of the pheromone [Kurjan and Herskowitz, 1982]. It should be possible to derive clones of genomic DNA containing the cPD1 pheromone sequence by making use of a synthetic DNA probe corresponding to part of the known amino acid sequence. This currently is being pursued in our laboratory.

VARIANTS IN PHEROMONE PRODUCTION

Development of a plating technique to detect production of sex pheromone by bacterial colonies has provided a means for isolating mutants [Ike et al., 1983]. *S. faecalis* 39-5 (harbors pPD1) was used as an indicator strain where overlays onto pheromone-producing colonies resulted in the appearance of a "halo" corresponding to aggregates of the responder cells around the colonies. Several mutants have been obtained by chromosomal insertion of the conjugative transposon Tn*916* [Franke and Clewell, 1981; Gawron-Burke and Clewell, 1982; Gawron-Burke and Clewell, 1984] or by using nitrosoguanidine [Ike et al., 1983; Clewell et al., 1984a]. These halo-negative mutants were found to excrete cPD1 at levels about 3–4% that of the parent strains. Interestingly, the level of cAD1 was reduced to a similar extent; similarly, six of eight clinical isolates exhibiting good responses to a filtrate of *S. faecalis* JH2-2 (plasmid-free) gave reduced CIA titers when exposed to filtrates from the mutants [Clewell et al., 1984b]. The mutants were reduced by orders of magnitude in their ability to behave as recipients in broth matings; however, those that did acquire plasmid DNA were able to efficiently donate plasmid DNA [Ike et al., 1983].

The fact that cAD1 also was reduced in variants with defects in cPD1 production is of particular interest and suggests that synthesis or subsequent processing or transport of these substances is linked. Derivation of the sequence of the pheromone structural genes (reflecting possible precursors?) should provide much insight into the relationship of the gene products to each other.

MODIFICATION OF PHEROMONES IN DONOR CELLS

Because donor cells do not produce sex pheromone, there must be a plasmid-encoded mechanism for shutting off production of the peptide (see Fig. 2). Recent studies have revealed that inactivation via a chemical addition to the pheromone is the mechanism by which this occurs [Ike et al., 1983]. Culture filtrates of donor cells harboring pPD1 or pAD1 derivatives contain a substance shown to act as a specific inhibitor of the related exogenous pheromone. In the microtiter dilution assay, pheromone activity was reduced significantly if the dilutions were made through a donor filtrate rather than through broth. (For example, the cPD1 titer in a late exponential culture of recipient cells was reduced from 64 to 8 when the dilutions were made through a filtrate of a similar culture of donors.) Like the pheromones cPD1 and cAD1, the related inhibitor substances were heat stable, sensitive to chymotrypsin, and resistant to trypsin. Estimates based on gel filtration chromatography indicated that the inhibitor substances are 350–400 daltons larger than the respective pheromones [Ike et al., 1983].

With the notion that the modification might involve addition of a nucleotide, phosphodiesterase treatment was performed to see if pheromone activity could be regenerated. Interestingly, phosphodiesterase II and not phosphodiesterase I regenerated activity [Ike et al., 1983]. Phosphodiesterase II generally gives rise to 3′ mononucleotides; thus, if a nucleotide addition indeed were involved, a 3′ phosphodiester bond would appear to be present. This is in contrast to the 5′ phosphodiester bonds involved in the ribonucleotide forms of glutamine synthetase [Shapiro et al., 1967; Bender et al., 1977], aminoglycosides [Davies and Smith, 1978], and clindamycin [Argoudelis et al., 1977]. In the latter cases, the original activity is regenerated (or such is expected) by exposure to phosphodiesterase I. It is noted that the pheromone activity regenerated from the corresponding inhibitor is an order of magnitude lower than the pheromone activity observed in a similarly prepared filtrate of recipient cells. Thus, the extracellular concentrations of the modified peptides would appear to be low. A unit of modified pheromone (referred to now as mcAD1 or mcPD1) is defined as the amount necessary for a 50% reduction in pheromone activity. Units of 4, 5, and 6 gave rise to pheromone activities (titers) of 4, 8, and 16 when exposed to phosphodiesterase II.

Although the exact nature of the modification remains to be determined, the evidence for a phosphodiester bond suggests the involvement of the single serine residue in cPD1 (see above). It is likely that the remaining structural similarity with the active unmodified peptide serves as the basis of inhibition via competition for a receptor site. From a physiological standpoint, the question arises as to the effect the modified pheromone has on the ability of donors to mate. If there is indeed a functional role, it may serve to desensitize the donor to "very low" concentrations of pheromone. Low pheromone levels would mean that the corresponding recipient is far away or in too low a concentration (relative to donors); a response under these conditions might not lead to productive matings or give rise to more donor-donor aggregates than donor-recipient aggregates. Consistent with this view, a derivative of strain Y11 (defective in pheromone production; see above section) harboring pAM714 (a pAD1: :Tn*917* derivative) and shown not to produce detectable mcAD1 was a better donor than the parent strain in matings with the recipient FA2-2. The difference in the number of transconjugants generated was one or more orders of magnitude, depending on the mating time and ratio of donors to recipients used (F. An and D. Clewell, unpublished data).

With regard to the kinetics of appearance in culture media, an interesting difference between mcDP1 and mcAD1 has been observed [Ike et al., 1983]. The production of both cPD1 and cAD1 from a plasmid-free strain parallels cell growth during most of exponential phase; both activities plateau a generation or so prior to stationary phase. A similar behavior is seen for mcPD1 in corresponding donor cell cultures. In contrast, mcAD1 appears as a burst (rising up to 3 units) in a freshly inoculated culture (lag phase) and then decreases to a very low and constant level (one unit) until stationary phase is reached, at which point appearance occurs again as a sudden burst to a relatively high level (6 units). The fact that activity decreases as cells enter exponential phase implies the existence of a specific removal or degradation process.

GENETIC ASPECTS OF THE pAD1 PHEROMONE RESPONSE

To analyze some of the molecular processes involved in the pAD1 mating response, insertion mutants were derived using the erythromycin-resistance transposon Tn*917* [Tomich et al., 1978, 1980]. In *S. faecalis* DS16, transposition of Tn*917* from the nonconjugative pAD2 to pAD1 is induced when cells are exposed to erythromycin; this phenomenon has been exploited in the generation of insertion mutants [Clewell et al., 1982a; Ike and Clewell, 1984]. Variants that exhibited constitutive clumping were derived easily [Ike and Clewell, 1984]. These strains transferred plasmid DNA in short (10 min) matings at relatively high frequency (10^{-4}) compared to the very low level ($< 10^{-7}$) normally observed (i.e., normally donors require 30 to 40 min to become fully induced by the pheromone of the recipient). The constitutively clumping mutants fell into two subclasses that exhibited colony morphologies that were "dry" or "normal." The "dry" types also exhibited significantly larger clumps in broth compared to "normal" types; the latter resembled the case more typical of pheromone induction. The Tn*917* insertions were mapped by restriction enzyme analysis to two separate clusters, designated *traA* and *traB* (Fig. 3). The "dry" colony subclass corresponded to *traA* and represented a span of 1.5 kb (12 independent insertions mapped), whereas the "normal" colony subclass corresponded to *traB* and spanned 1.3 kb (eight insertions mapped). The two clusters were separated by 1.7 kb where insertion of Tn*917* (seven mapped) did not affect the ability to respond normally to cAD1.

It was speculated that if *RcAD1* (Fig. 2) negatively controlled the synthesis of AS, mutations at this locus should result in constitutive aggregation. Similarly, mutations in *IcAD1* might be expected to aggregate constitutively due to a response to endogenous unmodified pheromone, and such mutants might be expected to excrete cAD1. None of the constitutive clumpers derived excreted cAD1; indeed, both *traA* and *traB* types excreted mcAD1. Thus, insertion mutants in *IcAD1* are very rare, or such derivatives still are incapable of responding to endogenous pheromone.

Conceivably, products determined by the two loci *traA* and *traB* (both equivalent to *RcAD1*) interact in some way in the repression of AS. The fact that *traA* mutants have a distinguishable colony texture and give rise to unusually large clumps suggests differences in surface properties of the cell

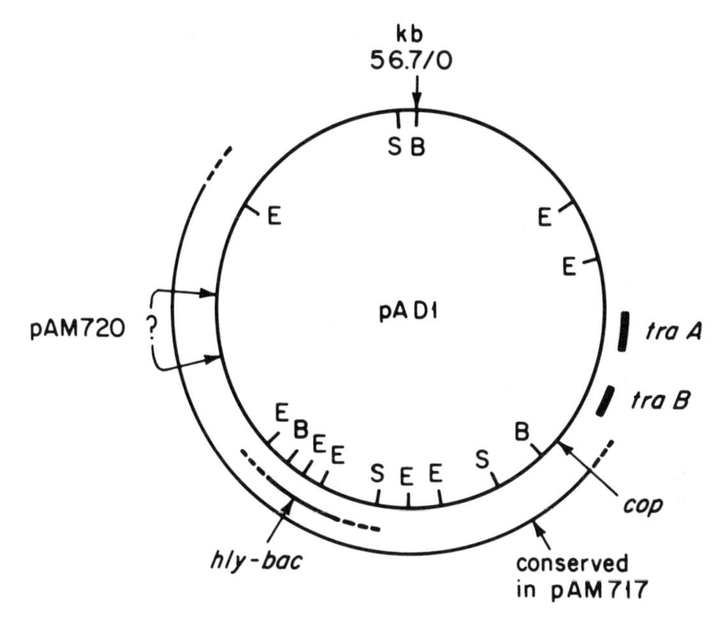

Fig. 3. Map of pAD1 based on analyses of Tn*917* insertions. *Hly-bac* refers to the hemolysin-bacteriocin determinant; *cop* refers to the location of the insertion of pAM710. *TraA* and *traB* are segments of DNA into which insertions give rise to constitutive clumping. The restriction sites were: E, *Eco*R1; S, *Sal* I; and B, *Bam*H1. (Reproduced from Ike and Clewell, [1984], with permission of the publisher.)

and that the AS in this case may be altered. Two different mutants with insertions mapping within *traA* were defective in their ability to aggregate, even upon exposure to cAD1. These variants were reduced greatly in transferability; however, the low level of transfer observed was of a "constitutive" nature (i.e., transfer occurred in short matings and could not be induced to higher levels). It is possible that AS structure or processing is related closely to or is influenced by *traA*.

Two mutants (pAM710 and pAM720) mapping outside of *traA* and *traB* (Fig. 3) were reduced in transfer (by 1–3 orders of magnitude) but could be induced to aggregate by cAD1 [Ike and Clewell, 1984]. With respect to donor potential, both were sluggish in responding to cAD1. The mutant pAM710 had an increased copy number and is designated *cop* on the map. A mutant designated pAM717 was incapable of transfer and sustained a large deletion that included *traA* and *traB* (Fig. 3). The cells harboring this derivative produced normal levels of cAD1.

Although not indicated on the pAD1 map, it is presumed that the plasmid encodes one or more proteins that serve as receptor sites for exogenous cAD1. As mentioned earlier, this site probably also has an affinity for mcAD1. Other plasmid encoded functions presumed to be present are those more directly involved in the transfer of plasmid DNA. Such functions, inducible by cAD1, have been inferred from experiments involving matings between isogenic donors with distinguishable derivatives of pAD1, i.e., pAD1: :Tn*916*(Tc) and pAD1: :Tn*917*(Em) [Clewell and Brown, 1980]. It was found that if one donor was induced by exposure to a filtrate containing cAD1 and then mixed for a short time (20 min) with the other donor, transfer occurred primarily in the direction from the induced to the uninduced strain. Transfer in the opposite direction was orders of magnitude lower. It was reasoned that, had the sole function of cAD1 been to generate aggregates, then transfer might have been expected to occur equally well in both directions. Because it did not, other functions (relating to transfer) also must be induced. When both donors were induced prior to mating, transfer occurred to a similar extent in both directions, although the frequencies were significantly lower than the case where transfer occurred from an induced donor to an uninduced donor. It is possible, therefore, that cAD1 also induces an entry (surface) exclusion function. Such a behavior would seem important in a natural environment where donors responding to recipients also are able to aggregate with other donors.

PHEROMONE-LIKE ACTIVITIES IN OTHER BACTERIAL SPECIES

In an effort to determine if different species of bacteria might excrete substances similar to the *S. faecalis* sex pheromones, 10 *S. faecalis* strains exhibiting high sensitivity to induced clumping were tested for their response to filtrates from a variety of bacterial sources, including eight species of streptococci, two species of staphylococci, *Bacillus subtilis* and *Escherichia coli* [Clewell et al., 1984a]. Nine of the 10 *S. faecalis* responders failed to respond to filtrates representing any of the other 12 species. One strain, *S. faecalis* RC73, responded to filtrates of *Streptococcus sanguis* Challis, *Streptococcus faecium* 9790, and *Staphylococcus aureus* 879R4S. All of an additional 22 strains of *S. aureus* also produced a CIA activity (titers of 16-64) that induced clumping of *S. faecalis* RC73. This was in contrast to 22 isolates of the less pathogenic coagulase negative staphylococci (e.g., *Staphylococcus epidermidis*), among which only two excreted such an activity. (In the two positive cases, the titers were only 2.) Among additional strains of *Streptococcus sanguis* tested, only 2 of 9 strains produced activity.

S. faecalis RC73 produces hemolysin, bacteriocin, and is resistant to tetracycline; sedimentation and electrophoretic analyses revealed at least five plasmids, the largest being about 57 kb. Studies have shown (Clewell et al., 1984a) that a 36 kb conjugative plasmid designated pAM373 encodes the response to the CIA activities in culture filtrates of *S. aureus, S. sanguis* Challis, and *S. faecium* 9790. The plasmid has no selectable markers; however, transposons such as Tn*917*, Tn*916*, and Tn*918*(Tc) (a transposon resembling Tn*916* discovered in strain RC73) have been inserted and used to select for transfer of drug resistance. Mating experiments showed that in *S. faecalis* filtrates, the CIA activity, designated cAM373, had all the typical properties of a sex pheromone. Interestingly, the plasmid would not establish in *S. aureus* or *S. sanguis,* even after matings on filter membranes.

A restriction map of pAM373 was derived [Clewell et al., 1984a] and showed no resemblance to previously derived maps of pAD1 and pPD1. At a size of 36 kb, pAM373 is significantly smaller than other plasmids (usually > 45 kb) encoding pheromone responses. The related pheromone cAM373 (from an *S. faecalis* filtrate) is about 850 daltons in size based on gel filtration chromatography; and, like cAD1 and cPD1, this pheromone is sensitive to chymotrypsin and resistant to trypsin [Clewell et al., 1984a]. In addition, filtrates of donors harboring the plasmid were found to have an inactivated (modified) pheromone.

The fact that members of a few other bacterial species can excrete a substance resembling a specific sex pheromone in *S. faecalis* does not necessarily imply a pheromonal role in those species. Indeed, the fact that pAM373 derivatives could not be established in *S. aureus* or *S. sanguis* suggests the possible absence of a relationship between cAM373 and mating in these organisms. The correlation of cAM373 production and pathogenicity in staphylococci raises the interesting possibility that the peptide could play a role in virulence.

CONCLUDING REMARKS

In *S. faecalis,* a pheromone-induced mating response can promote transfer of genetic information in addition to that borne by the related plasmid. The formation of mating aggregates can lead to the mobilization of nonconjugative elements such as the R-plasmids pAMα1 [Dunny and Clewell, 1975] and pAD2 [Tomich et al., 1979]. In the case of pAMα1, mobilization occurs at relatively high frequency much like the mobilization of *Col*E1 by F [Willetts and Skurray, 1980] in *E. coli,* whereas, in the case of pAD2 transfer occurs at low frequency and is probably dependent on cointegrate formation [Clewell et al., 1982a]. In addition, conjugative R-plasmids such as pAMβ1 [Clewell et al., 1974; LeBlanc et al., 1978], which transfer poorly if at all in broth, transfer at frequencies of 10^{-4} per donor if present in donors also harboring a pheromone-responding plasmid such as pAMγ2 or pAMγ3 [Clewell et al.,

1982b]. Interestingly, pAD1 and pAMγ1 both inhibit the transfer of pAMβ1 [Clewell et al., 1982b]. Pheromone-responding plasmids also have been shown to be capable of chromosomal mobilization (in filter matings) at frequencies of about 10^{-7} per donor [Franke et al., 1978; Franke and Clewell, 1981]. Further, the mobilization of chromosome-borne elements such as Tn*916* [Franke and Clewell, 1980, 1981] and the newly discovered Tn*918* [Clewell et al., 1984a] clearly is enhanced in the presence of such plasmids.

Analyses of 100 clinical isolates of *S. faecalis* showed that drug-resistant strains were significantly more likely to be both producers of and responders to sex pheromones [Dunny et al., 1979]. A role of pheromones in the evolution of drug resistance would be expected in that producers of these substances should be prime targets in the acquisition of drug-resistance determinants from strains harboring responding plasmids. Thus far, only one such naturally occurring plasmid, pCF10, has been shown to carry an antibiotic resistance determinant [Dunny et al., 1981]; most of the others encode hemolysins, bacteriocins, and/or UV-resistance [Clewell, 1981]. The hemolysin plasmids pAD1 and pAMγ1 appear almost identical and respond to the same pheromone despite their very different origins [Clewell et al., 1982b]. Other hemolysin plasmids have been shown to have significant homology with pAD1 and pAMγ1 [LeBlanc et al., 1983], implying these plasmids have a common evolutionary origin.

Whereas *S. faecalis* appears to excrete numerous sex pheromones, it is also likely that many isolates carry more than one pheromone-responding plasmid. Indeed, strain DS5 has been shown to carry three such plasmids (pAMγ1, pAMγ2, and pAMγ3), encoding responses to three distinct pheromones (Clewell et al., 1982b). DS5 is also a good example of the ability of pheromone-related plasmids to influence the transfer of co-resident R-plasmids (i.e., pAMα1 and pAMβ1).

Because plasmid-free strains of *S. faecalis* excrete multiple peptides that can serve as mating signals for a variety of plasmids that may have never been encountered by the recipient, it is conceivable that production of these substances may have preceded the evolution of the related conjugative systems. The latter simply may have evolved in such a way as to take advantage of the available extracellular compounds as mating signals. If this is true, the question arises as to the nature of the original, perhaps continuing, function of these peptides.

LeRoith, Roth and co-workers have reported that certain microorganisms (in some cases including bacteria) excrete hormones such as insulin, corticotropin, and β-endorphin-like substances [LeRoith et al., 1980, 1981; Roth et al., 1982]; and Acevedo et al. [1981] have observed a choriogonadotropin-like substance in certain streptococci and staphylococci isolated from cancer patients. The significance of these observations is unclear at the moment;

however, it raises the intriguing question of whether the peptide pheromones produced by *S. faecalis* could have activities related to substances occurring in higher systems. In this regard it is interesting that the yeast α-factor recently has been shown to have an activity related to gonadotropin releasing factor [Loumaye et al., 1982].

ACKNOWLEDGMENTS

I would like to thank B. White, Y. Ike, A. Suzuki, F. An, M. Smith, C. Gawron-Burke, R. Wirth, E. Ehrenfeld, J. Shaw, G. Fitzgerald, M. Yamamoto, Y. Yagi, R. Craig, Y. Sakagami, and R. Kessler for helpful discussions. The work discussed here was supported by National Institutes of Health grants DE02731 and AI10318.

REFERENCES

Aaronson S (1981): "Chemical Communication at the Microbial Level." Boca Raton, Florida: CRC Press, Vol. 1.

Acevedo H, Koide S, Slifkin M, Maruo T, Campbell-Acevedo E (1981): Choriogonadotropin-like antigen in a strain of *Streptococcus faecalis* and a strain of *Staphylococcus simulans*: detection, identification, and characterization. Infect and Imm 31:487–494.

Argoudelis J, Coates J, Mizak S (1977): Microbial transformation of antibiotics clindamycin ribonucleotides. J Antiobiot 30:474–487.

Bender R, Janssen K, Resnick A, Blumenberg M, Foor F, Magasanik B (1977): Biochemical parameters of glutamine synthetase from *Klebsiella aerogenes*. J Bacteriol 129:1001–1009.

Bezdek M, Soska J (1972): Sex determined chemotaxis in *Salmonella typhimurium* LT2. Folia Microbiol 17:366–369.

Biro S, Bekesi I, Vitalis S, Szabo G (1980): A substance affecting differentiation in *Streptomyces griseus*. Eur J Biochem 103:359–363.

Clewell D (1981): Plasmids, drug resistance, and gene transfer in the genus *Streptococcus*. Microbiol Rev 45:409–436.

Clewell D, An F, White B, Gawron-Burke C (1984a): Sex pheromones and plasmid transfer in *Streptococcus faecalis*. A pheromone, cAM373, which is also excreted by *Staphylococcus aureus*. In Helinski D, Cohen S, Clewell D, Jackson D, Hollaender A (eds): "Plasmids in Bacteria." New York: Plenum Press, pp 489–503.

Clewell D, Brown B (1980): Sex pheromone cAD1 in *Streptococcus faecalis*: indiction of a function related to plasmid transfer. J Bacteriol 143:1063–1065.

Clewell D, Fitzgerald G, Dempsey L, Pearce L, An F, White B, Yagi Y, Gawron-Burke C (1985): Streptococcal conjugation: plasmids, sex pheromones, and conjugative transposons. In Mergenhagen S, Rosan B (eds): "Molecular Basis of Oral Microbial Adhesion." Washington, D.C.: American Society for Microbiology, pp 194–203.

Clewell D, Tomich P, Gawron-Burke C, Franke A, Yagi Y, An F (1982): Mapping of *Streptococcus faecalis* plasmids pAD1 and pAD2 and studies relating to transposition of Tn*917*. J Bacteriol 152:1220–1230.

Clewell D, White B, Ike Y, An F (1984b): Sex pheromones and plasmid transfer in *Streptococcus*

faecalis. In Losick R, Shapiro L (eds): "Microbial Development." Cold Spring Harbor, New York: Cold Spring Harbor Press, pp 122–147.
Clewell D, Yagi Y, Dunny G, Schultz S (1974): Characterization of three plasmid DNA molecules in a strain of *Streptococcus faecalis*. Identification of a plasmid determining erythromycin resistance. J Bacteriol 117:283–289.
Clewell D, Yagi Y, Ike Y, Craig R, Brown B, An F (1982b): Sex pheromones in *Streptococcus faecalis*: multiple pheromone systems in strain DS5, similarities of pAD1 and pAMγ1, and mutants of pAD1 altered in conjugative properties. In Schlessinger D (ed): "Microbiology—1982." Washington, D.C.: American Society for Microbiology, pp 97–100.
Collins J, Broda P (1975): Motility, diffusion and cell concentrations affect pair formation in *Escherichia coli*. Nature 258:722–723.
Davies J, Smith D (1978): Plasmid-determined resistance to antimicrobiol angents. Annu Rev Microbiol 32:469–518.
Dunny G, Brown B, Clewell D (1978): Induced cell aggregation and mating in *Streptococcus faecalis*. Evidence for a bacterial sex pheromone. Proc Nat Acad Sci USA 75:3479–3483.
Dunny G, Clewell D (1975): Transmissible toxin (hemolysin) plasmid in *Streptococcus faecalis* and its mobilization of a noninfectious drug resistance plasmid. J Bacteriol 124:784–790.
Dunny G, Craig R, Carron R, Clewell D (1979): Plasmid transfer in *Streptococcus faecalis*. Production of multiple sex pheromones by recipients. Plasmid 2:454–465.
Dunny G, Funk C, Adsit J (1981): Direct stimulation of the transfer of antibiotic resistance by sex pheromones in *Streptococcus faecalis*. Plasmid 6:270–278.
Engel H, Soedirman N, Rost J, van Leeuwen W, van Embden J (1980): Transferability of macrolide, lincosamide, and streptogramin resistances between group A, B, and D streptococci, *Streptococcus pneumoniae,* and *Staphylococcus aureus*. J Bacteriol 142:404–413.
Ensign J (1978): Formation, properties, and germination of actinomycete spores. Annu Rev Microbiol 32:185–219.
Franke A, Clewell D (1980): Evidence for conjugal transfer of a *Streptococcus faecalis* transposon (Tn*916*) from a chromosomal site in the absence of plasmid DNA. Cold Spring Harbor Symp. Quant. Biol. 45:77–80.
Franke A, Clewell D (1981): Evidence for a chromososme-borne resistance transposon in *Streptococcus faecalis* capable of "conjugal" transfer in the absence of a conjugative plasmid. J Bacteriol 145:494–502.
Franke A, Dunny G, Brown B, An F, Oliver D, Damle S, Clewell D (1978): Gene transfer in *Streptococcus faecalis*. Evidence for the mobilization of chromosomal determinants by transmissible plasmids. In Schlessigner D (ed): "Microbiology—1978. Washington, D.C.: American Society for Microbiology, pp 45–47.
Gawron-Burke C, Clewell D (1982): A transposon in *Streptococcus faecalis* with fertility properties. Nature 300:281–284.
Gawron-Burke C, Clewell D (1984): Regeneration of insertionally inactivated streptococcal DNA fragments after excision of transposon Tn*916* in *Escherichia coli*: strategy for targeting and cloning of genes from gram-positive bacteria. J Bacteriol 159:214–221.
Hirosawa T, Wolk C (1979): Isolation and characterization of a substance which stimulates the formation of akinetes in the cyanobacterium *Cylindrospermum licheniforme* Kutz. J Gen Microbiol 114:433–441.
Ike Y, Craig R, White B, Yagi Y, Clewell D (1983): Modification of *Streptococcus faecalis* sex pheromones after acquisition of plasmid DNA. Proc Natl Acad Sci USA 80:5369–5373.
Ike Y, Clewell D (1984): Genetic analysis of the pAD1 pheromone response in *Streptococcus*

faecalis using Tn*917* as an insertional mutagen. J Bacteriol 158:777–783.
Kaiser D, Manoil C, Dworkin M (1979): Myxobacteria: cell interactions, genetics, and development. Annu Rev Microbiol 33:595–639.
Karlson P, Luscher M (1959): "Pheromones": a new term for a class of biologically active substances. Nature 183:55–56.
Kessler R, Yagi Y (1983): Identification and partial characterization of a pheromone-induced adhesive surface antigen of *Streptococcus faecalis*. J Bacteriol 155:714–721.
Kochert G (1978): Sexual pheromones in algae and fungi. Annu Rev Plant Physiol 29:461–486.
Kurjan J, Herskowitz I (1982): Structure of a yeast pheromone gene (MF): a putative α-factor precursor contains four tandem copies of mature α-factor. Cell 30:933–943.
Lacks S (1977): Binding and entry of DNA in bacterial transformation. In Reissig JL (ed): "Microbial Interactions, series B. Receptors and recognition." London: Chapman and Hall, Vol. 3, pp 177–232.
Landman O, Bodkin D, Finn C Jr., Pepin R (1981): Conjugal transfer of plasmid pAMβ1 from *Streptococcus anginosis* to *Bacillus subtilis* and plasmid-mobilized transfer of chromosomal markers between *B. subtilis* strains. In Polsinelli M (ed): "Transformation—1980." Oxford: Cotswold Press.
LeBlanc D, Hawley R, Lee L, St. Martin E (1978): "Conjugal" transfer of plasmid DNA among oral streptococci. Proc Nat Acad Sci USA 75:3484–3487.
LeBlanc D, Lee L, Clewell D, Behnke D (1983): Broad geographical distribution of a cytotoxin gene mediating beta-hemolysis and bacteriocin activity among *Streptococcus faecalis* strains. Infect Immun 40:1015–1022.
LeRoith D, Shiloach J, Roth J, Lesniak M (1980): Evolutionary origins of vertebrate hormones: substances similar to mammalian insulins are native to unicellular eukaryotes. Proc Natl Acad Sci USA 77:6184–6188.
LeRoith D, Shiloach J, Roth J, Lesniak M (1981): Insulin or a closely related molecule is native to *Escherichia coli*. J Biol Chem 256:6533–6536.
Loumaye E, Thorner J, Catt K (1982): Yeast mating pheromone activates mammalian gonadotrophs: evolutionary conservation of a reproductive hormone? Science 218:1323–1325.
Manney T, Duntze W, Betz R (1981): The isolation, characterization and physiological effects of the *Saccharomyces cerevisiae* sex pheromones. In O'Day D, Horgen P (eds): "Sexual Interactions in Eukaryotic Microbes." pp 21–51.
Nanney D (1977): Cell-cell interactions in ciliates: evolutionary and genetic constraints. In Reissig JL (ed): "Microbial Interactions, series B. Receptors and recognition." London: Chapman and Hall, Vol. 3.
Nester E, Kosuge T (1981): Plasmids specifying plant hyperplasias. Ann Rev Microbiol 35:531–565.
Roth J, LeRoith D, Shiloach J, Rosenzweig J, Lesniak M, Havrankova J (1982): The evolutionary origins of hormones, neurotransmitters, and other extracellular chemical messengers. Implications for mammalian biology. N Engl J Med 306:523–527.
Schaberg D, Clewell D, Glatzer L (1982): Conjugative transfer of R-Plasmids from *Streptococcus faecalis* to *Staphylococcus aureus*. Antimicrob Agents Chemother 22:204–207.
Shapiro B, Kingdon H, Stadtman E (1967): Regulation of glutamine synthetase, VII. Adenylyl glutamine synthetase: a new form of the enzyme with altered regulatory and kinetic properties. Proc Natl Acad Sci USA 58:642–649.
Sprague Jr., G, Blair L, Thorner J (1983): Cell interactions and regulation of cell type in the yeast *Saccharomyces cerovisiae*. Annu Rev Microbiol 37:623–660.
Stephens K, Hegeman G, White D (1982): Pheromone produced by the myxobacterium *Stig-*

matella aurantiaca. J Bacteriol 149:739–747.

Suzuki A, Mori M, Sakagami Y, Isogai A, Fujino M, Kitada C, Craig R, Clewell D (1984): Isolation and structure of bacterial sex pheromone, cPD1. Science 226:849–850.

Tomich P, An F, Clewell D (1978): A transposon (Tn*917*) in *Streptococcus faecalis* which exhibits enhanced transposition during induction of drug resistance. Cold Spring Harbor Symp Quant Biol 43:1217–1221.

Tomich P, An F, Clewell D (1980): Properties of erythromycin-inducible transposon Tn*917* in *Streptococcus faecalis*. J Bacteriol 141:1366–1374.

Tomich P, An F, Damle S, Clewell D (1979): Plasmid related transmissibility and multiple drug resistance in *Streptococcus faecalis* subsp. *zymogenes* strain DS16. Antimicrob Agents Chemother 15:828–830.

Willetts N, Skurray R (1980): The conjugation system of F-like plasmids. Annu Rev Genet 14:41–76.

Yagi Y, Kessler R, Shaw J, Lopatin D, An F, Clewell D (1983): Plasmid content of *Streptococcus faecalis* strain 39-5 and identification of a pheromone (cPD1)-induced surface antigen. J Gen Microbiol 129:1207–1215.

Yanagishima N (1978): Sexual cell agglutination in *Saccharomyces cerevisiae*: sexual cell recognition and its regulation. In Controlling Factors in Plant Development. Botan Mag Tokyo Special Issue. 1:61–81.

NOTE ADDED IN PROOF

Mori et al. (FEBS Letters, 178:97–100) have recently determined the structure of cAD1. Its sequence is H-Leu-Phe-Ser-Leu-Val-Leu-Ala-Gly-OH; its molecular weight is 818.

The Origin and Evolution of Sex, pages 29–45

DNA Repair and Complementation: The Major Factors in the Origin and Maintenance of Sex

Harris Bernstein, Henry Byerly, Frederick Hopf, and Richard E. Michod

Department of Microbiology and Immunology (H.B.), Department of Philosophy (H.B.), Optical Sciences Center (F.H.), and Department of Ecology and Evolutionary Biology (R.E.M.), University of Arizona, Tucson, Arizona 85721

INTRODUCTION

Overview

Sexual reproduction (or "sex") has two fundamental aspects: 1) recombination, in the sense of breakage and exchange of DNA (or RNA) between homologous chromosomes, and 2) outcrossing in the sense that the homologous genomes come from two different individuals. Section II presents an argument that repair of genome damage is necessary for survival and is the immediate selective force that gave rise to recombination, the first fundamental aspect of sex. Section III presents an argument that the need to cope with deleterious mutations through complementation explains the maintenance of outcrossing, the second fundamental aspect of sex. In section IV alternative theories on the evolution of sex are considered briefly in order to evaluate them in relation to the theory presented here. Finally, in section V the principal consequences of sex that follow from our point of view are considered.

Before developing these ideas, the distinction between genome damage and mutation and the mechanisms for dealing with these two basically different alterations in DNA are reviewed, as well as evidence that genome damage is an important and general problem in the struggle for survival.

The order of authors is strictly alphabetical and is not intended to imply seniority.

Information, Mutation, and Damage

The transmission of information from parent to offspring via genome replication is basic to life. Disruptions due to noise, i.e., random influences which change the information in unpredictable ways, are always a critical problem in the transfer of information which is encoded in a sequence of characters. Such disruptions can be of two major types: change of an allowed character to a disallowed character (damage) or change of one allowed character for another allowed character (mutation). There is an essential difference in the way a system can deal with these two types of changes. In the first case, the presence of a disallowed character can be recognized and the lost information can be recovered if a redundant copy of the information is present. However, in the case of a mutation, a comparison with a redundant copy can indicate that there is a mistake, but it is not possible to recognize which form is incorrect. Thus, a mutation cannot be corrected even if a redundant copy is present.

In the genetic material, damages are physical alterations (disallowed characters) in the structure of DNA (or RNA). Examples in DNA are depurinations, thymine dimers, crosslinks, various oxidatively altered bases, and single and double strand breaks. Damages usually interfere with replication [Rupp and Howard-Flanders, 1968; Cleaver, 1969] by blocking movement of the DNA polymerase, or transcription [Zieve, 1973; Hackett and Sauerbier, 1975; Nocentini, 1976; Leffler et al., 1977] by blocking RNA polymerase. Damages are neither replicated nor inherited. By contrast, mutations are changes in the sequence of base pairs (allowed characters) resulting from substitution, addition, deletion, or rearrangement of the standard base pairs. Mutations do not alter the physical regularity of the molecule. They are replicated and thus can be inherited.

One essential consequence of damages is that they can be recognized by enzymes and then be repaired, provided a redundant copy of the template is available to replace the lost information. Mispaired bases may be prevented from turning into mutations by enzymes which recognize them and correct the mispair. However, once the mispairing is resolved by replication and the mutation has actually occurred, there is no mechanism for determining which daughter strand is correct and carries the original information. Mutations can be removed only from a population of individuals on the basis of their effect on fitness, through natural selection. In general, damages can be repaired but mutations cannot.

By contrast, deleterious effects of mutation can usually be masked by having another correct copy of the homologous genome present in the same cytoplasm. When this masking effect is mutual between two genomes with different mutations, the phenomenon is called complementation.

Damage in Nature

In the current biota, a variety of gene functions whose only known role is to repair DNA have been described. Such functions occur in viruses, bacteria, fungi, and mammals [for numerous examples, see symposium volume edited by Hanawalt et al., 1978]. The prevalence of such gene functions implies that the damages they repair are also prevalent in nature and would reduce fitness if unrepaired. Damage which affects only one strand of DNA (single-strand damage) can often be repaired by excision repair. In such cases, the undamaged strand serves as a template for replacing the excised DNA. In contrast, when a single strand damage occurs in a genome composed of single stranded RNA (ssRNA) excision repair is not an option since there is no intact strand to serve as a template. Likewise, in double-stranded DNA (dsDNA) damages affecting both strands at nearby positions (double-strand damages) cannot be repaired by excision repair, again because there is no intact template. For single-strand damages in ssRNA and double strand damages in dsDNA, an intact template for repair is only available from another homologous genome. The process by which one homologue recovers information from another homologue is a process of exchange. These types of lesions, reparable only through exchange, are critical for our model of the origin and maintenance of sex.

To assess the importance of recombinational repair in present-day organisms, we need to ask how prevalent are double-strand damages in current organisms. The highly reactive superoxide radical (O_2^-) and hydrogen peroxide [H_2O_2] are apparently ubiquitous byproducts of cellular respiration [Fridovich, 1978]. These oxidizing agents are a particular hazard for DNA and have been linked to aging [Harmon, 1981] and cancer [Ames, 1983]. The relative rates of formation of the various types of damage produced by H_2O_2 have been estimated [Massie et al., 1972]. About 60 modified bases are produced for every double strand damage, and the two types of double-strand damages, crosslinks, and double-strand breaks are produced in roughly equal frequency. The modified bases produced by oxidative damages are removed from DNA by excision repair enzymes, three of which are known, that deal with specific types of altered bases. These are three repair glycosylases that remove hydroxymethyluracil [Hollstein et al., 1984], thymine glycol [Breimer, 1983; Breimer and Lindahl, 1984], and formamidopyrimidine [Margison and Pegg, 1981], respectively. Cathcart et al. [1984] have presented evidence that in humans an average of about 320 thymine glycols are removed by repair per cell, per day, and comparable levels of hydroxymethyluracil also may be removed. We therefore assume that about 600 oxidatively modified bases are removed per cell per day from DNA. Because one double strand lesion is formed in DNA per 60 altered bases, about 10 double strand damages also should be formed per cell per day; because at least 90% of oxidative reactions occur in mitochondria, the DNA of these organelles may incur most of the total damage. Yet if even a small fraction (e.g., 1%) of the oxidative DNA damages are in nuclear DNA, single-strand and double-strand damages would occur at frequencies of approximately 6 and 0.1 respectively per nuclear genome per day. Thus, from this one source alone, double-strand damages would be a substantial hazard to each cell in man. Presumably, similar problems confronted organisms throughout evolution.

ORIGIN OF SEX

Primitive RNA Replicators and Evolution

The capacities to encode information and to replicate may be regarded as the two most fundamental properties of life. We hypothesize, in accord with the extensive experimental and theoretical work of Eigen and collaborators [Eigen, 1971; Eigen and Schuster, 1979; Eigen et al., 1981] that life arose as a self-replicating heteropolymer similar to RNA. Variation arose simply as a result of errors of replication. Ribonucleotides, present in the early aqueous environment, were probably the first resources used for self-replication, and natural selection may have arisen from competition among RNA replicators for nucleotides.

The most primitive adaptations were likely folded configurations of RNA molecules [Kuhn, 1972; Bernstein et al., 1983]. Such conformations were the first phenotypes, and these were determined by the specific base sequences of RNA, the first genotypes [Michod, 1983]. Three classes of phenotypic adaptations likely arose in the earliest replicators. These are conformations of RNA that promoted 1) increased rate and accuracy of replication, 2) protection of the replicator against damage, and 3) increased ability to incorporate nucleotides from the environment. Fitness (per capita rate of increase) is determined by these three adaptive capacities and the availability of resources in the environment [Bernstein et al., 1983]. The three adaptations are intrinsic to the organism in the sense that they are encoded in the RNA, in contrast to the resources which are a property of the external environment.

As the RNA evolved the capacity to specify enzymes, the three types of adaptation became associated with increasingly complex structures. Here we will concentrate on the evolution of the second adaptation, protection of the replicator from damage.

The Vulnerability of the Earliest Protocells to Genome Damage

We do not know the principal source of damage when life first evolved. Sagan [1973] has argued that the flux of solar UV light penetrating the primitive reducing atmosphere of the earth posed a major problem for the early evolution of life. Damage in an RNA replictor would likely block its replication or interfere with expression of its encoded information. In a population of primitive, independently replicating RNA strands without repair, damaged ones would simply die. Although the early RNA replicators were probably initially independent, it has been proposed by Eigen [1971], Eigen and Schuster [1979], and Eigen et al. [1981] that they evolved mutual dependencies based on the benefits of joint uses of encoded products (primitive enzymes). These authors suggested a likely evolutionary sequence from free RNA replicators to "hypercycles" (sets of mutually dependent replicators) and then to encapsulated hypercycles. For additional discussion of the transition to encapsulation, see Bernstein et al. [1984b].

With the evolution of encapsulation within simple membranes, such as a lipid bilayer, the problem of damage to the RNA replicators comprising a hypercycle becomes more acute. The advantage of encapsulation is that it allows the set of interacting RNAs to localize their encoded products (e.g., primitive enzymes) for more efficient use. If each RNA strand within a protocell produces a distinct product that promotes the survival and duplication of the protocell, then each RNA molecule is equivalent to a gene in a haploid organism. Woese [1983] has proposed that the universal cellular ancestor contained an RNA genome which was "physically disaggregated, comprising a collection of gene size pieces." Simple haploid organisms such as this would be very vulnerable to damage, since a single lesion in any essential gene (ssRNA replicator) could be fatal to the protocell. In addition, when the genes replicate, at least one copy of each gene must be transmitted to each daughter protocell. Failure of this segregation leads to inviability. Encapsulation thus allows efficient product utilization, but at the cost of locking in genetic damages when they occur.

The Origin of Sex in Simple RNA Protocells

Vulnerability to loss of genetic information can be reduced by maintaining more than one copy of each gene in each protocell. This would greatly reduce lethality due to genome damage or failure of segregation, as a damaged or lost gene could be replaced by an extra replication of its homologue. However, to maintain two or more copies of each gene is costly in terms of resources expended and lengthened generation time. These costs of redundancy reduce fitness (per capita rate of increase). Thus, a basic problem facing early protocells was to cope with damaged or lost information while at the same time minimizing the costs of redundancy.

Bernstein et al. [1984a, 1984b] have shown that the selected strategy under a wide range of circumstances is for each protocell to be haploid, but to periodically fuse with another haploid protocell to form a transient diploid.

Periodic fusions between two lethally damaged haploid protocells allows them to mutually reactivate each other. To our knowledge, Dougherty [1955] was the first to propose that reactivation of damaged genomes was a primary factor in the origin of sex. Reactivation results simply from undamaged genes undergoing an extra replication which allows recovery of lost information. We consider this recovery to be the simplest form of repair. The cycle of haploid reproduction, fusion, and subsequent splitting to haploidy, we think, is the sexual cycle in its most primitive form. Haploid progeny would often have the genes of their two parents in new combinations. Thus, the reassortment of genes when passing from parents to offspring may have arisen as a byproduct of the recovery from genetic damage through sexual reproduction.

Without this sexual cycle, damaged protocells would simply die and be replaced in the population by replication of undamaged ones. Replication without repair may allow more rapid proliferation when damages are infrequent and resources are abundant, allowing maximum rate of growth. We refer to this as the "replication option." However, the sexual strategy is superior when damages are frequent and/or resources are limiting.

This proposal for the earliest stages of sexual reproduction is not merely hypothetical but corresponds to the known sexual behavior of the segmented RNA viruses, which are among the simplest of extant organisms. Influenza virus and reovirus are two well-studied examples. The genome of influenza virus is composed of ssRNA divided into eight physically separate segments [Lamb and Choppin, 1983]. Six of these segments encode only one polypeptide each and are thus equivalent to individual genes. The other two segments encode two or three polypeptides each. When two influenza viruses infect a single cell, the ssRNA segments of the two viruses are released into the cell. There they replicate, and progeny viruses are formed. These progeny contain a full genome composed of genes from both parents. Genetic recovery here is based on the replication and reassortment of undamaged ssRNA segments. Reovirus also has a genome composed of RNA segments, but in this case the segments are double stranded. It also forms progeny by replication and reassortment of RNA segments. Both influenza virus and reovirus undergo a form of recovery known as multiplicity reactivation [Barry, 1961; McClain and Spendlove, 1966]. Many viruses are capable of multiplicity reactivation [for reviews, see Bernstein, 1981; Bernstein, 1983]. Conventionally multiplicity reactivation is demonstrated by treating a population of viruses with an agent such as UV light that damages their genomic RNA or DNA. When these damaged viruses are allowed to infect susceptible cells, there is substantially greater virus survival if cells are infected by two or more damaged viruses then if they are infected by only one virus. Bernstein [1981] and Bernstein and Wallace [1983] summarized a considerable body of evidence, indicating that the enhanced survival results from exchange of genetic material between damaged viruses leading to progeny that are free of lethal damages. Multiplicity reactivation in segmented RNA viruses likely results simply from replication of undamaged RNA segments of the infecting virus, and the reassortment of these segments to form complete undamaged progeny viruses. Thus, the plausibility of our proposal for the origin of sexual reproduction in early protocells is supported by the existence of a similar process, i.e., multiplicity reactivation, in some of the simplist organisms known, the segmented RNA viruses.

Sex in Simple DNA-Containing Organisms: DNA Viruses

The genes of DNA viruses are generally linked end to end to form one continuous DNA molecule. This linkage allows reliable assortment of a complete set of genes to each progeny virus, which is an improvement over the unreliable assortment in the simple segmented RNA viruses. In the well-studied DNA viruses phage T4 and phage λ, both multiplicity reactivation and the associated recombination of genes depend on enzymes which catalyze breakage and physical exchange of DNA between chromosomes [for reviews see Bernstein, 1981; Bernstein and Wallace, 1983]. In phage T4, multiplicity reactivation is an efficient process for overcoming a wide variety of different types of DNA damages (e.g., those caused by UV, HNO_2, mitomycin C, psoralen-plus-light, x-rays and ^{32}P-decay).

In primitive RNA protocells, as in segmented RNA viruses, repair occurs simply by nonenzymatic reassortment of undamaged segments. This, we think, evolved into recombinational repair involving enzyme mediated breakage and exchange between DNA molecules from separate individuals, that occurs in most DNA viruses. In each step of the evolution of recombinational repair that we have envisioned, the adaptive advantage is the same, namely, the repair or replacement of damaged DNA.

Sex in Bacteria

Recombinational repair has been studied extensively in *E. coli*. However, most of this work was done on exchange between sister chromosomes before cell division. In *E. coli* there is efficient recombinational repair of gaps opposite pyrimidine dimers [Rupp and Howard-Flanders, 1968; Rupp et al., 1971; Hanawalt et al., 1979], x-ray induced double-strand breaks [Krasin and Hutchinson, 1977], and psoralen crosslinks [Cole et al., 1976, 1978]. This suggests that recombinational repair in bacteria is an effective mechanism for overcoming double strand DNA lesions. Each of these recombinational repair processes in *E. coli* depends on the *recA* protein whose mechanism of action has been extensively studied. The *recA* protein promotes pairing of homologous DNA molecules and strand exchange [Radding, 1981]. Sexual reproduction in bacteria can take several forms, including conjugation which probably is infrequent in nature [see Clark, this volume] and transformation which occurs among a wide range of bacteria. Unfortunately, little if any work has been done on the role of transformation in recombinational repair.

As ordinarily studied in the laboratory, bacteria are provided with optimal conditions which promote rapid exponential growth. Under these conditions sex appears to be dispensable. Lethally damaged bacteria are replaced in the population by replication of undamaged ones, which is basically the replication option, discussed above. Sexual processes, particularly transformation, occur in many different bacterial species. However, it is still not clear if these processes afford an advantage through recombinational repair. In the case of *E. coli*, sexual reproduction does not appear to be frequent in nature [Selander and Levin, 1980]. Because sex has high costs, it is not surprising that *E. coli* and perhaps other bacteria use sexual reproduction rarely, if at all, and thus deal with DNA damage in other ways. The identification of bacteria that may have abandoned sex does not warrant the conclusion that sex has arisen independently in procaryotes and eucaryotes.

COMPLEMENTATION AS A MAJOR FACTOR IN THE EVOLUTION OF OUTCROSSING SEX

The Basis of Outcrossing in Simple vs Complex Organisms

In simple protocells, we think, the costs of genome redundancy were decisive in the origin and maintenance of outcrossing [Bernstein et al., 1984b]. In more advanced organisms, resources involved in making DNA are a small fraction of the overall energy budget. Nevertheless, data suggest that these costs still play some, albeit a reduced, role [Bernstein et al., 1984b]. We argue below that complementation is the dominant factor in the maintenance of outcrossing in these organisms.

Sex in Diploid Organisms: The Origin and Maintenance of Diploidy

As primitive haploid organisms evolved genomes with increasing numbers of genes, they became more vulnerable to lethal mutation, whose frequency per genome is proportional to the number of essential genes. Probably the initial adaptive response of these haploid organisms was to improve the fidelity of replication. Bernstein et al. [1981] have proposed that at some point, when the genome had reached several thousand genes, a new strategy for dealing with this problem evolved. Although the diploid stage of the sexual cycle was probably transient in early haploid organisms, we think it eventually became cost effective to take advantage of the masking effect of diploidy which protects against expression of deleterious recessive mutations through complementation. Thus, in some lines of descent, the diploid stage became the dominant stage of the sexual cycle. This change may have come about as the direct costs of maintaining an extra genome became small relative to the costs of maintaining the total cellular metabolism. Once this strategy was adopted, deleterious mutations would accumulate in the diploid line, as they no longer could be weeded efficiently out of the population by natural selection. This accumulation would continue until a new balance is reached between selection against homozygous recessive alleles and new mutations [Muller, 1932; Crow and Kimura, 1970]. At this new balance point, many more deleterious recessive mutations are present in each individual. Should a diploid organism at this stage revert to predominant haploidy, all these deleterious recessive alleles are expressed immediately. Because of the accumulation of deleterious recessives, the advantage of diploidy is only transient. However, the transition back to haploidy is essentially blocked. We now argue that similar factors maintain outcrossing.

The Advantage of Outcrossing Sex

Closed and open systems. In diploid organisms, depending on the breeding system, germ line recombination can be achieved either by exchange between homologous chromosomes of different parental origin or between homologues from the same parent. We call a system *open* if the sources of the two chromosomes are different parents and *closed* if the source is a single parent. Examples of closed systems are self-fertilization and automixis (uniparental production of eggs through ordinary meiosis followed by fusion to restore diploidy).

The predominant type of system in nature outcrossing is an open one. In fact, outcrossing is commonly considered intrinsic to the concept of sex. If, as we have argued, the only selective advantage of sexual reproduction is the recombinational repair of genetic damage, the optimal strategy is some kind of closed system. Such systems avoid some of the major costs of sex: the cost of males [Maynard Smith, 1978], high recombinational load [Shields, 1982, especially Chapter 5], the lower genetic relatedness between parent and offspring [Williams, 1980; Uyenoyama, 1984] and the costs of finding a mate [Bernstein et al., 1985a, and references therein]. The question arises then, why is the open system of outcrossing the most common.

Maintenance of outcrossing. The key factor in maintaining outcrossing sex, we think, is the need to cope with deleterious recessive mutations. In a closed system, if there is recombination between nonsister chromatids, homozygosity will increase. The expression of deleterious recessive alleles clearly decreases fitness. Thus, we argue, that maintenance of outbreeding, like the maintenance of diploidy, depends on transient advantages of complementation.

At equilibrium, the rate at which lethal or deleterious mutations are eliminated by selection equals the rate at which they are produced [Haldane,

1937]. We have found [Hopf, et al., unpublished, in preparation] that at equilibrium the effect on survival of accumulated deleterious recessive alleles is independent of the breeding system, whether it is haploid or diploid or whether it is selfing, automictic, or outcrossing. If u is the rate of occurrence of deleterious mutations per haploid complement per generation, the equilibrium survival is approximately e^{-u} in all cases.

In Drosophila the rate of spontaneous deleterious mutation u is about 0.3 [Mukai et al., 1972], which gives an equilibrium survivorship of $e^{-.3} = .7$. We have argued [Bernstein et al., 1985b] that these values for Drosophila should be roughly typical for diploid organisms in general.

Although the effect of mutational load on survival at equilibrium gives no competitive advantage to any breeding system, when a new breeding system arises in an equilibrium population, there is a transient selective advantage or disadvantage corresponding to the masking ability of the new system. Consider the case of a new gene for outcrossing arising in a population of selfers in which each selfing individual has a small but finite number of deleterious recessive alleles. Each progeny of a pair of outcrossers will have two genomes, one from each parent. Like the selfer these two genomes will have a small number of deleterious recessive alleles. However, for the outcrosser these deleterious alleles will be statistically unrelated to each other so that there will be efficient complementation. Thus, the fitness of the outcrosser is not reduced at all by mutational load initially. However, the fitness of the outcrosser will be reduced by the costs of outcrossing listed above (e.g., cost of males, etc.). Individuals bearing a gene for outcrossing will expand in competition with the selfers if the costs of outcrossing are less than the costs associated with the mutational load in the population of selfers.

Descendents in the outcrossing line, in the above example, will accumulate increasing numbers of recessive mutations over many generations because of the masking effect of outcrossing. This will continue until equilibrium is reached. At this point, fitness is reduced by both the mutational load (which is equal for all breeding systems at equilibrium) and the additional costs of outcrossing. Thus, the *net effect* of the transition to outcrossing is ultimately a *reduction* in individual fitness. Nevertheless, a transition back to inbreeding, just as the transition back to haploidy, is strongly inhibited because unmasking the many accumulated deleterious recessive mutations would cause a precipitous decline in fitness. Our general argument, then, is that a closed breeding population would likely be at a competitive disadvantage to new mutant outcrossers, whereas an outcrossing population is not similarly vulnerable to closed strategies. Outcrossing will, however, give way to a closed system under conditions where the costs of outcrossing are very large (e.g., when a population is at very low density and finding a mate is costly). We have argued that parthenogens tend to be found in nature in precisely those situations where the costs of outcrossing, particularly the cost of mating, are large [Bernstein et al., 1985a].

Apomixis and vegetative reproduction. In apomixis, a form of parthenogenesis, meiosis is suppressed and diploid eggs are formed by a single mitotic division. The maternal genome passes intact from mother to daughter. This provides maximal masking of deleterious alleles. Vegetative reproduction, common in plants, also provides this advantage. However, because apomixis and vegetative reproduction bypass meiosis, they probably largely abandon recombinational repair of double strand damages. In these cases, the effects of genetic damage may be overcome by somatic selection for undamaged cells, the replicative option. This option is effective when damages are infrequent and/or growth is rapid, but not otherwise.

Endomitosis. A case of special interest is endomitosis, another parthenogenetic system, in which meiosis is preceded by two premeiotic replications; then, at the four chromatid stage, pairing is between chromatids derived from only one parental chromosome [Cole, 1984]. This mode of reproduction is used, for example, by whiptail lizards of the southwestern deserts of the U.S. Recombination, in this case, does not generate new homozygotes, and the maternal genome is maintained intact. Endomitosis might seem to reap the benefits of meiotic repair while avoiding the expression of deleterious recessives. However, recombinational repair in endomitosis cannot overcome double-strand damages occurring before the first premeiotic replication, as in that case all four potential sister chromatids would be damaged at the same site, with no intact template. This is not a problem in conventional meiosis, because recombination occurs between non-sister chromosomes. Hence, if damages are frequent, endomitosis is an unsatisfactory strategy.

Comparison of alternative reproductive systems. We have argued in this section that every transition from outcrossing to an alternative breeding system should reduce fecundity. The lowered fitness, we think, is transient for selfing and automixis, but permanent for apomixis and endomitosis because of the reduced ability to repair damage. In support of this argument, there are numerous observations indicating that parthenogens generally have lower reproductive rates than their sexual relatives, often less than 50% [see Lynch, 1984, for review].

ALTERNATIVE THEORIES OF SEX

Sex as a Parasitic Process

Hickey [1982] proposed that "sex itself, and especially outbreeding, is a product of parasitic genes." Dawkins [1982] has suggested that there are replicating "engineers" of meiosis which "achieve their own replication success as a byproduct of forcing meiosis upon the organism." Could sexual reproduction have arisen and been maintained, not as an adaptation but rather as an unavoidable parasitic disease? We think this idea is implausible primarily because it attributes no advantage to sexual reproduction to balance its costs which are usually substantial (see previous section). If parasitic genes imposed these costs on organisms to the organism's detriment, it seems likely that some effective method would have evolved to eliminate these genes. For instance, bacteria can deal with foreign parasitic DNA through restriction enzymes that recognize and degrade this DNA [Kornberg, 1980]. Other mechanisms also exist in microorganisms for inhibiting gene expression of competing DNA. It seems likely that higher organisms would have evolved similar methods of avoiding the high costs of sex if this process was maintained solely for the benefit of parasitic genes.

The Variation Hypothesis

We categorize under the title "variation hypothesis" all theories on the evolution of sex based primarily on the selective advantage of the variation produced by sexual reproduction. Such theories have been critically reviewed by Williams [1975], Maynard Smith [1978], and Bell [1982]. Our position is that sexually produced variation is largely a byproduct of recombinational repair.

Indeed, it is the tendency of recombination to randomize genetic information that produces variation, and this, we think, generally reduces short-term fitness. In this respect, recombination is similar to mutation, as in both cases most variants produced are deleterious. However, in both cases infrequent beneficial variants may provide raw material for long-term evolution. Mutational and recombinational variation are also similar in that they may be byproducts of other fundamental life processes, mutation being a byproduct of genome replication and recombinational variation being a byproduct of genome repair. Generally, the variation produced by recombinational load contributes to the immediate cost of sex, and we consider that any short-term advantages to variation *per se* are limited to special cases.

Bernstein et al. [1985b] have argued that the existence of premeiotic replication, a general feature of meiosis, implies that recombinational repair is the primary process and variation is a consequence. Studies in *E. coli* show that replication of DNA with single-strand damages leads to gaps in the new strands opposite the damages [Rupp and Howard Flanders, 1968] and that these gaps promote recombinational repair [West et al., 1982]. Such gaps have a molecular structure that is likely independent of the molecular structure of the original damage and can serve as a universal initiator of recombinational repair. Thus, on the repair hypothesis, premeiotic replication can serve a generally useful function.

Because of premeiotic replication, sister chromatid exchanges (SCEs) become possible. SCE does not generate variation. Under a hypothesis that attributes no function to recombination other than variation, there are two alternatives: If SCE is suppressed, a selective advantage of variation is indicated, but premeiotic doubling is a wasteful excess. If SCE is not suppressed, it is counter-productive from the standpoint of producing variation.

NEW PERSPECTIVES ON OTHER ISSUES IN BIOLOGY

Aging of the Soma Versus Immortality of the Germ Line

An essential feature of sexual reproduction is the contrast between the potential immortality of the germ line and the apparently inevitable aging of the individual organism. Under the repair hypothesis, recombinational repair of genome damage in the germ line can explain the potential immortality of this line. In contrast, the somatic line, which probably has little, if any, capacity for recombinational repair, should accumulate damages and be mortal [for further discussion see Bernstein, 1977; Gensler and Bernstein, 1981; Medvedev, 1981]. A specific possibility, in line with our previous discussion, is that accumulation of double-strand damages due to oxygen radicals in long-lived nondividing cells such as neurons may be the primary cause of aging in humans. Such double-strand damages might be irreversible in somatic cells but reparable by meiotic recombination in germ cells.

Sexual Reproduction Leads to Formation of Species

Darwin, viewing evolution as a gradual process, wondered why we find the biota segregated into distinct species rather than being "blended together in an inextricable chaos" [Darwin, 1859]. As discussed below, we think Darwin's expectation is logical if populations obey a geometric, or Malthusian, law of increase as he postulated. Asexual organisms such as parthenogens should follow such a law so that when nutrients are not limiting their growth is exponential. This type of growth is referred to as linear, because it can be plotted as a straight line on semilog coordinates. Sexual reproduction, however, has a different, non-linear, non-Malthusian dynamic. Eigen and Schuster [1979] have shown that these nonlinearities affect the qualitative outcome of natural selection. Non-linearities arise from the necessity for two individuals to come together for reproduction. This fact leads to density dependent fitness (per capita rate of increase) with an intrinsic low fitness at low population density. Costs associated with finding a mate result in a "cost of rarity" with fitness approaching zero as population density approaches zero.

A cost-benefit analysis of a sexual population splitting into species each more finely adapted to a continuous resource but with smaller numbers of individuals, shows that with a cost of rarity splitting does not proceed indefinitely, but leads, rather, to segregation into distinct species [Bernstein et al., 1985a; Hopf and Hopf, 1984]. Also, a large number of very finely adapted, densely packed species are found by this analysis to evolve into a few large distinct species. These results were obtained when various different ecological conditions were superimposed and even when the cost of rarity was small (in this case, segregation into distinct species was slower). Asexual organisms, in contrast, have no cost of rarity due to sex; thus, no limitations on fine scale adaptation and proliferation of clonal types. Such proliferation is observable in nature [Bell, 1982; Parker, 1979].

The concept of species is tied intimately to sexual reproduction because individual organisms, by convention, are allocated to species on the basis of sexual compatibility. We believe our analysis of the consequences of sexual reproduction makes clear why this convention is natural. The concept of niche is also basic to most definitions of species [e.g., Mayr, 1982]. Two views of the existence of niches are possible. First, one could argue that distinct niches exist in the world and species adapt to them, thus becoming themselves distinct. However, there are an infinite number of possible niches in the world as Lewontin [1978] has pointed out. Also, our analysis shows that even on an environmental continuum the cost of rarity will create, distinct species distributions which will then be recognized as distinct niches [Bernstein et al., 1985a]. According to our view, sex is primary and distinct niches are constructed, even out of an environmental continuum, by the sexual dynamic of natural selection. Consequently, we believe that the answer to Darwin's dilemma as to why the biota is not in "an inextricable chaos" lies in the fact that most organisms are sexual and that sex results in an intrinsic cost of rarity which creates distinct species.

The Quantum Character of Evolution of Species

The non-linear character of the rate of population increase introduced by sexual reproduction decouples the adaptedness (appropriateness of design) of the phenotype from its fitness (per capita rate of increase) at low population densities [Michod, 1984; Eigen and Schuster, 1979]. When replication is linear as for parthenogens, a more adapted variant will tend to increase by natural selection at all relative densities. However, sexual reproduction inhibits the expansion of a better adapted species if its members are few. Because new species generally will arise with relatively few members, the cost of rarity tends to stabilize an already adapted species against competition with a new competing species, unless the new species is substantially, rather than moderately, better adapted. Thus, evolutionary advance will tend to be held in check over long periods until a substantially better adapted new species emerges. Then evolution will spurt ahead. This gives species displacement a quantum rather than a gradual character, which depends on the commonness of the established, but less well-adapted, species. A similar argument can be given for the expansion of a rare coadapted genotype within a species [Bernstein et al., 1984a, 1985a]. These conclusions are consistent with the view of Eldridge and Gould [1972] based on the fossil record that evolution is marked by long periods of species stasis punctuated by rapid (in geological time) changes of species in the ecological community.

ACKNOWLEDGMENTS

We thank Carol Bernstein for her critical comments on the manuscript. This work was supported by grants NIH GM 27219 (HB), NSF DEB 81-18248 (REM), and NIH RCDA 1 K04 HD00583 (REM).

REFERENCES

Ames BN (1983): Dietary carcinogens and anticarcinogens. Science 221:1256–1264.

Barry RD (1961): The multiplication of influenza virus II. Multiplicity reactivation of ultraviolet irradiated virus. Virology 14:398–405.

Bell G (1982): "The Masterpiece of Nature: The Evolution and Genetics of Sexuality." Berkeley: University of California Press.

Bernstein C (1981): Deoxyribonucleic acid repair in bacteriophage. Microbiol Rev 45:72–98.

Bernstein C, Wallace SS (1983): DNA repair. In Mathews CK, Kutter EM, Mosig G, Berget PB (eds): "Bacteriophage T4." Washington, D.C.: American Society for Microbiology, pp 138–151.

Bernstein H (1977): Germ line recombination may be primarily a manifestation of DNA repair processes. J Theor Biol 69:371–380.

Bernstein H, Byers GS, Michod RE (1981): Evolution of sexual reproduction: Importance of DNA repair, complementation and variation. Am Nat 117:537–549.

Bernstein H (1983): Recombinational repair may be an important function of sexual reproduction. BioScience 33:326–331.

Bernstein H, Byerly HC, Hopf FA, Michod RE, Vemulapalli GK (1983): The Darwinian dynamic. Q Rev Biol 58:185–207.

Bernstein H, Byerly HC, Hopf FA, Michod RE (1984a): The evolutionary role of recombinational repair and sex. Int Rev Cytol (in press).

Bernstein H, Byerly HC, Hopf FA, Michod RE (1984b): Origin of sex. J Theor Biol 110:323–351.

Bernstein H, Byerly HC, Hopf FA, Michod RE (1985a): Sex and the emergence of species (submitted).

Bernstein H, Byerly HC, Hopf FA, Michod RE (1985b): Genetic damage, mutation and the evolution of sex. Science (in press).

Breimer LH (1983): Urea-DNA glycosylase in mammalian cells. Biochemistry 2:4192–4203.

Breimer LH , Lindahl T (1984): DNA glycosylase activities for thymine residues damaged by ring saturation, fragmentation, or ring contraction are functions of endonuclease III in *Escherichia coli*. J Biol Chem 259:5543–5548.

Cathcart R, Schwiers E, Saul RL , Ames BN (1984): Thymine glycol and thymidine glycol in human and rat urine: A possible assay for oxidative DNA damage. Proc Natl Acad Sci USA 81:5633–5637.

Cleaver JE (1969): DNA repair in Chinese hamster cells of different sensitivities to ultraviolet light. Int J Radiat Biol 16:277–285.

Cole CJ (1984): Unisexual lizards. Sci Am 250:94–100.

Cole RS, Levitan D, Sinden RR (1976): Removal of psoralen interstrand cross-links from DNA of *Escherichia coli:* Mechanism and genetic control. J Mol Biol 103:39–59.

Cole RS, Sinden RR, Yoakum GH, Broyles S (1978): On the mechanism for repair of cross-linked DNA in *E. coli* treated with psoralen and light. In Hanawalt PC, Friedberg ED, Fox FC (eds): "DNA Repair Mechanisms, ICN-UCLA Symposia on Molecular and Cellular Biology." New York: Academic Press, pp 287–290.

Crow JF, Kimura M (1970): An introduction to population genetics theory. New York: Harper and Row.

Darwin C (1859, 6th ed): "The Origin of Species by Means of Natural Selection." New York: Avenel Books, (reprinted, 1979).

Dawkins R (1982): "The Extended Phenotype." Oxford: WH Freeman.

Dougherty EC (1955): Comparative evolution and the origin of sexuality. Syst Zool 4:145–190.

Eigen M (1971): Self-organization of matter and the evolution of biological macromolecules. Naturwissenschaften 58:465–523.

Eigen M, Schuster P (1979): "The Hypercycle, a Principle of Natural Self-Organization." Berlin: Springer-Verlag.

Eigen M, Gardiner W, Schuster P, Winkler-Oswatitsch R (1981): The origin of genetic information. Sci Am 244(4):88–118.

Eldridge N, Gould SJ (1972): Punctuated equilibria: an alternative to phyletic gradualism. In Schopf TJM (ed): "Models in Paleobiology." San Francisco: Freeman, Cooper and Company pp 82–115.

Fridovich I (1978): The biology of oxygen radicals. Science 201:875–880.

Gensler HL, Bernstein H (1981): DNA damage as the primary cause of aging. Q Rev Biol 56:279–303.

Hackett PB, Sauerbier W (1975): The transcriptional organization of the ribosomal RNA genes in mouse L cells. J Mol Biol 91:235–256.

Haldane JBS (1937): The effect of variation on fitness. Am Natur 71:337–349.

Hanawalt PC, Cooper PK, Ganesan AK, Smith CA, (1979): DNA repair in bacteria and mammalian cells. Ann Rev Biochem 48:783–836.

Hanawalt PC, Friedberg EC, Fox FC (1978): "DNA Repair Reports." ICN-UCLA Symposia on Molecular and Cellular Biology, IX. New York: Academic Press.

Harmon D (1981): The aging process. Proc Natl Acad Sci USA 78:7124–7128.

Hickey DA (1982): Selfish DNA: A sexually-transmitted nuclear parasite. Genetics 101:519–531.

Hollstein MC, Brooks P, Linn S, Ames BN (1984): Hydroxymethyluracil DNA glycosylase in mammalian cells. Proc Natl Acad Sci USA 81:4003–4007.

Hopf FA, Hopf FW (1985): The role of the Allee effect on species packing. Theor Popul Biol 27:27–50.

Kornberg A (1980): "DNA Replication." San Francisco: W.H. Freeman and Co.

Krasin F, Hutchinson F (1977): Repair of DNA double-strand breaks in *Escherichia coli* which requires *recA* function and the presence of a duplicate genome. J Mol Biol 116:81–89.

Kuhn H (1972): Self-organization of molecular systems and evolution of the genetic apparatus. Angew Chem Internat Edit 11:798–820.

Lamb RA, Choppin PW (1983): The gene structure and replication of influenza virus. Ann Rev Biochem 52:467–506.

Leffler S, Pulkrabak P, Grunberger D, Weinstein IB (1977): Template activity of calf thymus DNA modified by a dihydrodiol epoxide derivative of benzo(a)pyrene. Biochemistry 16:3133–3136.

Lewontin RC (1978): Adaptation. Sci Amer 239:212–230.

Lynch M (1984): Destabilizing hybridization, general purpose genotypes and geographic parthenogenesis. Q Rev Biol 59:257–290.

Margison GP, Pegg AE (1981): Enzymatic release of 7-methylguanine from methylated DNA by rodent liver extracts. Proc Natl Acad Sci USA 78:861–865.
Massie HR, Samis HV, Baird MB (1972): The kinetics of degradation of DNA and RNA by H_2O_2. Biochim Biophys Acta 272:539–548.
Maynard Smith J (1978): "The Evolution of Sex." Cambridge: Cambridge University Press.
Mayr E (1982): "The Growth of Biological Thought: Diversity, Evolution and Inheritance." Cambridge: Harvard University Press.
McClain ME, Spendlove RS (1966): Multiplicity reactivation of reovirus particles after exposure to ultraviolet light. J Bacteriol 92:1422–1429.
Medvedev ZA (1981): On the immortality of the germ line: genetic and biochemical mechanisms. A review. Mech Ageing Dev 17:331–359.
Michod RE (1983): Population biology of the first replicators. Am Zool 23:5–14.
Michod RE (1984): Genetic constraints on adaptation, with special reference to social behavior. In Price PW, Slobodchikoff CN, Gaud WS (eds): "The New Ecology: Novel Approaches to Interactive Systems." New York: Wiley.
Mukai T, Chigusa SI, Mettler LE, Crow JF (1972): Mutation rate and dominance of genes affecting viability in Drosophila melanogaster. Genetics 72:335–355.
Muller HJ (1932): Some genetic aspects of sex. Am Natur 66:118–138.
Nocentini S (1976): Inhibition and recovery of ribosomal RNA synthesis in ultraviolet irradiated mammalian cells. Biochim Biophys Acta 454:114–128.
Parker ED Jr (1979): Ecological implications of clonal diversity in parthenogenic morphospecies. Amer Zool 19:753–762.
Radding CM (1981): Recombination activities of *E. coli RecA* protein. Cell 25:3–4.
Rupp WD, Howard-Flanders P (1968): Discontinuities in the DNA synthesized in an excision-defective strain of *Escherichia coli* following ultraviolet irradiation. J Mol Biol 31:291–304.
Rupp WD, Wilde CE, Reno DL, Howard-Flanders P (1971): Exchanges between DNA strands in ultraviolet-irradiated *Escherichia coli*. J Mol Biol 61:25–44.
Sagan C (1973): Ultraviolet selection pressure on the earliest organisms. J Theor Biol 39:195–200.
Selander RK, Levin BR (1980): Genetic diversity and structure in *Escherichia coli* populations. Science 31:545–547.
Shields WM (1982): "Philopatry, Inbreeding and the Evolution of Sex." Albany: State University of New York Press.
Uyenoyama MK (1984): On the evolution of parthenogenesis: A genetics representation of the "cost of meiosis." Evolution 38:87–102.
West SC, Cassuto E, Howard-Flanders P (1982): Postreplication repair in *E. coli:* Strand exchange reactions of gapped DNA by recA protein. Mol Gen Genet 187:209–217.
Williams GC (1975): "Sex and Evolution." Princeton: Princeton University Press.
Williams GC (1980): Kin selection and the paradox of sexuality. In Barlow GW, Silverman J (eds): "Sociobiology: Beyond Nature/Nurture? Reports, Definitions and Debate." Boulder: Westview Press, pp 371–384.
Woese CR (1983): The primary lines of descent and the universal ancestor. In Bendall DS (ed): "Evolution from Molecules to Men." Cambridge: Cambridge University Press, pp 209–233.
Zeive FJ (1973): Effects of the carcinogen N-acetoxy-2-fluorenylacetamide on the template properties of deoxyribonucleic acid. Mol Pharmacol 9:658–669.

The Origin and Evolution of Sex, pages 47–68

Conjugation and Its Aftereffects in *E. coli*

Alvin J. Clark

Department of Molecular Biology, University of California, Berkeley, California 94720

OVERVIEW OF CONJUGATION IN *E. COLI*

Conjugation in *E. coli* is a coordinated series of cell surface and DNA transactional events that results in the direct transfer of DNA between cells. There is a donor and a recipient cell in this polarized one-way process. While engaged, the cells are called conjugants and, when they have finished, they are called exconjugants. Conjugational offspring are called transconjugants and are descended from a merozygote, which is a special name for a recipient exconjugant that has received DNA from the donor.

The cell surface changes of conjugation are relatively poorly understood. The following is an oversimplified scenario that ignores alternatives and points of dispute. Proteinaceous projections from the surface of the donor, called sex pili, contact specific receptors in the outer membrane of the recipient. Retraction of the attached sex pili draws the conjugants together until their outer envelopes make contact. This contact first increases in stability and then decreases until it eventually ruptures. Early events in this process initiate the DNA transactions of transfer, and late events may result from transfer.

The DNA transactions associated with transfer require a nucleotide sequence at which to originate called the origin of transfer and symbolized *oriT*. At this sequence, DNA is mobilized (i.e., prepared) for transfer. A single strand exits the donor, passes between the cells, and enters the recipient. A specialized channel is thought to be required for this transfer but remains to be demonstrated.

Accompanying transfer, DNA synthesis occurs in both cells. In the donor, the transferred strand is replaced, and in the recipient a complementary strand is added to the transferred strand. Such synthesis is not required for transfer and so can be considered an aftereffect. The processes of inheritance that occur in the merozygote are also aftereffects, as they apparently do not contribute to transfer itself. All or part of the transferred DNA can be inherited as a plasmid by a process called repliconation. Recombination of the transferred DNA with the recipient chromosome can lead to replacement of recipient genes by their donor alleles. If the transferred DNA has circularized, recombination could result in its inheritance by addition to the chromosome.

A diagrammatic summary of this overview appeared originally in a review by Clark and Warren [1979]. More recent review articles provide other views of the overall process and its components [Manning and Achtman, 1979; Willetts and Skurray, 1980; Willetts and Wilkins, 1984].

Although study of this subject is almost 40 years old, evidence on the molecular details is still only fragmentary. This makes it necessary to use extrapolation and interpolation to fill out the framework and flesh out the form of conjugation in *E. coli,* much as paleontologists do to recreate an individual from bone fragments. As a result, informed speculation uniting disparate observations will be found in this article, freely mixed with experimentally demonstrated facts. Researchers studying conjugation may consider these speculations to be working hypotheses, as the author does.

THE F PLASMID

A conjugational donor is differentiated from a recipient strain by the presence of a plasmid in the donor that contains all the genes necessary for transfer, including *oriT*. Such a plasmid is called a conjugative plasmid. The first conjugative plasmid recognized was called F for fertility; it is the prototype for the many other conjugative plasmids since discovered.

Plasmid F consists of about 100 kb of DNA whose circular map (Fig. 1) corresponds to the circular state of the DNA in donor cells. The map can be broken into three functional regions: 1) conduction region 0–20 kb, 2) replication-partition region 37–67 kb, and 3) transfer region 67–100 kb. This classification is not meant to imply that all genes in a region contribute to, nor that genes outside of a region cannot participate in, the function mentioned. The region from 20–37 kb has been little studied and so can be called an undetermined region.

As shown in Figure 1, *oriT* is the dividing line between the transfer and replication-partition regions. Whether or not *oriT* itself is divided by the transfer process is not clear, but transfer occurs so that the replication-partition region enters the recipient first and the transfer region last. This unidirectional process is accomplished by transferring, in a 5′ to 3′ manner, only one of the two plasmid strands.

The replication-partition region is the first region transferred by conjugation (Fig. 1). The key components relating to replication are the origins of rep-

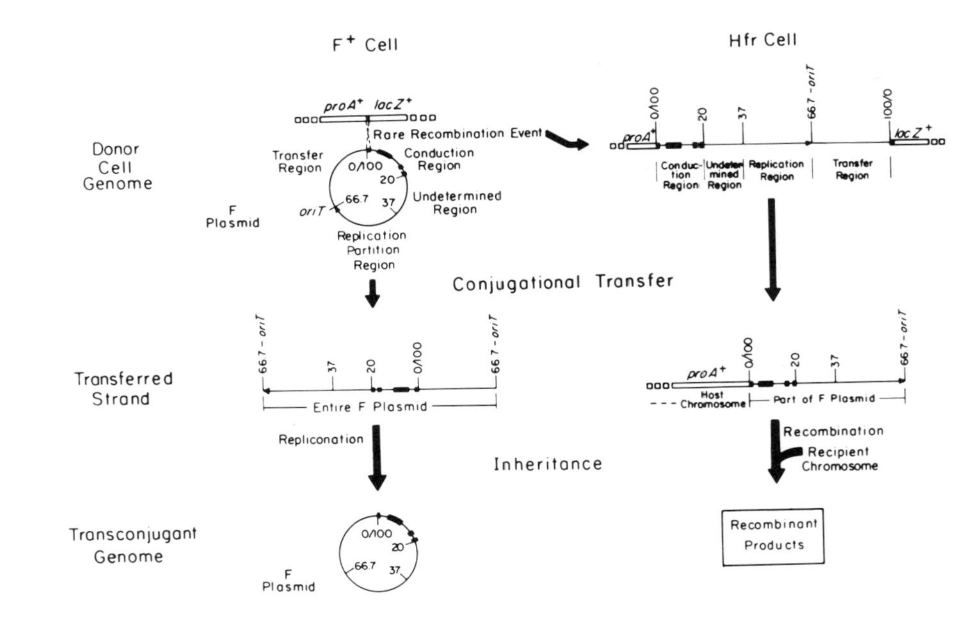

Fig. 1. Conjugational transfer and inheritance of F plasmid DNA from F^+ and Hfr cells. Thin lines represent F plasmid DNA, open rectangles represent *E. coli* chromosomal DNA, and filled rectangles represent transposable elements. Coordinates in kb are marked on the F plasmid maps and the genetic regions marked off are named. No attempt is made to distinguish double- and single-stranded DNA.

lication (there are two at least) and the genes whose products allow initiation at these origins [Scott, 1984]. Other genes in this region encode products that are analogues of host replication products, e.g., *ssf*, which encodes a single strand DNA binding protein similar in activity to that of the host gene *ssb* [Kolodkin et al., 1983; Chase et al., 1983].

The partition components of F, by analogy with those of other systems, regulate initiation of replication and prevent or facilitate binding of plasmid DNA to cell components [Scott, 1984]. Partition refers to the way plasmid DNA copies are distributed to sister cells at division. One way involves random distribution. To ensure a high probability that each sister will obtain at least one copy, the plasmid copy number must be high. Another way depends on plasmid copies binding to cell components that are partitioned regularly. Strong binding can ensure a high probability that each sister will obtain at least one plasmid copy, even if the copy number is low. Interference with binding or, in the absence of binding, reduction of copy number leads to increased probability of obtaining a plasmid-free sister cell at each division. When such reduction or interference is caused by plasmid-plasmid interactions, it is called incompatibility. Where it has been studied, incompatibility involves highly sequence-specific interactions between RNA molecules, between protein and RNA molecules, or between protein and DNA molecules [Scott, 1984]. Thus, incompatibility is considered to be diagnostic of close relatedness, and plasmids are classified into incompatibility groups [Datta, 1979]. Members of different groups cohabit cells stably; members of the same group cohabit cells unstably and can be maintained only by killing or inhibiting the cells that have not obtained copies of both plasmids at division. There are at least four F genes that contribute to incompatibility among F plasmids [Scott, 1984], and there is at least one gene that confers stability to a plasmid whose copy number is not high enough to ensure stability [Ray and Skurray, 1984].

The undetermined and conduction regions are the second and third to be transferred during conjugation. It is the conduction region where recombination events with the *E. coli* chromosome occur, producing Hfr cells (Fig. 1). Conduction is the process whereby DNA can be conjugationally transferred even though it may lack an *oriT* type sequence or may contain one that is nonfunctional [Clark and Adelberg, 1962; Clark and Warren, 1979]. The operative portions of the conduction region are transposons and insertion sequences. There are three different ones on F: IS*2*, IS*3* (two copies), and Tn*1000*. These can recombine with homologous regions in a general or site-specific manner [Deonier and Hadley, 1980]. Depending on the location of such homologous regions on the chromosome and their orientation relative to a reference point, different origins and directions of chromosomal transfer are produced. This was first understood by Jacob and Wollman [1961] and resulted in their proposal of a circular map of the *E. coli* chromosome. Another mechanism leading to conduction has become apparent more recently due to the existence of a cointegrate state in the transposition of Tn*1000* to the chromosome or to another plasmid [Reed, 1981].

The transfer region comprises one third of F. It is the last region to enter the recipient during plasmid transfer (Fig. 1). When conduction of the 4000 kb *E. coli* chromosome occurs from an Hfr cell, entry of the transfer region is delayed by about 100 min. Before this time has elapsed, most specific contacts have weakened and ruptured, thus excluding the transfer region from the recipients. As a result, repliconation is greatly reduced in frequency and recombination with the recipient chromosome is required for inheritance of donor genes.

tra GENE FUNCTIONS

The transfer region extends from about position 67 to 100 on the map of F. In this 33 kb of DNA 24 translated genes, four (possibly five) transcriptional promoters and one transcriptional operator have been identified, not to mention *oriT* [Willetts and Skurray, 1980; Willetts and Wilkins, 1984]. The translated genes can be divided into six functional groups, as indicated in Table 1. Four of these groups are concerned directly with the transfer process and two are involved indirectly. Each group of genes will be discussed separately.

Pilin is the structural component of sex pili. Evidence now suggests that the 14 kDa *traA* gene product is a prepropilin and is cleaved in two steps to 7 kDa pilin [Ippen-Ihler et al., 1984; K. Ippen-Ihler, personal communication]. The *traQ* gene product, which is here called prepropilin hydrolase, presumably carries out the first cleavage, and the *E. coli* signal peptidase may carry out the second [K. Ippen-Ihler, personal communication].

Pilus assembly requires at least 12 gene products. At one time, it was thought that pilin was modified by the addition of phosphate and sugar residues [Brinton, 1971; Armstrong et al., 1980] and that some of these products would be kinases or phosphosugar transferases. Now, however, it is thought that these residues were contributed by contaminating lipopolysaccharide and phosphotidyl choline [Armstrong et al., 1981; W. Paranchych, personal communication]. Therefore, most of the 12 gene products are thought to form a structure to nucleate assembly, although no structure has so far been identified. Pili also are thought to be retracted, presumably by disassembly of subunits [Folkhard et al., 1979]. How this is accomplished and what determines the transition from assembly to disassembly and vice versa are unknown.

The *traG* protein (TraGp) is involved in assembly, and preliminary evidence suggests that it acetylates the N terminus of pilin [K. Ippen-Ihler,

TABLE 1. Genes of the Transfer Region[a]

A. Translated genes

1. Transfer genes

a. Pilin synthesis		b. Pilus assembly	c. Mating pair stabilization	d. DNA transactions	
traA(13.2)	Prepropilin	*traB*(63)	*traG*(112)	*traD*(89)	
TraQ	Prepropilin hydrolase	*traC*(98)	*traN*	*traI*(180)	Helicase I
		traE(21.2)		*traM*(15)	
		traF(25.5)		*traY*(17.5)	*oriT*-nuclease
		traG(112)		*traZ*(87)	*oriT*-nuclease
		traH(44)			
		traK(25.5)			
		traL(10.3)			
		traP(25.5)			
		traU			
		traV			
		traW			

2. Transfer-related genes

a. Surface exclusion	b. Regulation
traS(19.5)	*finP*
traT(26.8)	*traJ*(25.5)

B. Nontranslated genes

1. Transfer gene (DNA transaction)	*oriT*	
2. Transfer-related genes		
a. Transcriptional promoters	*traMp*	*traTp(?)*
	traJp	*traIp*
	traYp	
b. Transcriptional operator *traJo*		

[a]Following the gene name the number in parenthesis refers to the mass in kDa of the gene product as determined by polyacrylamide gel electrophoresis [A. Ray, PhD Thesis, 1984, from Monash University, Melbourne, Australia, supervised by R. S. Skurray]. In the cases of *traA*, *traE*, and *traL* the molecular masses were calculated from the putative nucleotide sequence of each gene [W. Paranchych, personal communication]. A name has been invented in two cases to reflect the specific function of the product.

personal communication]. *TraGp* also is involved in mating pair stabilization [Kingsman and Willetts, 1978]. At 112 kDa it is one of the largest *tra* gene products. The *traN* gene with a presently unidentified product also is required for stabilization. How stabilization is accomplished and what leads to destabilization are unknown.

The DNA transactions associated with transfer are carried out by five gene products. Cleavage of DNA occurs at *oriT* and requires the *traY* and *traZ* products [Everett and Willetts, 1980; Thompson et al., 1984]. Presumably, these constitute a sequence-specific endonuclease. They probably also constitute a sequence-specific ligase, as only a small amount of *oriT* DNA can be found cleaved at any one time, implying that the relative rates of cleavage and ligation are such that there is a steady state level of cleaved DNA. Because cleaved *oriT* DNA can be found even when no transfer is occurring [Everett and Willetts, 1980], something must initiate the transfer. Willetts and Wilkins [1984] hypothesize that this is achieved by attaching the 3′ terminus of cleaved *oriT* to a cell membrane component, thus inhibiting ligation; *traD* and traM proteins may be involved in this. To accomplish transfer, the cleaved strand must be unwound unidirectionally from its complement. An oligomeric form of DNA helicase I, the *traI* product [Abdel-Monem et al., 1983], can unwind DNA and is available for this part of the process [Willetts and Wilkins, 1984]. Unpublished results suggest that traD protein may anchor helicase I to the cell membrane [E. Minkley, personal communication].

Other translated genes in the transfer region are not involved directly in transfer. Two of these, *traS* and *traT*, are involved in sexual differentiation [Achtman et al., 1980]. The products they encode (TraSp and TraTp) make it virtually impossible for an F-containing cell to act as a recipient in conjugation. TraTp is a lipoprotein [Perumal and Minkley, 1984] that acts by inhibiting specific contacts with the pili from another F-containing cell [Achtman et al., 1977]. TraTp may act as a high affinity unproductive pilus binding site competing with the productive lower affinity site, *ompA* protein, which occurs in the outer membranes of cells of both sexes [Minkley and Willetts, 1984]. TraSp prevents triggering transfer DNA transactions if any specific contacts are formed [Manning and Achtman, 1979]. Physiologically, *traS* and *traT* are of great selective advantage. This is evidenced by the necessity to grow *traS* or *traT* single mutant strains in the presence of a detergent that inhibits specific contact formation [Achtman et al., 1980]. In the absence of the detergent, cells carrying multiple mutant F plasmids are enriched, and cells carrying the single mutant plasmids are lost eventually to the investigator [M. Achtman, personal communication]. On this basis, it seems likely that donor cells with inefficient surface exclusion spend so much of their metabolism being recipients that their growth rate or survivability is reduced relative to their nonmating mutant offspring.

Two other transfer-related genes, *traJ* and *finP*, are involved in regulation; traJ encodes a product that is necessary for transcription of the rest of the *tra* genes into a single mRNA [Gaffney et al., 1983]. Because expression of *traJ* has such a general effect, its regulation is the focal point for overall regulation of the transfer genes; *finP* encodes a specificity product, which brings about negative regulation of *traJ* expression in the presence of the *finO* encoded aporepressor [Finnegan and Willetts, 1971, 1972]. F does not contain a functional *finO* gene, so expression of the *tra* genes is constitutive. Other compatible plasmids do carry *finO* genes, however; thus, their presence leads to repression of most of the F *tra* genes [Meynell et al., 1968].

The most important untranslated gene of the transfer region (*oriT*) has been discussed already. One operator, *traJo*, is presumed to exist as the sequence at which the *finP* and *finO* products act to block transcription from initiating at the promoter *traJp* [Thompson and Taylor, 1982]; traJ protein (TraJp) acts to stimulate transcription of all other *tra* genes except *traM* at the promoter *traYp* [Gaffney et al., 1983]; *traM* is transcribed from its own TraJp-regulated promoter, *traMp* [Gaffney et al., 1983]. Likewise, *traI* and *traZ* are thought to be transcribed from a subsidiary promoter, *traIp*, [Achtman et al., 1978; Willetts and Maule, 1979]. Another subsidiary promoter, *traTp*, is thought to lead to *traJ*-independent transcription of the surface exclusion and possibly other downstream genes [Rashtchian et al., 1983].

AFTEREFFECTS

Only one strand of DNA is transferred from the donor; the remaining strand is used as a template for synthesis, which accompanies, but is not required, for transfer [Vapnek and Rupp, 1970, 1971; Sarathy and Siddiqi, 1973; Kingsman and Willetts, 1978]. Because rifampicin blocks this synthesis, the host RNA polymerase, which is rifampicin sensitive, presumably initiates synthesis of a strand to replace the one transferred [Kingsman and Willetts, 1978]. In the recipient, DNA synthesis also accompanies and is not necessary for transfer [Wilkins and Hollom, 1974]. In this case, synthesis provides a complement for the transferred strand [Ohki and Tomizawa, 1968; Vapnek and Rupp, 1970]. This synthesis is presumably similar to lagging strand synthesis and thus involves multiple initiations by host *dnaG* primase and synthesis of short Okazaki fragments [Willetts and Wilkins, 1984].

If the transfer region of F is transferred, then repliconation can occur presumably by rejoining the 5′ and 3′ termini, which were separated to initiate the process [Willetts and Wilkins, 1984]. If too much DNA intervenes and the mating connection ruptures before the transfer region has been transferred, high frequency repliconation is blocked. This is ordinarily the case with Hfr crosses, and recombination with the recipient chromosome is the only high frequency way donor genes are inherited.

Recombination between conjugationally transferred Hfr and DNA and the recipient chromosome normally proceeds by a pathway called the RecBC pathway after two of the genes whose products are involved: *recB* and *recC* [Clark, 1973, 1974]. These genes contribute subunits to a multifunctional enzyme called DNA exonuclease V or ExoV [Smith, 1983], whose role seems both positive and negative. In the presence of recA protein, ExoV is thought to enable recombination by separating strands from the distal terminus, where the formation of the last Okazaki fragment will have left a 3′ single strand terminus [Smith, 1983]. Then *recA* protein would use the single strands so produced to search for homology and catalyze synapsis, as explained by C. Radding in this symposium. In the absence of *recA* protein, ExoV degrades donor DNA to a form unusable for recombination [Itoh and Tomizawa, 1971; Bresler et al., 1981]. Competing reactions thus give a narrow window for recombination to occur.

Pathways of recombination that do not use ExoV have been revealed by the discovery of mutations that increase the low recombination frequency of $ExoV^-$ mutants [Clark, 1973, 1974]. These genes, which suppress *recB* and *recC* mutations and are therefore called *sbc* genes, seem to reduce inhibition of the RecF pathway, so-called because *recF* was the first gene in this pathway to be genetically characterized [Horii and Clark, 1973]. At least four genes that participate in this pathway and not in the RecBC pathway (*recF, recJ, recN,* and *ruv*) have been discovered to date [Horii and Clark, 1973; Lovett and Clark, 1984; Lloyd et al., 1983, 1984], but nothing is known yet about their biochemical functions. This pathway is thought to participate in the repair of some types of damage to DNA, e.g., gaps left by incomplete replication of UV irradiated DNA [Rothman and Clark, 1977; Wang and Smith, 1983] and double strand breaks caused by gamma rays [Picksley et al., 1984].

In order to inherit donor genes by recombination, a minimum of two recombination events are required. There are, however, three main classes of products that can be formed, as shown in Figure 2. In the case of single strand replacement, only one initiation event is required, and this could happen at the distal terminus. Likewise, one initiation event could occur at the distal terminus for double strand replacement and the circularization event leading to double strand addition. The second event, however, must be initiated internally, as the first 60 kb or so of DNA transferred is derived from F and has no counterpart on the recipient chromosome (Fig. 1). One way for such initiation to occur would be for single-stranded donor DNA to interact with recA protein and the chromosome before its complement is synthesized [R. Lloyd and A. Thomas, personal communication]. Another way is for additional double strand termini to be provided by nuclease action within portions of the duplex exogenote homologous to the chromosome [Bresler et al., 1978]. Alternatively single-stranded regions of the recipient chromosome may initiate the second recombination event. A few experiments on the capacity of RecBC and RecF pathways to carry out these events have been done [e.g., Siddiqi and Fox, 1973; Mahajan and Datta, 1979; Bresler et al., 1981; R. Lloyd and A. Thomas, personal communication], but a discussion of these is more appropriate in a specialized review article.

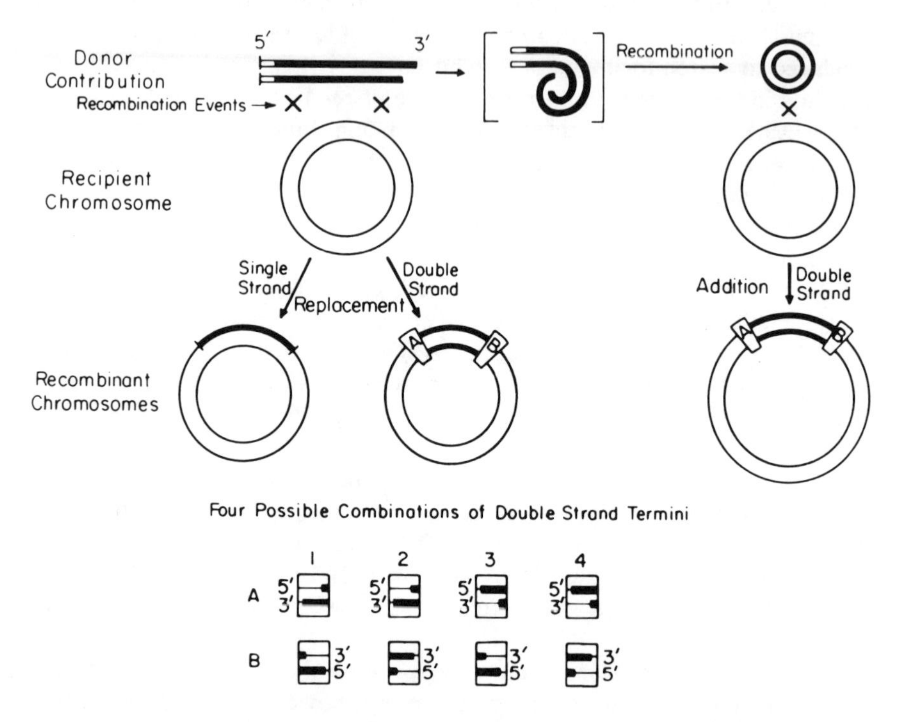

Fig. 2. Possible conjugational recombinant products from an Hfr by F⁻ cross. This is a continuation of Figure 1. Concentric circles and parallel lines represent duplex DNA. Thin lines represent the recipient chromosome. Heavy lines represent the donor strand transferred and its complement synthesized in the recipient. Open rectangles represent F plasmid DNA. Rectangles labeled A and B hide the heteroduplex junctions between a recipient and donor duplex DNAs. Four possible combinations of heteroduplex joints are shown below. No attempt was made to draw the DNA molecules to scale or to distinguish newly replicated strands from those existing before conjugation. Recombination of the duplex donor molecule with itself forming a circular molecule and eliminating F DNA is shown to proceed through a hypothetical intermediate. Such recombination is presumably rare because homology is limited.

EVOLUTIONARY ASPECTS

Conjugative Plasmid Evolution

Besides F, a wide variety of other conjugative plasmids have been discovered, primarily by their antibiotic and heavy metal resistance genes [Meynell et al., 1968]. The plasmids are classified into incompatibility groups, usually by adding an upper-case letter to the prefix "Inc," e.g., IncF, IncI,

IncP, etc. [Datta, 1979]. The exceptions include certain plasmids judged similar to F by two criteria: 1) they carry a *finO* aporepressor gene whose product can interfere with F *tra* gene expression when combined with the F *finP* specificity subunit; and 2) most of the *tra* region and part of the replication-partition region are similar enough to those of F that the two plasmid DNAs hybridize in these regions. Such plasmids are called IncFII, IncFIII, etc., plasmids to indicate their compatibility with, but similarity to, F and other IncFI plasmids.

Comparing the IncFII plasmids R1, R6-5, and R100 to one another and to F has revealed some sources of variation that contribute to plasmid evolution. Heteroduplexes formed by annealing DNA strands of different plasmids show regions of gross difference and similarity [Sharp et al., 1973]. R6-5 and R100 hybridize over their entire length except for three small regions [1.4, 1.7, and 3.8 kb], where DNA appears to be inserted in R6-5 or deleted from R100; thus, these two plasmids are very closely related. When R6-5 is hybridized with F, however, there are 15 places where the DNAs do not hybridize. Five of these are locations of insertions or deletions from one or the other plasmid. Ten are regions where DNA exists in both plasmids but does not hybridize. These are called substitution loops. The many differences between R6-5 and F indicate that these plasmids are related more distantly to each other than are R6-5 and R100. Plasmid R1 shows 10 differences with F, 12 with R6-5, and 11 with R100. This would appear to indicate that R1 is related distantly to all three plasmids; however, one substitution loop with F involves 33 kb of R1 DNA and 47 kb of F DNA, whereas no substitution loops with the other plasmids exceed 1 kb of either partner. This, coupled with the compatibility of R1 with F and its incompatibility with R6-5 and R100, indicates the closer relationship of the R plasmids to one another than to F [Meynell et al., 1968; Datta, 1979].

All three R factors show the large substitution loop with F. Figure 3 shows that this corresponds to part of the replication-partition region, all of the undetermined region, and part of the conduction region of F. Within the large region of noncomplementarity are most of the resistance genes and the replication genes of the R plasmids. Like F, these plasmids possess multiple replication systems [Scott, 1984]. This probably indicates a multiplasmid ancestry. Because only one system of each seems to be preferred in *E. coli* (the *oriV* system of F and the *repA* system of the R plasmids), it seems plausible to consider that each aggregate gives broader host range by allowing replication in hosts with different biochemical preferences. Detailed studies, however, have shown no similarity between the *oriV* system of F and the *repA* system of the three R plasmids [Scott, 1984; K-I Arai, personal communication]. Thus, it seems likely that they were derived separately and came together by transposition to form a plasmid cointegrate (pProF::pRepA), which

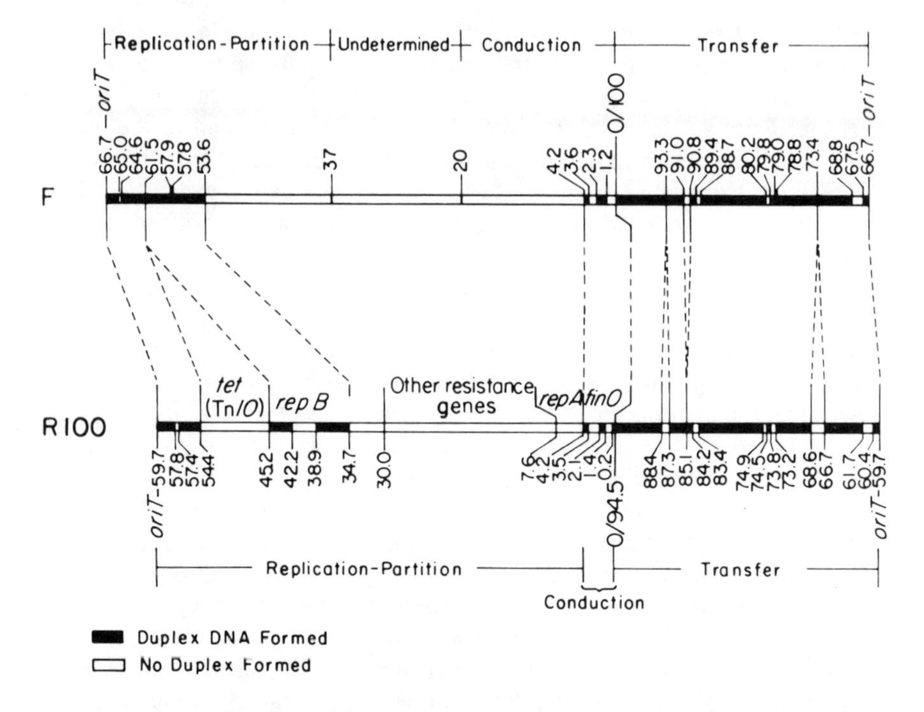

Fig. 3. Representation of a heteroduplex map of plasmids R100 and F. Kilobase coordinates of Sharp et al. [1973] were increased by a factor of 100 ÷ 94.5 to compensate for a change in the coordinates of F [Willetts and Skurray, 1980]. Heteroduplex DNA of R100 and F actually was not examined by Sharp et al. [1973], but inferences can be made by comparing their maps of R6-5 with F and R6-5 with R100-1. Two errors in the map of R6-5 with F were corrected: 1) the 0/94.5 mark was misplaced and 2) F coordinate 71.2 should have been 68.7. R100-1 was assumed to be a point mutant of R100. The zero point of R100 was chosen to be the same as the zero point for F of Willetts and Skurray [1980]. The coordinates for *repA* and *repB* regions were deduced from the maps of R1 [Uhlin et al., 1983], R6-5 [Timmis et al., 1978], and R100 [Ryder et al., 1981, 1982]. Dempsey and McIntire [1983] have mapped *finO*. Dashed lines show the positions of insertion-deletion loops. Other unfilled regions represent substitution loops.

broke down to retain the transfer genes of the F plasmid ancestor (pProF) coupled to the replicative genes of the *repA* ancestor (pRepA). The breakdown product (pProR) could then have given rise to two lines of descent, one leading to R1 and the other to R100 and R6-5. Meanwhile, the ancestor of F (pProF) could have gone through changes to give us the F plasmid we have today.

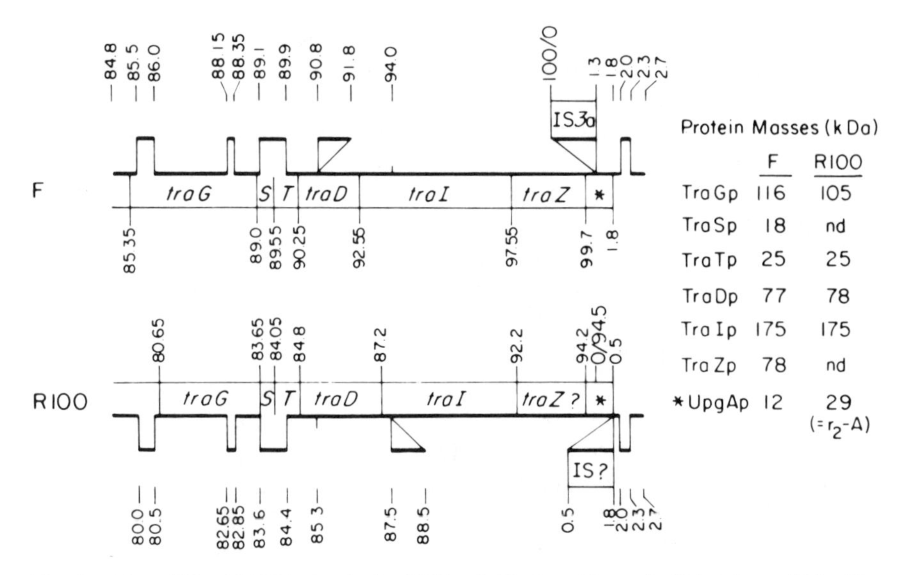

Fig. 4. Map of F and R100 heteroduplex DNA redrawn from Figure 5 of Manning and Morelli [1982]. The renatured strands of F and R100 are drawn as thickened lines. Where duplex DNA was formed the strands are closer together. Projections above or below the duplex regions represent insertion-deletion and substitution loops. Coordinates for these features are found outside, whereas estimated gene coordinates are found inside the duplex. Distances in kb were obtained by measurement of the details of the figure. Coordinates were obtained from these by setting the left side of the IS3a insertion-deletion loop as the 0/100 kb point for F [Willetts and Skurray, 1980]; the 0/94.5 kb point on R100 is the location of the IS3a loop. Protein mass data are taken from the results of Manning et al. [1982] and Hansen et al. [1982]. The asterisk indicates hypothetical unsuspected protein gene *upgA*. Judging from the data of Dempsey and McIntire [1983], it seems possible that this gene is actually *finO* and that F carries an IS*3* insertion in this gene, thereby rendering it constitutive for *tra* gene expression. The 12 kDa protein described by Manning et al. [1982] may then be a truncated *finO* product.

Plasmid Gene Evolution

Whereas cointegrate formation by transposition and breakdown may have led to plasmid variety, other processes seem to be at work which lead to gene evolution. These processes can be inferred from a fine structure analysis of heteroduplex regions of related plasmids and from nucleotide sequences. Figure 4 shows a heteroduplex analysis of a portion of the transfer regions of F and R100 [Manning and Morelli, 1982]. There are four substitution and four insertion-deletion loops. Superimposed is an indication of where the translational genes (i.e., coding sequences) may lie if they do not overlap [Manning and Morelli, 1982].

One of the insertion-deletion loops corresponds with the location in F of IS*3*a and marks the zero reference point for F. This insertion may have mutated a gene adjacent to *traZ*, causing the production of a 12 kDa truncated protein. From the identical region in R100, a 29 kDa protein called r2-A is produced [Hansen et al., 1982]. Two other insertion-deletion loops lie in translational genes *traD* of F and *traI* of R100. Although these may have been derived from insertion sequences, they do not interrupt but seem to be part of the coding sequence. The *traD* products are interchangeable, as each plasmid complements a *traD* mutation in the other but the *traI* products are not interchangeable.

Most of the substitution loops lie within translational genes. This implies there are blocks of amino acid changes differentiating the gene products. A case of plasmid specificity correlated with such a block of DNA sequence difference has been studied in great detail by Ryder et al. [1981, 1982]. The nucleotide sequence of R1 and R100 differ so much in the *copB* region that one can infer that a small substitution loop would be seen if heteroduplexes were examined (Fig. 5). The first 11 codons of *copB* are identical, but then the sequence diverges so that there are 44 amino acid substitutions and two additional amino acids in the *copB* protein ($CopB_p$) of R1 as compared to that of R100. Although seven of the substitutions are conservative and the overall conformational profiles of the proteins are similar, the *copB* proteins are plasmid specific in their action. The reason is that the site of action of CopBp is included within the heterogeneous region [Light and Molin, 1982]. CopBp is a negative regulator of transcription of RNAII, which has its 5′ terminus 17 nucleotides from the end of the heterogeneous region. Because promoters are at least 35 nucleotides long [Rosenberg and Court, 1979] the promoter for RNAII must overlap the heterogeneous region. The operator at which CopBp acts is located within a 60 nucleotide region that includes the promoter. Presumably, the operator lies in the heterogeneous region, and differences in DNA binding specificity result from the amino acid differences in *copB* proteins.

Because Ryder et al. [1982] could detect similarities in the heterogeneous region of *copB*, they hypothesized that it originated from a similar plasmid that had diverged from the common ancestor of R1 and R100, called (pProR). It would seem, however, that the origin of such a high degree of difference has to be set further in the past to the common ancestor of two pRepA plasmids pRepA1, giving rise to pProR and pRepA2. Ryder et al. [1982] further hypothesized that a gene-conversion-type mechanism might have been responsible for the substitution. An alternative would be the recruitment of similar sequences from a transposable element through sequential deletions of the termini.

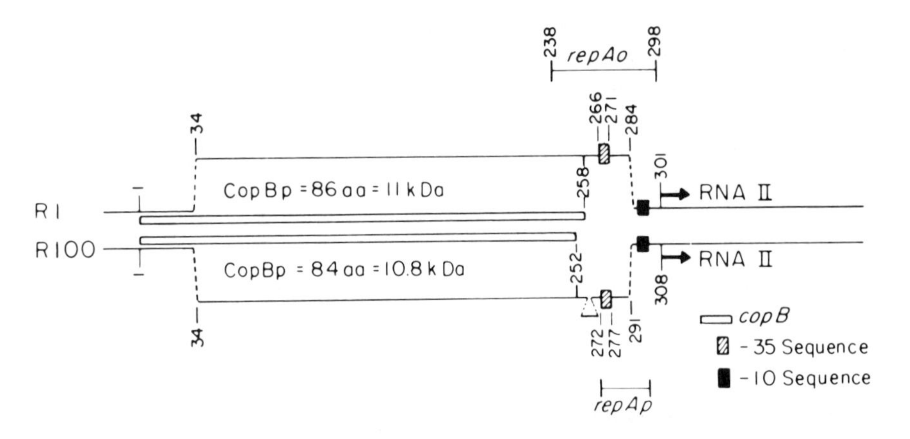

Fig. 5. Hypothetical heteroduplex map of *copB* regions of R1 and R100. The data used was obtained by Ryder et al. [1981, 1982]. Nucleotides are numbered from the beginning of the first codon of *copB*. Close parallel lines indicate where duplex DNA would form. Distant parallel lines joined by dashed lines to the duplex represent a substitution loop. There are seven more nucleotides in the R100 than in the R1 loop; these are represented as a small triangle as done by Ryder et al. [1981]. The *repA* promoter (*repAp*) has been located on both R1 and R100 [Ryder et al., 1981] while the *repA* operator (*repAo*) has been localized only in R1 [Light and Molin, 1982].

Host Dependency of Conjugative Plasmids

Conjugative plasmids differ in the range of Gram-negative hosts by which they can be inherited [Datta and Hedges, 1972]. Most plasmids labeled IncP have a broad host range, whereas those labeled IncFI or IncFII have a narrow host range. For one IncP plasmid (RP4) part of this difference is attributed to a gene encoding two forms of a DNA primase [Lanka et al., 1984]. A primase-defective mutant form of RP4 showed reduced inheritance in about half the intergeneric crosses tried by Lanka and Barth [1981]. Apparently, some host primases inefficiently initiate synthesis of the complementary strand following conjugational transfer of RP4, and this leads to lower inheritance [Willetts and Wilkins, 1984].

Plasmids also depend on their hosts for efficient gene expression. For example, expression of the *tra* genes of F depends on the function of at least four genes in *E. coli*. One of these (sfrA) may be involved in peptide transport, and F may depend on it to produce sufficient *traJ* product to perform its positive regulatory function [Beutin et al., 1981; Gaffney et al., 1983]. Mutations of this gene that affect expression of F *tra* genes were discovered independently by four groups, and so the gene was given the names *cpxC*, *fexA*, and *msp*, besides *sfrA*. The gene also was studied for effects of its mutant alleles on sensitivity to dyes [Buxton and Drury, 1984]; and so it is important for the physiological integrity of *E. coli*. Another gene (*sfrB* = *fexB*) interacts with *sfrA* so that disfunction of *sfrA* alone will not inhibit the expression of the *tra* genes [Lerner and Zinder, 1982]. Another pleiotropic epistatic host gene pair (*cpxA* and *cpxB*) also contribute to full expression of F *tra* gene functions [McEwen and Silverman, 1980] and affect the protein composition of the outer and inner membranes of *E. coli* [McEwen and Silverman, 1982; McEwen et al., 1983].

Other Evolutionary Aspects

Two major evolutionary problems can be broached only briefly and speculatively: 1) the evolutionary origin of the transfer genes and 2) the roots of the transfer system in an anoxic lifestyle. Evolutionarily speaking, it seems logical to trace enzymes specialized for conjugational DNA transactions back to enzymes with a role in replication and repair of DNA damage. Two crucial specialized enzymes of the F transfer system are the *oriT*-endonuclease and helicase I. It can be suggested that, on the basis of their similar activities, that *oriT*-endonuclease is related to *E. coli* DNA gyrase and that helicase I is related to *E. coli recA* protein. Although in vivo tests have shown only nicking by *oriT*-endonuclease, transient double strand breaks similar to those made by DNA gyrase may be made. This is an inference based on the stimulation of RecBC pathway recombination by the *oriT* region and the *traY* and *traZ* genes [Porter, 1983; R. D. Porter, personal communication]. Because the *recB recC* enzyme ExoV requires a double strand break to perform helicase and double-strand exonuclease action [reviewed by Smith, 1983], it seems warranted to hypothesize the occurrence of transient double-strand breaks. The similarities of helicase I and *recA* protein are their ATPase activities, their action as oligomeric aggregates on single-stranded DNA, and their similar directionality of action [Geider and Hoffmann-Berling, 1981; Radding, 1982].

Studies by Fisher [1957a, 1957b, 1961] showed that transconjugant formation required aerobic metabolism and that DNA transfer could be inhibited by dinitrophenol. It is also known that F pili disappear from donor cells in the presence of cyanide [W. Paranchych, personal communication]. These observations seem to indicate that the F transfer system is adapted to the aerobic lifestyle. To what extent it has its roots in a transfer system that originated under anoxic early earth conditions remains to be probed experimentally.

CONCLUSION

Conjugation in *E. coli* reveals a situation of cell differentiation completely dependent on the specialized chromosome called a conjugative plasmid. Such a plasmid encodes what is necessary to make contacts with a recipient cell, draw the cells together, deliver a DNA strand to the recipient, and close that strand into a circle once it has been transferred. Some conjugative plasmids even encode a key enzyme to initiate complementary strand synthesis, thus improving the probability of inheritance. A conjugative plasmid also prevents cells carrying it from mating with themselves, thus making donor and recipient cells into essentially exclusive categories. There is at present no evidence that the recipient cell chromosome encodes a protein specialized for DNA uptake. Even the *ompA* protein, which is the conjugation receptor site protein for F pili, contributes to normal viability by maintaining outer membrane integrity [Hall and Silhavy, 1981]. Thus, sexual differentiation of *E. coli* seems to be entirely a function of the plasmid.

Besides their transfer genes, conjugative plasmids carry a variety of genes that enhance cell viability in special environments. Antibiotic and heavy metal resistance genes are common. The *traT* product of R100 provides resistance to serum complement [Ogata et al., 1982]. Numerous cases also exist in which bacterial isolates from hospital patients carry virulence genes on conjugative plasmids or large nonconjugative plasmids that integrate with conjugative plasmids [Gaster and deGraaf, 1982]. Thus, it seems that conjugational transfer of these plasmids leads to niche specialization and adaptation to hostile environments.

By contrast, *E. coli* chromosome transfer and inheritance of donor genes by homologous genetic recombination, which was an extraordinarily germinal discovery in genetics [Lederberg and Tatum, 1946], seems to be mainly of laboratory importance, with little selective advantage in nature. What makes this opinion compelling are two main arguments. First, there is no evidence that an *oriT*-type sequence and the genes for its mobilization exist on the wild-type *E. coli* chromosome. Instead, conduction of the chromosome depends on low frequency events involving transposable elements. Thus, the capacity for chromosome conduction seems derivative from two sources with obvious selective advantages: 1) transposition of regulatory elements and advantageous translational genes and 2) conjugational transmission of transposable elements via plasmids. Second, all of the genes of homologous recombination thus far found in bacteria are involved in DNA damage repair and replication [Clark and Margulies, 1965; Clark and Volkert, 1978; Little and Mount, 1982]. Homologous recombination thus appears to derive from an enzymatic capacity to remove damage, repair it, or completely replicate damaged templates. A denial of the natural occurrence of chromosome conduction and reshuffling of linked genes is not included in the author's opinion. Rather, these are viewed as epiphenomenal capacities that contribute little to natural population variation [e.g., Ochman and Selander, 1984] and waiting for the appropriate circumstances to provide selective advantage in their own right.

ACKNOWLEDGMENTS

I am very grateful to K-I Arai, W. Dempsey, H. Masai, S. McIntire, E. Minkley, and W. Paranchych for helpful discussions and for providing published and unpublished information on the F transfer system and a comparison of IncFI and IncFII plasmids. This work was supported by grants #NP-237 from the American Cancer Society and by NIH AI 05371 from the National Institute of Allergy and Infectious Diseases.

REFERENCES

Abdel-Monan M, Taucher-Scholz G, Klinkert M (1983): Identification of *Escherichia coli* DNA helicase I as the *traI* gene product of the F sex factor. Proc Natl Acad Sci USA 80:4659–4663.

Achtman M, Kennedy N, Skurray R (1977): Cell-cell interactions in conjugating *Escherichia coli:* Role of *traT* protein in surface exclusion. Proc Natl Acad Sci USA 74:5104–5108.

Achtman M, Manning PA, Kusecek B, Schwuchow S, Willetts N (1980): A genetic analysis of F sex factor cistrons needed for surface exclusion in *Escherichia coli*. J Mol Biol 138:779–795.

Achtman M, Skurray RA, Thompson R, Helmuth R, Hall S, Beutin L, Clark AJ (1978): The assignment of *tra* cistrons to *Eco*RI fragments of F sex factor DNA. J Bacteriol 133:1383–1392.

Armstrong GD, Frost LS, Sastry PA, Paranchych W (1980): Comparative biochemical studies on F and EDP208 conjugative pili. J Bacteriol 141:333–341.

Armstrong GP, Frost LS, Vogel HJ, Paranchych W (1981): Nature of the carbohydrate and phosphate associated with Co1B2 and EDP208 pilin. J Bacteriol 145:1167–1176.

Beutin L, Achtman M (1979): Two *Escherichia coli* chromosomal cistrons *sfrA* and *sfrB* which are needed for expression of F factor *tra* functions. J Bacteriol 139:730–737.

Beutin L, Manning PA, Achtman M, Willetts N (1981): *sfrA* and *sfrB* products of *Escherichia coli* K-12 are transcriptional control factors. J Bacteriol 145:840–844.

Bresler SE, Goryshin IY, Lanzov VA (1981): The process of general recombination in *Escherichia coli* K-12: Structure of intermediate products. Mol G Genet 183:139–143.

Bresler SE, Krivonogov SV, Lanzov VA (1978): Scale of the genetic map and genetic control of recombination after conjugation in *Escherichia coli* K-12: Hot spots of recombination. Mol G Genet 166:337–346.

Brinton Jr CC (1971): The properties of sex pili, the viral nature of "Conjugal" genetic transfer systems, and some possible approaches to the control of bacterial drug resistance. CRC Crit Rev Microbiol 1:105–160.

Buxton RS, Drury LS (1984): Identification of the dye gene product mutational loss of which alters envelope protein composition and also affects sex factor F expression in *Escherichia coli* K-12. Molec Gen Genet 194:241–247.

Chase JW, Merrill BM, Williams KR (1983): The F sex factor encodes a ssDNA binding protein with extensive homology to *Escherichia coli* SSB: comparison of their DNA and protein sequences. Proc Natl Acad Sci USA 80:5480–5484.

Clark AJ (1973): Recombination deficient mutants of *E. coli* and other bacteria. Annu Rev Genet 7:67–86.

Clark AJ (1974): Progress toward a metabolic interpretation of genetic recombination of *Escherichia coli* and bacteriophage lambda. Genetics 78:259–271.

Clark AJ, Adelberg EA (1962): Bacterial conjugation. Ann Rev Microbiol 16:289–319.

Clark AJ, Margulies AD (1965): Isolation and characterization of recombination-deficient mutants of *Escherichia coli* K-12. Proc Natl Acad Sci USA 53:451–459.

Clark AJ, Volkert MR (1978): A new classification of pathways repairing pyrimidine dimer damage in DNA. In Hanawalt PC, Friedberg EC, Fox CF (eds): "DNA Repair Mechanisms." New York: Academic Press, pp 57–72.

Clark AJ, Warren GJ (1979): Conjugal transmission of plasmids. Ann Rev Genet 13:99–126.

Datta N (1979): Plasmid classification: Incompatibility grouping. In Timmis KN, Pühler A (eds): "Plasmids of medical, environmental and commercial importance." Amsterdam: Elsevier/North-Holland Biomedical Press, pp 3–12.

Datta N, Hedges RW (1972): Host range of R-factors. J Gen Microbiol 70:453–460.

Dempsey WB, McIntire SA (1983): The *finO* gene of antibiotic resistance plasmid R100. Mol G Genet 190:444–451.

Deonier RC, Hadley RG (1980): IS2-IS2 and IS3-IS3 relative recombination frequencies in F integration. Plasmid 3:48–64.

Everett R, Willetts N (1980): Characterization of an in vivo system for nicking at the origin of conjugal DNA transfer of the sex factor F. J Mol Biol 136:129–150.

Finnegan DJ, Willetts NS (1971): Two classes of F*lac* mutants insensitive to transfer inhibition by an F-like R factor. Mol Gen Genet 111:256–264.

Finnegan D, Willetts N (1972): The nature of the transfer inhibitor of several F-like plasmids. Mol G Genet 119:57–66.

Fisher KW (1957a): The role of the Krebs cycle in conjugation in *Escherichia coli* K-12. J Gen Microbiol 16:120–135.

Fisher KW (1957b): The nature of the endergonic processes in conjugation in *Escherichia coli* K-12. J Gen Microbiol 16:136–145.

Fisher KW (1961): Environmental influence on genetic recombination in bacteria and their viruses. In Meynell GG, Gooder H (eds): "Microbial reaction to environment." Symp Soc Gen Microbiol 11, Cambridge, England: Cambridge University Press, pp 272–295.

Folkhard W, Leonard KR, Malsey S, Marvin DA, Dubochet J, Engel A, Achtman M, Helmuth R (1979:) X-ray diffraction and electron microscope studies on the structure of bacterial F-pili. J Mol Biol 130:145–160.

Gaastr W, de Graaf FK (1982): Host specific fimbrial adhesins of noninvasive enterotoxigenic *E. coli* strains. Microbiol Rev 46:129–161.

Gaffney D, Skurray R, Willetts N (1983): Regulation of the F conjugation genes studied by hybridization and *tra-lacZ* fusion. J Mol Biol 168:103–122.

Geider K, Hoffmann-Berling H (1981): Proteins controlling the helical structure of DNA. Ann Rev Biochem 50:133–260.

Givskov M, Molin S (1984): Copy mutants of plasmid R1: Effects of base pair substitutions in the *copA* gene on the replication control system. Mol G Genet 194:286–292.

Hall MN, Silhavy TJ (1981): Genetic analysis of the major outer membrane proteins of *Escherichia coli*. Annu Rev Genet 15:91–142.

Hansen BS, Manning PA, Achtman M (1982): Promoter-distal region of the *tra* operon of F-like sex factor R100 in *Escherichia coli* K-12. J Bacteriol 150:89–99.

Horii ZI, Clark AJ (1973): Genetic analysis of the RecF pathway of genetic recombination in *Escherichia coli* K12. Isolation and characterization of mutants. J Mol Biol 80:327–344.

Ippen-Ihler K, Moore D, Laine S, Johnson DA, Willetts NS (1984): Synthesis of F-pilin polypeptide in the absence of the F *traJ* product. Plasmid 11:116–129.

Itoh T and Tomizawa JI (1971): Inactivation of chromosomal fragments transferred from Hfr strains. Genetics 68:1–11.

Jacob F, Wollman EL (1961): "Sexuality and the genetics of bacteria." New York: Academic Press.

Kingsman A, Willetts N (1978): The requirements for conjugal DNA synthesis in the donor strain during F*lac* transfer. J Mol Biol 122:287–300.

Kolodkin AL, Capage MA, Golub EI, Low KB (1983): F sex factor of *Escherichia coli* K-12 codes for a single-stranded DNA binding protein. Proc Natl Acad Sci USA 80:4422–4426.

Kumar S, Srivastava S (1983): Cyclic AMP and its receptor protein are reequired for expression of transfer genes of conjugative plasmid F in *Escherichia coli*. Mol G Genet 190:27–34.

Lanka E, Barth PT (1981): Plasmid RP4 specifies a deoxyribonucleic acid primase specified by I-like plasmids. Proc Natl Acad Sci USA 76:3632–3636.

Lanka E, Lurz R, Kröger M, Furste JP (1984): Plasmid RP4 encodes two forms of a DNA primase. Mol G Genet 194:65–72.

Lederberg J, Tatum EL (1946): Novel genotypes in mixed cultures of biochemical mutants of bacteria. Cold Spring Harbor Symp Quant Biol 11:113–114.

Lerner TJ, Zinder ND (1982): Another gene affecting sexual expression of *Escherichia coli*. J Bacteriol 150:156–160.

Light J, Molin S (1982): The sites of action of the two copy number control functions of plasmid R1. Mol Gen Genet 187:486–493.

Little JW, Mount DW (1982): The SOS-Regulatory system of *Escherichia coli*. Cell 29:11–22.

Lloyd RG, Benson FE, Shurvinton CE (1984): Effect of *ruv* mutations on recombination and DNA repair in *Escherichia coli* K12. Mol Gen Genet 194:303–309.

Lloyd RG, Picksley SM, Prescott C (1983): Inducible expression of a gene specific to the RecF pathway for recombination in *Escherichia coli* K12. Mol Gen Genet 190:162–167.

Lovett ST, Clark AJ (1984): Genetic analysis of the *recJ* gene of *Escherichia coli* K-12. J Bacteriol 157:190–196.

Mahajan SK, Datta AR (1979): Mechanisms of recombination by the RecBC and the RecF pathways following conjugation in *Escherichia coli* K12. Mol Gen Genet 169:67–78.

Manning PA, Achtman M (1979): Cell to cell interactions in conjugating *Escherichia coli:* the involvement of the cell envelope. In Inouye M (ed): "Bacterial Outer Membranes." Wiley & Sons, pp 409–447.

Manning PA, Morelli G (1982): DNA homology of the promoter-distal regions of the *tra* operons of sex factors F and R100 in *Escherichia coli* K-12. J Bacteriol 150:389–394.

McEwen J, Sambucetti L, Silverman PM (1983): Synthesis of outer membrane proteins in *cpxA cpxB* mutants of *Escherichia coli*. J Bacteriol 154:375–382.

McEwen J, Silverman P (1980): Genetic analysis of *Escherichia coli* K-12 chromosomal mutants defective in expression of F-plasmid functions: Identification of genes *cpxA* and *cpxB*. J Bacteriol 144:60–67.

Meynell E, Meynell GG, Datta N (1968): Phylogenetic relationships of drug resistance factors and other transmissible bacterial plasmids. Bacteriol Rev 32:55–83.

Minkley Jr EG, Willetts NS (1984): Overproduction, purification and characterization of the F *traT* protein. Mol G Genet (in press).

Ochman H, Selander RK (1984): Evidence for clonal population structure in *Escherichia coli*. Proc Natl Acad Sci USA 81:198–201.

Ogata RT, Winters C, Levine RP (1982): Nucleotide sequence analysis of the complement

resistance gene from plasmid R100. J Bacteriol 151:819–827.
Ohki M, Tomizawa J (1968): Asymmetric transfer of DNA strands in bacterial conjugation. Cold Spring Harbor Symp Quant Biol 33:651–658.
Perumal NB, Minkley Jr EG (1984): The product of the F sex factor *traT* surface exclusion genes is a lipoprotein. J Biol Chem 259:5357–5360.
Picksley SM, Attfield PV, Lloyd RG (1984): Repair of DNA double strand breaks in *Escherichia coli* K12 requires a functional *recN* product. Mol G Genet 195:267–274.
Porter RD (1983): Specialized transduction with λplac5: involvement of the RecE and RecF recombination pathways. Genetics 105:247–257.
Radding CM (1982): Homologous pairing and strand exchange in genetic recombination. Ann Rev Genet 16:405–437.
Rashtchian A, Crooks JH, Levy SB (1983): *traJ* independence in expression of *traT* on F. J Bacteriol 154:1009–1012.
Ray A, Skurray R (1984): Stabilization of the cloning vector pACYC184 by insertion of F plasmid leading region sequences. Plasmid 11:272–275.
Reed RR (1981): Transposon-mediated site-specific recombination: a defined in vitro system. Cell 25:713–720.
Rosenberg M, Court D (1979): Regulatory sequences involved in the promotion and termination of RNA transcription. Ann Rev Genet 13:319–353.
Rothman RH, Clark AJ (1977): The dependence of postreplication repair on *uvrB* in a *recF* mutant of *Escherichia coli* K-12. Mol G Genet 155:279–286.
Ryder TB, Davison DB, Rosen JI, Ohtsubo E, Ohtsubo H (1982): Analysis of plasmid genome evolution based on nucleotide-sequence comparison of two related plasmids of *Escherichia coli*. Gene 17:299–310.
Ryder T, Rosen J, Armstrong KA, Davison D, Ohtsubo E, Ohtsubo H (1981): Dissection of the replication region controlling incompatibility, copy number and initiation of DNA synthesis in the resistance plasmids R100 and R1. ICN-UCLA Symp Mol Cell Biol 22:91–111.
Sarathy PV, Siddiqi O (1973): DNA synthesis during bacterial conjugation II. Is DNA replication in the Hfr obligatory for chromosome transfer. J Mol Biol 78:443–451.
Schmidt F (1984): The role of insertions deletions and substitutions in the evolution of R6 related plasmids encoding aminoglycoside transferase. Mol G Genet 194:248–259.
Scott JR (1984): Regulation of plasmid replication. Microbiol Revs 48:1–23.
Sharp PA, Cohen SN, Davidson N (1973): Electron microscope heteroduplex studies of sequence relations among plasmids of *Escherichia coli* II. Structure of drug resistance (R) factors and F factors. J Mol Biol 75:235–255.
Siddiqi O, Fox MS (1973): Integration of donor DNA in bacterial conjugation. J Mol Biol 77:101–123.
Smith GR (1983): General recombination. In Hendrix RW, Roberts JW, Stahl FW, Weisberg RA (eds): "Lambda II." Cold Spring Harbor, New York: Cold Spring Harbor Laboratory, pp 175–209.
Thompson R, Taylor L (1982): Promoter mapping and DNA sequencing of the F plasmid transfer genes *traM* and *traJ*. Mol G Genet 188:513–518.
Thompson R, Taylor L, Kelly K, Everett R, Willetts N (1984): The F plasmid origin of transfer: DNA sequence of wild-type and mutant origins and location of origin-specific nicks. EMBO Journal 3:1175–1180.
Timmis KN, Cabello F, Andrés I, Nordheim A, Burkhardt HJ, Cohen SN (1978): Instability of plasmid DNA sequences: Macro and micro evolution of the antibiotic resistance plasmid R6-5. Mol G Genet 167:11–19.
Vapnek D, Rupp D (1970): Asymmetric segregation of the complementary sex-factor DNA

strands during conjugation in *Escherichia coli*. J Mol Biol 53:287–303.

Vapnek D, Rupp WD (1971): Identification of individual sex-factor DNA strands and their replication during conjugation in thermosensitive DNA mutants of *Escherichia coli*. J Mol Biol 60:413–424.

Wang TV, Smith KC (1983): Mechanisms for recF-dependent and recB-dependent pathways of postreplication repair in UV-irradiated *Escherichia coli uvrB*. J Bacteriol 156:1093–1098.

Wilkins BM, Hollom SE (1974): Conjugational synthesis of F*lac*$^+$ and ColI DNA in the presence of rifampicin and in *Escherichia coli* K12 mutants defective in DNA synthesis. Mol G Genet 134:143–156.

Willetts N, Maule J (1979): Investigations of the F conjugation gene *traI*: *traI* mutants and λ*traI* transducing phage. Mol G Genet 169:325–336.

Willetts N, Skurray R (1980): The conjugation system of F-like plasmids. Ann Rev Genet 14:41–76.

Willetts N, Wilkins B (1984): Processing of plasmid DNA during bacterial conjugation. Microbiol Rev 48:24–41.

The Origin and Evolution of Sex, pages 69–85

What Is Sex?

Lynn Margulis, Dorion Sagan, and Lorraine Olendzenski

Boston University, Department of Biology, Boston, Massachusetts 02215

THE NATURE OF AT LEAST TWO PROBLEMS

The term "sex" evokes many different images in different contexts for different people, especially biologists and biochemists. This paper attempts to clarify the nature of the evolutionary problems pertaining to the "origins of sex." We define sex as a process always involving at least one autopoietic parent that forms an individual with a genetic constitution that differs from both of the parents. Autopoiesis is metabolic self-maintenance at the expense of carbon and energy sources. It is characteristic of all cells but not of viruses or plasmids [Varela and Maturana, 1974].

We sharply distinguish prokaryotic sex from eukaryotic sex. Prokaryotic sex may involve genetic recombination of autopoietic entities (like bacterial cells) and non-autopoietic entities (such as viruses and plasmids). Eukaryotic or meiotic sex which leads to the alternation of haploid and diploid cells requires fertilization and meiosis. It evolved in protoctists, those eukaryotic microorganisms and their multicellular descendants that are not animals (lack blastulas), plants (lack plant embryos), or fungi (are not haploid or dikaryotic osmotrophs that develop from spores) [Margulis and Schwartz, 1982]. Prokaryotic sex probably evolved in the Archean Aeon more than 3000 million years ago. Bernstein et al. [1984, 1985] even argue that sex predated the origin of cells, although we tend to doubt this [Margulis and Sagan, 1985]. In any case, prokaryotic sex evolved in bacteria and viruses long before eukaryotic organisms even appeared on the earth. Meiotic sex, on the other hand, evolved in protoctists in the late Proterozoic Aeon, probably about 1,000 million years ago [Vidal, 1984]. Certainly meiotic sex evolved prior to the fossil animals of the Ediacaran fauna, approximately 700 million years ago [Glaessner, 1984].

The vast differences between prokaryotic and meiotic sexuality suggest independent origins for the two types of processes. However, the chemistry of DNA repair and recombination suggests that prokaryotic sex preadapted meiotic eukaryotes for the crossing over aspect of meiosis [Bernstein et al., 1984, 1985].

The origin of sex must be distinguished from its maintenance. Biologists have long assumed that biparental sexuality itself (mixis) is selected for in populations of animals and plants because it generates genetic variation, but there is no direct evidence for this assumption. Rather, animal and plant sexuality, we believe, is maintained because it is directly tied to tissue differentiation and histogenesis. It is the products of these processes, namely morphologically complex organisms, not sexuality itself, that have been subjected to powerful natural selection. Of course, in morphologically complex organisms, such as mammals, in which meiosis and fertilization are inextricably bound to reproduction, there is no way for meiotic sex to be selected against. Beyond this, however, we hypothesize that meiosis, in particular, the processes usually found in meiotic prophase I, controls complex tissue development in eukaryotes. Meiosis is a physiological process that insures the transmission of accurate, undamaged DNA representative of the community of former microbes that we perceive as the eukaryotic cell [Margulis, 1981]. As such, then, meiosis, whether in mictic or apomictic organisms, is invulnerable—not open to the vicissitudes of natural selection in animals and plants. We have argued these points of view in detail in a book devoted to the origins of sex [Margulis and Sagan, 1985].

In the case of fungi, where the selection on differentiated tissue is far less rigorous than it is in animals and plants, sexuality is absent in over ten thousand species of Hyphomycetes alone [Alexopoulos, 1962]. Although more than one thousand genera of Deuteromycotina (fungi that lack a sexual stage and that are thought to be derived from ascomycotes or basidiomycotes) are known, not all of these genera (of hypho- or coelomycetes) are valid [Kendrick and Carmichael, 1973]. In any case, complex morphology and tissue differentiation is clearly dispensable in large numbers of fungal taxa that display immense variation and superb adaptability in the total absence of biparental sexuality. It is clear that the loss of biparental sex (mixis) certainly has occurred not only in fungi but independently in many species in all four eukaryotic kingdoms. The loss of sex has been the subject of much observation and speculation [Maynard-Smith, 1978; Bell, 1982]. No entirely acceptable explanation for the loss of mixis, nor indeed for its maintenance has been proffered. We have a limited approach to these problems. Rather, it is the first evolutionary appearance of both prokaryotic sex and the biparental fertilization-meiosis cycle that we wish to discuss. The origin of meiotic sex has been the subject of almost no work except that of Cleveland [1947] and

Raikov [1982]. Because the details of our analysis, based on Cleveland's impetus, have appeared elsewhere [Margulis, 1982; Margulis and Sagan, 1984; Margulis and Sagan, 1985], we only summarize our conclusions and provide references here.

WHAT IS SEX?

Sex, as the formation of a new live individual that contains genes from more than a single parent, includes all of the processes listed in Table I. The major themes of sexuality and the order in which they appeared are indicated on a phylogeny of all life in Figure 1.

WHY DOES MEIOTIC SEX SURVIVE?

What keeps sexual organisms competitive with asexual organisms that do not require two organisms for the production of new offspring but simply bud or divide in far shorter times? Biparental sexual organisms which leave fewer organisms per unit time are clearly less fit than those capable of by-passing sexuality. To put it in the insidious economic language still in vogue in evolutionary biology, the "cost" of sex—the finding of mates, the production of sex cells with half the number of chromosomes such as eggs or sperm, and the time invested in these and like activities—seems warranted by the acquisition of increased potential for variation and thus adaptation to changing environments. In fact, a recent mammoth compilation of sexual studies reveals that there is no evidence that organisms with a single parent show less heritable variation than organisms with two parents [Bell, 1982].

Ramets, the asexual growing parts of animals, such as the gemmae of sponges or the buds of hydras, clearly reproduce more rapidly and more efficiently than the sexual products of these same organisms. Why wasn't biparental sexuality eliminated long ago by natural selection? When this idea is tested by comparing asexually and sexually reproducing rotifers, lizards, and grasshoppers, it is found that the asexually produced animals (those having only a single parent) are found just as frequently, if not more so, in varying environments [Bell, 1982].

We do not believe there is any evidence that meiotic sexuality is selected for independently. Rather, the continued existence of sexual organisms is a product of their entire evolutionary history. We suggest that "sex" is maintained by natural selection not because of sexuality (mixis), but because of the connection of meiosis with morphogenesis—or, in the case of the mammals, because of the obligatory association of meiotic sexuality with reproduction [Margulis and Sagan, 1985].

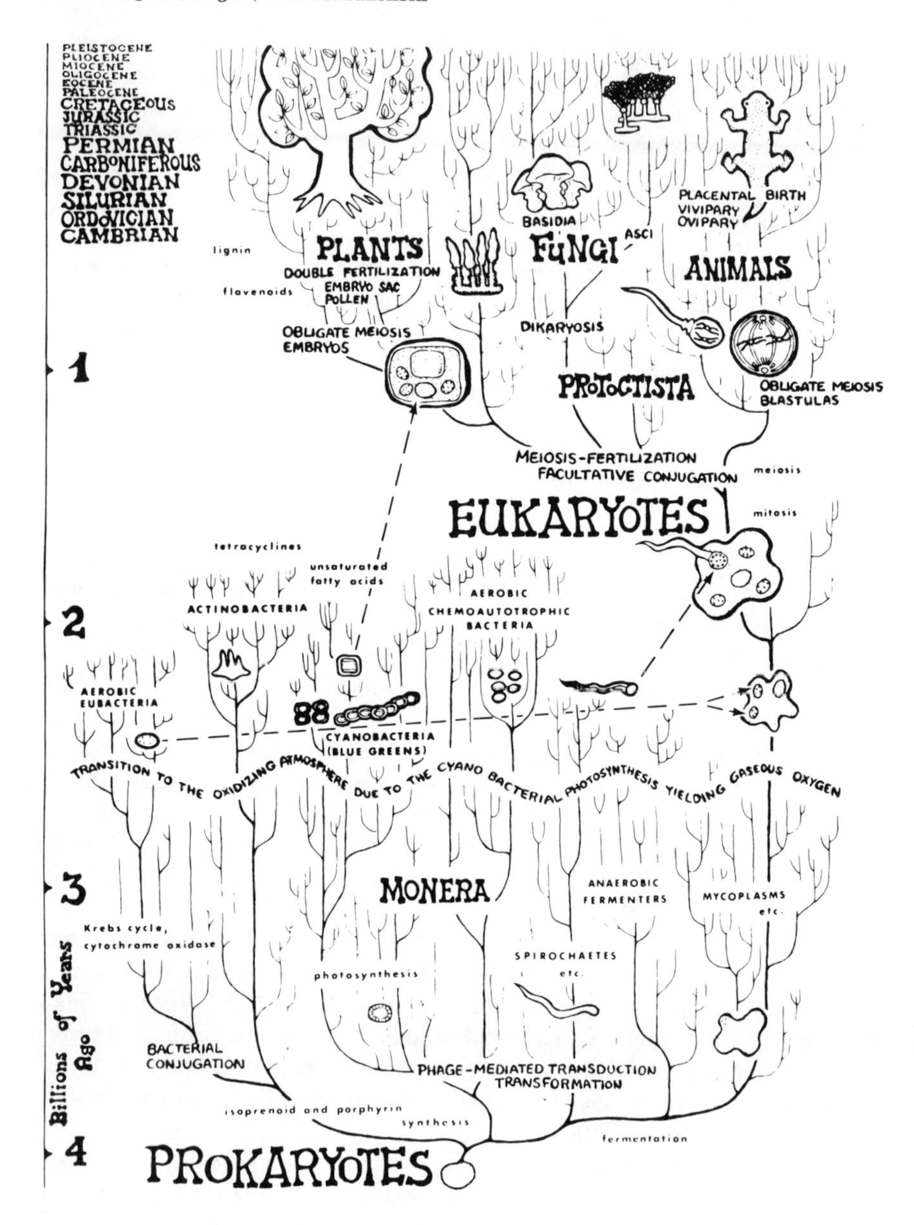

Fig. 1. Sexual processes (phage-mediated transduction and transformation, bacterial conjugation, facultative conjugation, meiosis fertilization, etc.) superimposed on a phylogeny of all life.

If sex in most vertebrate animals as a result of evolutionary history is inseparable from reproduction, and reproduction is of course imperative, there is little point in mathematical descriptions of the maintenance of sex in these organisms. This is a little like wondering about the nature of the selective force that keeps water faucets periodically wet.

TABLE I. Terminology: Sexual Processes

Name of process	Description
Reproduction	The process that augments the number of cells or organisms
Sex	Any process that unites genes (DNA) in an individual cell or organism from more than a single source
Recombination	Breakage and reunion of DNA molecules
Syngamy	Contact or fusion of gametes (cells)
Conjugation	Contact or fusion of gametes or gamonts (cells or organisms)
Gametogamy	Fusion of gametes (cells or nuclei)
Gamontogamy	Aggregation or union of gamonts (organisms) (mating, conjugation)
Karyogamy	Fusion of gamete nuclei
Crossing over	Breakage and reunion of DNA of homologous non-sister chromatids in meiosis (or mitosis)
Amphimixis (Mixis)	Syngamy or karyogamy leading to fertilization to form an individual with two different parents; outcrossing, outbreeding
Amixis	Absence of meiosis and fertilization at all stages in the life cycle (asexual reproduction)
Apomixis	Altered meiosis or fertilization such that amphimixis is bypassed (e.g., parthenogenesis)
Automixis	Syngamy or karyogamy of nuclei or cells deriving from the same parent (selfing; extreme inbreeding; autogamy)
Parthenogenesis	Development of eggs or macrogametes in the absence of amphimixis
Arrhenotoky	Parthenogenesis producing haploid males and amphimictically produced diploid females (incomplete fertilization event, unfertilized eggs become males)
Heterogony	Parthenogenesis in the absence of karyogamy stimulated by sperm or a second species
Thelytoky	Parthenogenesis in which diploid individuals are formed by karyogamy of the egg with its own female pronucleus
Tychoparthenogenesis	Occasional parthenogenesis
Gametic Meiosis	Meiosis immediately preceding gametogenesis (characteristic of members of Animalia kingdom)
Zygotic Meiosis	Meiosis immediately following zygote formation (characteristic of members of Fungi kingdom)
Haplodiplomeiosis	Meiosis preceding an extensive haploid life cycle in organisms which also have extensive diploid life cycle phases (characteristic of members of Plantae kingdom)

Sex, in a deep sense, has nothing necessarily to do with reproduction. Neither is sex always related to gender. Sex is a genetic mixing feature of organisms that operates at a variety of levels. In some organisms, it occurs at more than a single level simultaneously. The smallest sort of sexual event, here considered to be any union of genetic material within a single cell to produce an individual from more than a single parental source, is at the nucleic acid level. By this definition an influenza virus causing illness in humans is sexual because the viral genes enter human cells. The simplest form of sex is the entry of heterologous DNA into a cell. Genetic integration via the recombination of sequences of genetic material (DNA or RNA) from two sources, to form a new and different sequence of DNA or RNA, is the second form of sexuality probably both in simplicity and order of evolutionary appearance. This kind of genetic recombination—the breaking of the integrity of a DNA sequence and the reforming of a new sequence—occurs during virus and plasmid incorporation into bacteria and in the chromatid exchanges of plants and animals in DNA crossing over. Crossing over, a supplementary form or recombination that occurs during certain meioses (for example, in the gonadal tissue to make sperm or egg cells) is not homologous to the union of chromosomes from the parental male and female cells in fertilization that for the most part produces the recombined offspring. Crossing over is a DNA-recombinational process that probably evolved by utilization of enzyme systems already present in ancestral bacteria.

Meiotic sex always involves two reciprocal processes—the reduction by half of the number of chromosomes (prior to the differentiation of vegetative or gamete nuclei, sperm, eggs, or spores) and the fertilization that reestablishes the original chromosomal number. Meiotic sex has a separate traceable history from that of DNA recombinational sex. We believe that meiosis itself is constantly selected for because it is imperative for histogenesis in members of the animal and plant kingdoms because the evolution of members of these kingdoms meiosis was associated with mixis. In cases where mixis can be separated from meiosis and histogenesis still retained (for example, in over 100 species of angiosperms and in parthenogenetic grasshoppers), it has been.

THE HISTORY OF BACTERIAL SEX

Bacterial sex most likely began early in the Archean Aeon, when the earth's atmosphere was devoid of free oxygen. About three and a half billion years ago, less than one part in a million of free oxygen is estimated to have been present in the atmosphere. Consequently, no ozone layer protected genetic material from the ultraviolet radiation emitted by the sun. Oxygen did not become prevalent as a free gas until about two thousand million years ago [Cloud, 1983]. The ozone layer did not form until photosynthetic bacteria released significant amounts of gaseous oxygen. From observations of sun-like stars by the Explorer X satellite, it has been estimated that the probable output of light energy of the early sun was so great that had there been no atmosphere life would never have developed at all [Canuto et al., 1982]. Yet life did develop under constant bombardment from both visible radiation and wavelengths of ultraviolet light. Bacterial and viral sex, we believe, evolved in response to nucleic acid-threatening radiation and led to the spread of genes as "small genetic entities" in protected packets [Sonea and Panisett, 1983].

Bacterial sexuality most likely evolved from genetic repair systems which restored DNA damaged by ultraviolet radiation. Those bacteria that detected the ultraviolet-damaged DNA and excised it survived; as time went on, DNA repair systems became refined. Bacteria today still respond dramatically to ultraviolet radiation. For example, the "S.O.S." system, which is induced by short lengths of DNA such as those that are produced by ultraviolet lesions, is still intact today. When the S.O.S. system is activated, the bacteria stop dividing and they form long chains equivalent to N cell lengths and each containing N copies of the entire genophore. Oxygen uptake is inhibited immediately and growth is dramatically slowed. The irradiated cells, as part of the S.O.S. phenomenon, then undergo mutagenesis producing many new mutants by error-prone DNA synthesis. If they harbor them, the irradiated cells also release phage particles. "S.O.S." is a syndrome, a coordinated response assuring that the threatened bacteria's descendants—or their genes at least, in the non-autopoietic, miniaturized form of viruses—survive.

Standard DNA repair involves copying an undamaged strand to produce a healthy double-stranded DNA molecule. Damaged bacteria must have at least one intact functional DNA strand or, incapable of DNA replication, they will die. Involving a number of splitting and splicing procedures, DNA repair, which requires the protein product of the rec A gene in *E. coli,* an ATPase, is closely related to mechanisms governing the acceptance of similar DNA from foreign sources namely bacterial sex. Thus, it is logical to assert that the methods of survival against ultraviolet irradiation of single cells were used later for the bacterial sex that involved recombination of DNA from more than a single source.

Bacterial sex promoted diversity; the giving, receiving and trading of genes increased the spread of variation through the biosphere [Sonea and Panissett, 1984]. Here, it is again worth noting the rather large difference between prokaryotic and eukaryotic sex: whereas eukaryotic sex (as mixis) has not been consistently shown to produce variety, prokaryotic sex, in the sensu lato of genetic transfer meant here, was indispensable to the production of diversity. Any novelty resulting from recombination was propagated directly to succeeding generations by cell division.

The salient fact is that the loss of ability for repair of ultaviolet light-induced damage is often accompanied by the loss of the genetic recombination system. For example, some *Escherichia coli* known as "rec minus" mutants are far more sensitive to death by ultraviolet radiation than their sexual relatives. The observation that sexless rec minus mutants are unable to perform DNA repair while their sexual relatives are able to shows clearly that bacterial sex and DNA repair are very closely related. The most plausible explanation of this phenomenon is that the repair of damage induced by ultraviolet light preadapted bacteria to recombinational sex. By the formation of "thymine dimers," ultraviolet radiation put selection pressure on the development of repair systems. Those bacteria that developed systems that could copy damaged DNA sequences by using heterologous DNA, by definition engaged in the first "sex." No matter the source of the undamaged DNA, whether from other bacteria directly or through intermediary agents such as viruses or plasmids, bacteria using an external source of DNA for repair became sexual. Little more than sequences of DNA capable of being copied within cells, such viruses and plasmids could have themselves originated when ultraviolet light irradiated bacteria and excised portions of bacterial genetic material. Viruses, plasmids, and other replicating entities were released into the environment. The enzymes used by bacteria to repair ultraviolet damage are in some cases exactly those necessary to splice and ligate DNA in viral and conjugational recombination [Clark, this volume]. Thus the same enzymes were sometimes employed to add nearby genes to already functioning bacteria.

Today's rise in "genetic engineering" merely mimics bacterial sex in nature. When the DNA sequence used as a template is heterologous, that is, when it comes from a different parental source, the act that produces a new cell is by definition sex. People now mediate bacterial sexuality. Heterologous DNA, because it was discovered by many different researchers in a variety of contexts, has many names ("transforming principle," "fertility factors," "episomes," "plasmids," and "viral DNA"). There is, however, very little fundamental difference between infection, conjugation, transformation, and transduction. In all cases a recombination of DNA sequences from separate sources produces an offspring with more than a single parent.

The first fossil bacteria appear in rocks dated at about 3,500 million years old on continents now recognized as Africa and Australia [Schopf, 1983]. These first cells presumably already had evolved ultraviolet defense mechanisms involving the enzymatic splicing of DNA. Early photosynthetic life especially faced the dilemma of the requirement to be near surface sunlight and yet avoid ultraviolet radiation. Whereas heterotrophic organisms may have avoided the problem of ultraviolet radiation by moving deeper underwater, underground, or into another microbe or microbial community, surface organisms were exposed to the ultraviolet threat until enough ozone accumulated to block out the light. This was estimated to occur some 2,000 million years ago [Cloud, 1983; Margulis et al., 1976].

The argument that ultraviolet hardship led to the first sex can be summarized. DNA, RNA, and protein—the essential components of the first cells—all strongly absorb ultraviolet radiation at 260–280 nanometers. The Archean atmosphere was composed mostly of nitrogen, water vapor, and carbon dioxide, none of which absorb ultraviolet light in the 260–280 nanometer wavelength part of the spectrum. All microbes in the photic zone prior to the appearance of the absorbent and therefore safeguarding ozone layer were in danger of macromolecular destruction. Photosynthetic bacteria especially, as they require visible light, developed ultraviolet defense mechanisms or they died. Many modes of protection evolved, including enzymes that directly repaired ultraviolet damage to DNA integrity. Some methods of DNA repair act by removing damaged stretches of nucleic acid and resynthesizing DNA using available intact DNA from neighboring sources. If these sources were heterologous, that is, from other cells, plasmids, viruses or even DNA in solution, the phenomenon of recombination becomes one of sex according to our definition.

The appearance of the ozone layer made repair mechanisms less important for immediate survival. However, it is likely that by the time of the early Proterozoic Aeon, repair mechanisms had become integrated into the life cycles of cells. A series of bacterial symbioses have been hypothesized in which communities of bacteria formed larger nucleated cells [Margulis, 1981] (Fig. 2). In such communities, the recombinational sexuality of the component bacteria was utilized inside the emerging entity: the eukaryotic cell. This statement is deduced from many observations. Genes originating in the mitochondria have been found in the nucleus, whereas nonfunctional genes from chloroplasts have been found inside the mitochondria of corn cells [Stern and Lonsdale, 1982; Ellis, 1982]. Furthermore, the small protein component of ribulosebisphosphocarboxylase (RuBPCase) the CO_2-fixing enzyme of microbes and plants, has been found inside mitochondria [Lacoste-Royal and Gibbs, 1985]. Eukaryotic cells clearly have been "having sex with themselves." A vast quantity of genetic material in plant and animal cells are nonfunctional in the sense that they contain no information that codes for protein. These nonfunctional sequences of DNA are interpreted to be meaningless "junk" or "selfish DNA." Some of this DNA appears to code for proteins but in fact does not ("pseudogenes"). The discoveries of excess DNA point to a widespread tendency to retain molecular mechanisms that generate both accurate and defective copies of DNA, largely gratuitous with respect to protein synthesis. The essential propensity to mix and match genes, to

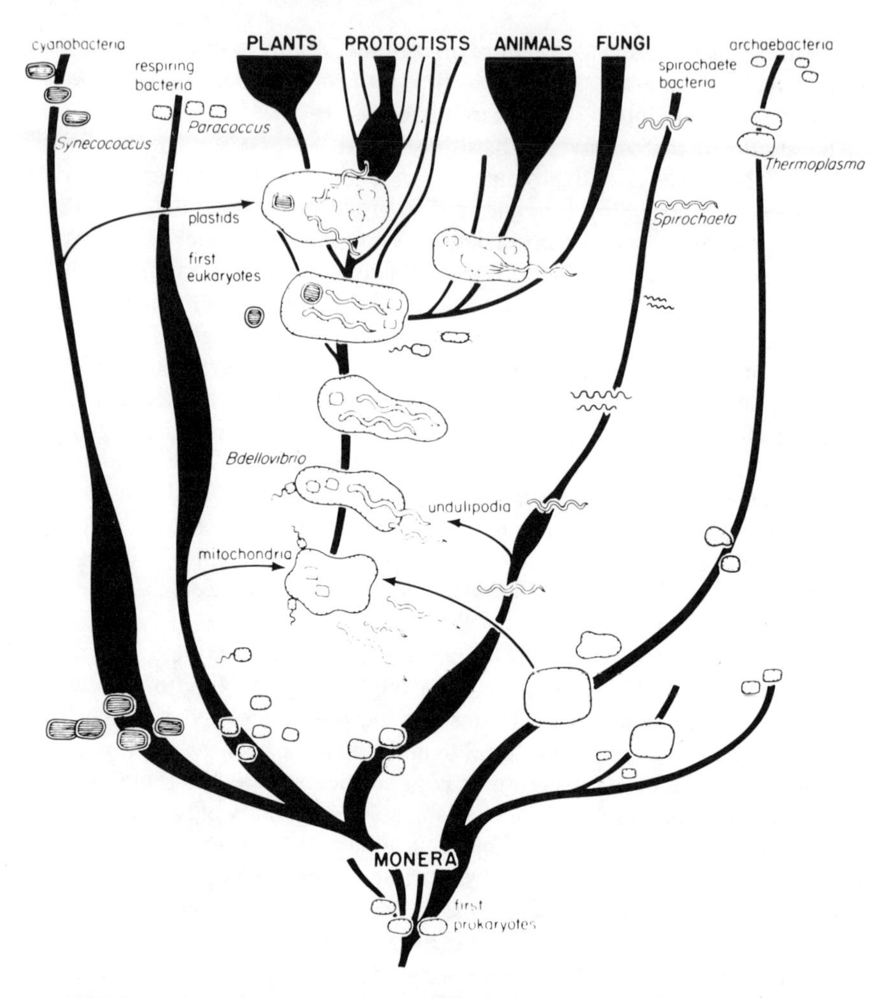

Fig. 2. Eukaryotic cells as microbial communities [see Margulis, 1981 for details].

accept foreign stretches of similar nucleic acid, probably was preserved along with the organisms that survived intense solar radiation and evolved bacterial sex.

THE ORIGINS OF MEIOTIC SEX

The origin of sex in plants and animals is not homologous to the origin of bacterial sex. Although meiotic sex brings separate genetic source materials together into one cell or organism, meiotic sex differs from bacterial sex in its origin, its details, and its complexity. A fundamental difference between meiotic and bacterial forms of sex is, of course, the obligate correlation of meiosis to organismal reproduction in many lineages. The bacterial cell does not require sexuality for reproduction as all bacteria reproduce by direct cell division or budding.

How did meiotic sex evolve, and how did it become tied to reproduction? Meiotic sex clearly evolved from mitotic cell division, the standard mechanism by which all animal, plant, and fungal cells reproduce. The origin of mitosis in Proterozoic protists is an evolutionary puzzle which has been discussed elsewhere [Margulis, 1981; Roos, 1984]. Although the details of the origin of mitosis are questions of conjecture, there is a general consensus that the ultimate advantage of mitosis was to distribute packaged DNA-protein complexes (chromosomes) evenly to offspring cells. By carrying just-reproduced chromosomes to opposite poles of a dividing cell, mitosis acts as a meticulously accurate genetic delivery system of eukaryotes that handles hundreds of times more DNA than the direct segregation of DNA on membrane that is characteristic of bacterial cells. Mitosis became the standard mechanism of cell division (and therefore of reproduction) in the first protoctists, that is, in nucleated cells prior to the appearance of animals, plants, and fungi in the fossil record.

Meiosis, the cell division that reduces by half the number of chromosomes in a nucleated cell, is very similar to mitosis and derived from it [Margulis, 1982]. The only major difference is that during meiosis first the kinetochores (also called spindle fiber attachments or centromeres) and later the chromosomes fail to double before the cell divides. The result is that only and exactly half the chromosomes of a diploid are incorporated into offspring cells. These cells with half the chromosome complement—often called sex cells (sperm, eggs, spores, or pollen)—sometime later became fertilized by counterpart sex cells. Fertilization restores the number of chromosomes and, in the case of an animal, begins the process of development with the fertilized egg or zygote.

Meiosis probably evolved with mitosis in protists such as the hypermastigotes, as Cleveland [1947] first argued. Members of the Kingdom Protoctista, which may have as many as 60,000 to 120,000 species [Corliss, 1984], show remarkable variations on the theme of mitosis and meiotic sex [Margulis and Schwartz, 1982; Margulis et al., 1985]. Meiotic sex evolved separately in at least three different lineages ("one-step" or "one-division" meiosis in apicomplexa and parabasalids like *Holomastigotoides* and "two-step" or "two-division" meiosis in heliozoans, foraminiferans, volvocaleans, opalinids, labyrinthulans, and ciliates [Raikov, 1982].

All organisms that undergo meiosis at some time during their life cycle always undergo mitosis during growth. No organisms are known that undergo meiosis in the absence of mitosis. Neither mitosis nor meiosis are universal methods of cell division in protists and mitosis is a prerequisite of meiosis in the sense that neither a haploid to diploid transformation (fertilization) nor a diploid to haploid transformation (meiosis) is possible in organisms unable to distribute regularly their chromosomes each cell division by mitosis. Meiotic organisms, therefore, are thought to have derived from mitotic ones.

One obligate aspect of meiotic sex is fertilization and the second is meiosis or chromosome reduction. Although either could have originally occurred by chance, these two processes had to become interlocked into a ritualized feedback cycle for meiotic sex to evolve. The exact environmental pressures which insured that the doubling of the chromosome number by fertilization always would be followed by the halving of chromosome members is the most difficult aspect of the evolutionary origin of sex to imagine. By themselves, however, each of the two processes can be visualized: Fertilization first occurred as a form of cannibalism, whereas meiosis originally came about as a disturbance in timing of the spindle attachment organelles, the kinetochores.

Protists ingest each other often, and yet digestion does not necessarily follow. L.R. Cleveland, the late Harvard professor, first proposed incomplete cannibalism as a precursor to fertilization when he saw a protist (a parabasalid zoomastiginid hypermastigote) living in the very dense communities of the hindguts of termites ingest another without digesting the genetic material of its prey's nucleus. Protists have no immune systems to prevent cell fusion, and such protistan cannibalism has been reported for stentors, amoebae, *Blepharisma,* and other protists as well. In some of Cleveland's observations, the nuclei of the undigested cannibalized cell fused with that of the cannibal. The cell membranes combined and new, fused, double hypermastigote protists were formed. Cleveland reasoned that in times of starvation such fusion of protists could have led to diploidy, which he considered a doubled state of chromosomes that had to be "relieved" by meiosis. Diploidy was originally accomplished in some cell lineages by a premature nuclear division (karyokinesis) and then fusion of the resulting cell nuclei in the absence of cell division (cytokinesis).

The second aspect of the origin of meiotic sex, the meiotic cell division itself, is somewhat more subtle. Also observed by Cleveland [1947] were cells in which the kinetochores that generally reproduce with the chromosomes and connect the chromosomes to the mitotic spindle during mitosis failed to divide on time. Thus, instead of moving one chromosome to each of the two offspring protist cells, each kinetochore moved two. The probability of frequent tardiness of the reproduction of these kinetochores is increased if we accept the aspect of the theory of serial endosymbiosis that regard kinetochores as remnants of spirochetes living in the chimera of the modern nucleated cell

[Margulis, 1981]. If the spirochete origin of kinetochores is vindicated, it will be relatively easy to comprehend their intracellular autonomy including the tendency of kinetochores to reproduce more rapidly than the chromosomes in which they are embedded. Such a process of kinetochoric reproduction (≡ centromeric fissioning) leads to karyotypic fissioning and has been an important mechanism in mammalian evolution [Todd, 1970; Guisto and Margulis, 1981]. When the kinetochores reproduce more slowly, the consequence is meiosis-like reduction division [Margulis, 1982].

Single-celled organisms doubled by fertilization-like cannibalism would be saved by meiosis-like reduction division, as this would return them to their original state, presumably the haploid condition to which they were optimally adapted. The problem is that in order that the meiotic life cycle be perpetuated, selection pressure on the haploid cell which favored diploidy would have to reappear. The establishment of a cyclical alteration of double (and single) chromosome states might have had something to do with an alternation of "good" and "bad" times, in other words, seasonal change. In times of food or water scarcity, for example, diploid cells with less surface area per volume might be favored. It is absurd to argue that homozygous lethals were masked in the first diploid organisms because those lethal genes would have been expressed, killing the haploid parents before the diploids were even formed. The original diploids were most likely selected for because, relative to the haploids, they tolerated environmental extremes such as starvation and desiccation. In times of plenty, selection pressure would restore the more rapidly reproducing haploids to prominence. Then again, if cannibalism itself provided sufficient nutrition in times of starvation, natural selection would favor those doubled organisms which were capable of cyclically ridding themselves of their extra genetic baggage.

However meiotic sex became established, once it did, it certainly flourished. Yet there is no evidence that obligately sexual organisms are more varied or better adapted to changing environments [Bell, 1982]. Indeed, a gamut of forced views on the maintenance of sex have been repeated, almost by rote, in popular science literature [Maranto and Brownlee, 1984]. One prevalent idea is the rejuvenation hypothesis, which suggests that meiotic sex is required to preclude aging. It is based on the observation that certain paramecia growing by cell division in the absence of sexual encounters survive for only months, whereas if conjugation occurs such that conjugating partners exchange nuclei the paramecia clones survive indefinitely; but this is not because their genes are "freshened up" with new genetic material. The generation of new variation is clearly totally irrelevant to the "rejuvenating" aspect of sex. This can be proved by study of "autogamy" in these same paramecia. Autogamy is a process in which a paramecium having undergone meiosis in preparation for conjugation, self-fertilizes. Paramecia undergoing autogamy also do not age; thus, autogamy is just as effective as a means of rejuvenation as sex.

Paramecium biaurelia has one large macronucleus containing many copies of genes and two small diploid micronuclei. The macro-nucleus synthesizes messenger RNA, whereas the diploid micronuclei, which reproduce only themselves, are inactive throughout cell division. During autogamy each diploid micronucleus divides twice meiotically, forming four haploid micronuclei. These four then divide mitotically once, to form eight haploid micronuclei. Then, in nature's typically abstruse style, all but one of the haploid micronuclei die. Seven nuclei are digested inside the paramecium. The eighth resistant nucleus divides mitotically to form two micronuclei with exactly the same genes. If a partner is present, conjugation will occur and one of the two micronuclei will travel to the partner as another one is received from the partner. However—and this is the crux of the matter— if no partner is available for conjugation, the two haploid nuclei, each from the same parent, will fuse. Not only have no new genes entered the conjugant, but the newly formed diploid nucleus is homozygous for all loci; all nuclear genetic variation has been removed. Yet the former conjugant is recharged, "rejuvenated," able to reproduce again for generations before the next conjugation or autogamy.

The fact that autogamy, which includes the process of meiosis but in which no new genes are received, is just as beneficial to the paramecia as sexual conjugation shows that mixis itself, the formation of an individual by two parents, is not the source of rejuvenation. Meiosis, which often is accompanied by mixis (but need not be) is apparently crucial for the continued health of the clone. We conclude that meiosis, which during its origin became correlated with biparental sex, and not biparental sex itself, was probably a prerequisite for the evolution of complex organisms. Thus, the general perception in the scientific community of a causative relation between complex morphological development and biparental sexuality (and its genetic variation) is in fact only a correlation. Mixis or outbreeding is an epiphenomenon of meiotic sexuality. It has been selected for because of the advantages it confers as the only known pathway to extensive tissue differentiation (in most animals and many plants).

We believe that meiotic sex flourished not because of its tendency to mix genes from separate sources and to generate genetic variation, but because it became fixed in the life cycle of a rapidly evolving group—the first animals. In animals there seems to be a direct relationship between the development of meiotic sex and that of tissue differentiation. Whereas there is no evidence that biparental sex was selected for because it permitted organisms rapidly to adapt to changing environments, there is a suggestion that meiotic sex evolved as part of the process of differentiation. We have developed the hypothesis that meiosis itself, with the formation of synaptonemal complexes, is a process somewhat like a roll call or game plan ensuring that all genes are in formation prior to the unfolding process of development of the multicellular animal or the plant embryo. Meiosis is hypothesized to be indispensable to tissue differentiation whereas mixis is only a nearly inextricable component of meiotic sex. Biparental meiotic sex was maintained because it was obligatorily associated with physiological necessities in the differentiated organisms in which it evolved. When these organisms survived, so did the process.

One of the many pieces of evidence to support our conclusion are the observations of Haskins and Therrien [1978]. Although differentiation and synaptonemal complexes are well known in *Echinostelium minutum,* there is no evidence for alternating haplo- diploidy changes, mixis fertilization or meiosis of any kind. On the contrary, both amoeba and plasmodial stages in the life cycle display the same ploidy levels. In such cases it is difficult to argue that mixis is required for variation or that synaptonemal complexes form to allow crossing over in meiosis.

CONCLUSION

We are developing a new view of sexuality. Bacterial sexuality is a strategy for immediate survival by which these microorganisms received new genetic components as easily as people catch a cold. The evolution of bacterial sex was a prerequisite to the origin of eukaryotic cells. Mechanisms of gene transfer by recombination were necessary among evolving microbial communities that led to the origin of the first eukaryotes. These mechanism also are employed in the crossover process of meiosis in modern eukaryotic cells. Only after eukaryotic cells evolved microtubule-based mitotic cell division could meiosis appear as a variation on mitosis. Subsequently, in protoctists and their multicellular descendants, meiosis and fertilization evolved together, first as part of the tissue differentiation process and later (in some lineages) intrinsic to reproduction.

We assert that the sexual, or mictic, aspect of meiotic sex has been misunderstood. Meiosis is obligate for reproduction and development in animals and plants with highly differentiated tissues and organs and, because of evolutionary history, this meiotic process tends to be associated with biparental sex. Since we, as mammals, obligately reproduce by meiotic sex, and since reproduction is imperative, it is counterintuitive to insist that biparental sexuality itself is not maintained by natural selection. Sex is so deeply rooted in the human thought process that we even hypostatize gender, referring to neutral objects as "he" or "she." As biologists, we invented the just-so story of enhanced variety to explain the existence of sex in a world where ramets so clearly reproduce more quickly; but it is the complex multicellular animals and plants themselves, not biparental sex, that has been preserved by natural selection. We require reproduction by way of meiotic sex because our ancient, single-celled ancestors survived by way of meiotic sex. Sexual reproduction is still a waste of time and energy. If the evolutionary process can circumvent biparental sex and still preserve complex multicellularity, it will. If human beings can devise a way to clone themselves from single parents and if the work of the cloners is reinforced by pleasure, this surely will occur. Cumbersome biparental sex will be bypassed. If our analysis is correct, however, meiosis cannot be dispensed with in any animal or plant even after two-parent sex disappears.

REFERENCES

Alexopoulos CJ (1962): "Introductory Mycology." New York: John Wiley & Sons, Inc.

Bell G (1982): "The Masterpiece of Nature: The Evolution and Genetics of Sexuality." Berkeley University of California Press.

Bernstein H, Byerly HC, Hopf FA, Michod R (1984): Origin of sex. J Theor Biol 102 (in press).

Bernstein H, Byerly HC, Hopf FA, Michod R (1985): The evolutionary role of recombinational repair and sex. Int Rev Cytol (in press).

Canuto VM, Levine JS, Augustsson TR, Imhoff CL (1982): UV radiation from the young sun and oxygen and ozone levels in the pre-biological palaeoatmosphere. Nature 295:816–820.

Cleveland LR (1947): Origin and evolution of meiosis. Science 105:287–288.

Cloud PE Jr (1983): The biosphere. Sci Am 249:176–189.

Corliss JO (1984): The kingdom Protista and its 45 phyla. Biosystems 17:87–126.

Ellis J (1982): Promiscuous DNA—chloroplast genes inside a plant mitochondrion. Nature 299:678–679.

Giusto J, Margulis L (1981): Karyotypic fissioning theory and the evolution of old world monkeys and apes. BioSystems 13:267–302.

Glaessner MF (1984): "The Dawn of Animal Life." Cambridge: Cambridge University Press.

Haskins EF, Therrien CD (1978): The nuclear cycle of the myxomycete *Echinostelium minutum*. 1. Cytophotometric analysis of the nuclear DNA content of the amoebal and plasmodial phases. Exp Mycol 2:32–40.

Kendrick B, Carmichael JW (1973): Hyphomycetes. In Ainsworth GC, Sparrow FK, Sussman AS (eds): "The Fungi." New York: Academic Press, Vol. IVA, pp 323–509.

Lacoste-Royal G, Gibbs S (1985): *Ochromonas* mitochondria contain a specific chloroplast protein. Proc Natl Acad Sci USA 82:1456–1459.

Maranto G, Brownlee S (1984): Why sex. "Discover" magazine, February. New York: Time, Inc., pp 24–28.

Margulis L (1981): "Symbiosis in Cell Evolution." San Francisco: W.H. Freeman & Company.

Margulis L (1982): Microtubules in microorganisms and the origins of sex. In Cappucinelli P, Morris NR (eds): "Microtubules in Microorganisms.." New York: Marcel Dekker Inc, pp 341–349.

Margulis L, Chapman D, Corliss J (1986): "Protoctista—A Guide to the Algae, Protozoa, Slime Molds, Slime Nets, Water Molds, Sporozoa, and the Other Protoctists." Boston: Jones & Bartlett Publishers, Inc. (in press).

Margulis L, Sagan D (1984): Evolutionary origins of sex. In Dawkins R, Ridley M (eds): "Oxford Surveys in Evolutionary Biology." London: Oxford University Press, 1:16–47.
Margulis L, Sagan D (1985): "Origins of Sex." New Haven: Yale University Press (in press).
Margulis L, Schwartz KV (1982): "Five Kingdoms." San Francisco: W.H. Freeman & Company.
Margulis L, Walker JCG, Rambler MB (1976): A reassessment of the roles of oxygen and ultraviolet light in Precambrian evolution. Nature 264:620–624.
Maynard-Smith J (1978): "The Evolution of Sex." Cambridge: Cambridge University Press.
Raikov IB (1982): "The Protozoan Nucleus." Heidelberg: Springer-Verlag.
Roos UP (1984): From proto-mitosis to mitosis—an alternative hypothesis on the origin and evolution of the mitotic spindle. Origins of Life 13:183–193.
Schopf WJ (1983): "Earth's Earliest Biosphere." Princeton: Princeton University Press.
Sonea S, Panisset M (1983): "A New Bacteriology." Boston: Jones & Bartlett Publishers, Inc.
Stern D, Lonsdale D (1982): Mitochondrial and chloroplast genomes of maize have a 12 kilobase DNA sequence in common. Nature 299:698–702.
Todd NB (1970): Karyotypic fissioning and carnivore evolution. J Theor Biol 26:445–480.
Varela F, Maturana H (1974): "Autopoiesis and Cognition." Dordrecht: Reidel Publishing Co.
Vidal G (1984): The oldest eukaryotic cells. Sci Am 250:48–57.

The Origin and Evolution of Sex, pages 87–88

General Overview

Edward A. Adelberg

Department of Human Genetics, School of Medicine, Yale University, New Haven, Connecticut 06510

In order to speculate intelligently about the origin and evolution of sex, we must first accomplish the following: 1) define sex; 2) describe it as it occurs in the lowest to the highest forms of life; 3) search for relationships that might represent true evolutionary progressions; and 4) consider the selective forces that might have driven the process.

I will start, then, with a definition: Sex is the process whereby a cell containing a new combination of genes is produced from two genetically different parent cells. In this book are presented many descriptions of the sexual process in prokaryotes (bacteria), lower eukaryotes (fungi, algae, protozoa), and higher eukaryotes (metazoans, from Cnidiaria to *Homo sapiens*). The challenge will be to recognize specific evolutionary developments; I will begin this process by asking if the sexual mechanisms that operate in the lower eukaryotes have any antecedents in the prokaryotes.

The pertinent features of sexuality in prokaryotes and lower eukaryotes are presented in Table 1. The differences are so great (intrachromosomal DNA exchange being the only common feature) that it seems to me likely that sexuality as defined above arose independently in the two groups; I can see no relation between the prokaryotic system of partial transfer and the eukaryotic system of genetic fusion, nor between prokaryotic partition systems and mitosis/meiosis.

An analogue of genetic fusion does occur, however, in prokaryotes: namely, *protoplast fusion*. It is a laboratory artefact in present-day forms, but it is conceivable that it served as a primitive sexual process prior to the emergence of eukaryotes, which retained it and modified it to produce the fusion process seen today in these forms.

In looking for possible transitions between the bacterial systems and the meiotic systems of eukaryotes, I am intrigued by the process (discovered by

TABLE I. Features of Sexuality in Prokaryotes and Lower Eukaryotes

Structure or process	Prokaryotes		Lower eukaryotes
Genome	Single, relatively simple chromosome		Several, relatively complex chromosomes
Vegetative partition	Membrane attachment		Mitosis
Zygosis	Transformation Transduction Conjugation	Partial transfer	Gamete fusion complete diploidy
Intrachromosomal DNA exchange	Yes		Yes
Chromosomal reduction	Dilution out of non-replicating, incomplete genote		Meiosis

Guido Pontecorvo) called *parasexuality,* which occurs in certain fungi (Aspergillus) and slime molds (Dictyostelium). In these organisms, haploid heterokaryons are formed by cell fusion. As a rare event, two genetically dissimilar nuclei fuse to produce a diploid, heterozygous nucleus, which proliferates mitotically in the coenocytic cytoplasm. From time to time, random nondisjunctions occur, eventually restoring the haploid condition; intrachromosomal exchanges (mitotic recombination events) take place throughout the process, so that the final haploid cell is a true recombinant. Parasexuality may have preceded meiosis during the evolution of sexuality in eukaryotes.

During the further evolution of eukaryotes, sexuality became a more and more complex process. There emerged such features as persistent diploidy, sex chromosomes, and all of the structures and activities associated with the differentiation of specialized gonadal tissues. These features will be described and discussed within the succeeding sections of this book.

Finally, there is the question of the selective forces that have driven the evolution of sex. As Dr. Zinder has pointed out, selection may at first have operated at the DNA level, favoring DNA molecules that gained the ability to spread from host to host in the form of plasmids and phages; only later, when these molecular parasites began to transfer host genes, might selection have begun to operate on new host gene combinations and on the systems that produced them. Dr. Kimball Atwood will give us his views on this subject.

In summary, then, many descriptions of sexual systems will follow in this text; let us keep in mind the challenge of understanding how these systems have originated and evolved.

The Origin and Evolution of Sex, pages 89–90

Bacterial Sexuality: Beginnings of Sexuality in Prokaryotes: Discussion

Leader: Edward A. Adelberg

Department of Human Genetics, Yale University School of Medicine, New Haven, Connecticut 06510

The discussion leader opened the session by asking Dr. Kimball Atwood to present his views on the selective forces that might have contributed to the evolution of sexuality. Dr. Atwood began by pointing out numerous examples of organisms that have evolved complex mechanisms for *preventing* sexuality, reproducing instead by parthenogenesis—as though sexuality is no longer an advantage in these species. He then addressed two issues raised by Dr. Norton Zinder: first, with respect to group *versus* individual selection, he reminded the audience that group diversity inevitably arose as a result of mutation and did not depend on horizontal gene transmission; second, with respect to periodic selection, he emphasized that the effect of this process is to make the population in question always appear to be about 200 generations old. Although the rate of accumulation of neutral mutations is not affected, periodic selection greatly reduces the number of different combinations of neutral mutations that ever arise; i.e., diversity is greatly reduced. The rate of horizontal gene transfer is sufficiently low to prevent its interfering with periodic selection. Finally, Dr. Atwood pointed out that persistent diploidy, which is believed to be advantageous because it prevents the expression of (recessive) deleterious mutations, is of recent evolutionary origin (at least in plants).

In the open discussion that followed, Dr. Charles Radding was asked whether an evolutionary trend could be discerned in the enzymatic mechanisms of DNA recombination. Radding responded by describing the Recl protein of the eukaryotic fungus Ustilago, which resembles the prokaryotic recA protein in its function but has several differences, including its ability to function at a much lower protein:DNA ratio and to promote the pairing of DNA duplexes. Dr. Alvin Clark then commented that the recA protein shares several structural properties with other nucleotide-binding proteins, from one or another of which it may have evolved, and that a good place to look for evolutionary antecedents of conjugation-promoting proteins would be the enzymes of replication.

The discussion then turned to two possible origins of mitosis (and thus of meiosis). Dr. Zinder considered the chromosomal transfer by bacterial Hfr donors to be a likely beginning and rejected Dr. Adelberg's nomination of protoplast fusion as the antecedent of mitosis on the ground that, in the bacterial diploids formed by protoplast fusion, only one parental set of genes appears to function. Dr. Adelberg then proposed that fungal parasexuality, as first described in Aspergillus, might have been the forerunner of meiosis; Dr. Maurice Sussman suggested that parasexuality in Dictyostelium would be an even better candidate, as it involves the fusion of haploid cells to form diploid cells rather than nuclear fusion within a coenocytic mycelium.

Dr. Ruth Sager called the audience's attention to the greater complexity of the eukaryotic chromosome: it is a differentiated structure, with a centromere and specialized termini that are made up of long runs of amplified (repeated) short DNA sequences. She proposed that such amplification might have been a novel process in the evolution of the eukaryotic chromosome, but Dr. Clark pointed out that such amplification is seen also in the replication origins and partition regions of plasmids, where they appear to be involved in binding to replication and membrane proteins.

The discussion then turned back to the question of the selective forces acting on sexuality. Dr. Harlyn Halvorson pointed out that in nature, bacteria are more likely to grow in aggregates than as free-swimming individuals, and he asked about the evolution of cell–cell interaction mechanisms. Dr. Zinder repeated a point that he had made in the opening talk of the session: that sexuality provides a rate of evolution that is two times faster than would be possible on the basis of asexual processes, as first proposed by R.A. Fisher, and this is sufficient to explain the selection for sexuality once it had originated. Dr. Bernard Davis stated that selection must operate on whole genomes, not on individual genes, and that sexuality, which recombines many genes, should therefore have more than a twofold effect on the supply of variation and hence on the rate of evolution. Dr. Atwood commented that it is difficult to make any quantitative prediction, since we do not know what the selective forces actually are in nature.

The evening's discussion ended with an exchange of views on the role of genetic transposition in the evolutionary process. Dr. Sager cited Werner Arber, who found over 50% of spontaneous mutation in certain phages to result from transposition; further, Dr. Sager suggested that the duplications generated by transposable elements may have been an important source of raw material for the evolution of new gene functions, including those necessary for sexuality. Dr. Atwood disagreed: he pointed out that transposable elements cause mutations principally by disrupting the function of existing genes, and he doubted that transposition would cause a significant number of beneficial changes. Dr. Zinder also doubted the significance of transposable elements in producing the raw material of evolution, stating 1) that the transposable elements discerned in *Drosophila* appear to be of very recent origin and 2) that Dr. Arber's finding (of transposable elements as the major source of spontaneous mutation) probably could not be generalized.

THE DNA OF SEX IN UNICELLULAR EUKARYOTES

The Origin and Evolution of Sex, pages 93–96

Introduction to the Second Day: Microbial Sexual Eukaryotes

Ruth Sager

Division of Cancer Genetics, Dana-Farber Cancer Institute, and Department of Microbiology and Molecular Genetics, Harvard Medical School, Boston, Massachusetts 02115

One way to identify the eukaryotic sexual process is by the life cycle. The centerpiece of the sexual life cycle is meiosis, in which diploids are reduced back to haploids by a process that includes recombination of chromosomal genes. To complete the cycle, haploid gametes fuse to form the diploid zygote or fertilized egg.

In the single-celled organisms we will be discussing—yeast, *Chlamydomonas*, and the ciliates—the sex difference is determined by a single gene locus: the mating type locus. In yeast the two mating types are called *a* and *alpha*, in Tetrahymena they are *A* and *a*, and in *Chlamydomonas* + and −. The disarming simplicity of a single genetic locus controlling sexual differentiation is, however, quite misleading. Mating type genes regulate many complex reactions at the molecular level. Indeed, today's research is focussed on new understandings of fundamental sexual processes as they are being worked out in molecular terms. This book will present new lines of research and new technologies that are updating our knowledge of molecular mechanisms involved in the regulation of sex in three organisms: yeast, Tetrahymena and *Chlamydomonas*. As an introduction, I wish to call attention to a new topic in the evolution of sex, a subject only about 2 years old, born of the new technology, and with considerable evolutionary significance—namely, the subject of *Promiscuous DNA*.

Promiscuous DNA, the term coined by Ellis [1982], refers to a DNA sequence found in more than one of the three genetic compartments of the eukaryotic plant cells: nucleus, chloroplast, and mitochondria, or in both nucleus and mitochondria of animal cells. The existence of promiscuous DNA is itself remarkable. It should be recalled that chloroplast and mitochondrial DNAs were both identified initially on the basis of their genetic and physical uniqueness: Each is distinctive in DNA sequence. Furthermore, in mitochondria, even the genetic code for translation of nucleic acid sequence into amino acid sequences is different from that in the nucleus.

Nonetheless, examples have recently been reported of chlDNA sequences that hybridize with mitDNA and others that hybridize with nuclear DNA and of mitDNA sequences also present in the nuclear DNA fraction. The chlDNA studies have been carried out primarily with maize [Lonsdale et al., 1983], but examples are also being found in other higher plants including pea, mung bean, and spinach [Stern et al., 1984]. In spinach, sequences of chlDNA have also been found in nuclear DNA [Timmis et al., 1983]. Because the nuclear sequences are incomplete copies of sequences known to be translated in the chloroplast, it seems likely that the direction of transposition was from the chloroplast to either the mitochondria or the nucleus. Examples have also been reported of transposition of mitDNA sequences into the nucleus in the sea urchin [Jacobs et al., 1983], in yeast [Farrelly et al., 1983], and in the fungus Podospora [Wright et al., 1983]. Of particular importance is the fact that, in all of these transpositions from mitDNA to the nucleus, the transposed sequences are found to be flanked by terminal inverted repeats, analogous to those of retroviruses and classical transposons. None of the transpositions so far identified appear to produce functional transcripts, although the fact that they exist at all is very important. Their presence alerts us to the fact that intracellular, interorganellar transpositions occur and thereby provide the potential for transcriptional activity.

In the studies in maize, for example, one 12 kb sequence in mitDNA was shown to be fully homologous to a region in chlDNA that contains coding sequences for $tRNA^{ile}$, and $tRNA^{val}$, and the 16S rRNA [Stern et al., 1982]. Two other mitDNA fragments from maize contain sequences that hybridize to chlDNA: one to the 3′ end of 23S rRNA and the other to a part of the large subunit of ribulose 1-5 bisphosphate carboxylase (rbcL) [Lonsdale et al., 1983]. These sequences are all within the *inverted repeat* regions of plant chloroplast DNA, regions that may be undergoing continuous rearrangement by homologous recombination [Palmer, 1983]. If fragments of DNA were released during recombination events, they could become available for incorporation into other cellular DNAs. Transfer from chloroplast to mitochondrion could be facilitated by transient membrane fusions of the kind described for several species [Stern et al., 1984].

Other suggested mechanisms [Stern et al., 1984] include DNA transformation per se and transposition by attachment to a transposable element containing suitable sequences for integration.

Cytoplasmic male sterility in maize provides a fascinating example of plasmidlike structure and behavior of segments of mitochondrial DNA [Kemble et al., 1983]. Plants with "S" cytoplasm transmit male (pollen) sterility from the female parent but not from the male, as in maternal inheritance of chloroplast genes. "S" cytoplasm contains two linear molecules, S-1 (6 kb) and S-2 (5 kb), which have 200 bp inverted repeats at their ends, a structure typical of transposons and retroviruses. The S-1 and S-2 sequences are also present in integrated form in mitDNA. When cytoplasmic male sterility is lost, these linear DNA molecules disappear from the cytoplasm, but S-1 is then found in nuclear DNA and S-2 in mitDNA. The fact that S-1 and S-2 have the structural features of transposons, i.e., terminal inverted repeats, suggests that transposition might be a common feature of organellar as well as nuclear DNAs, occurring not only between but also, perhaps more commonly, within each DNA compartment.

With respect to the topic of this book, the phenomenon of promiscuous DNA introduces evidence of a novel evolutionary mechanism. Genes present on organelle DNAs are maternally transmitted in the sexual life cycle, whereas those on nuclear chromosomes are subject to classical meiotic reshuffling. Thus if a functional gene is transposed from organelle to nucleus its pattern of inheritance changes. Precisely this effect, the switching of male sterility from maternal to biparental inheritance [Laughnan et al., 1973], led to the discovery that the mitochondrial S-1 had been transposed into the nucleus in maize.

In summary, the available evidence suggests that promiscuous DNA is functional only in a single location and that the direction of transfer has been from the organelles to the nucleus and from chloroplast to mitochondrion. One mechanism of transfer is probably analogous to that of transposons, but other mechanisms might also be operative. The occurrence of this phenomenon, however, represents a functional link between eukaryotic and prokaryotic sexual mechanisms. As discussed earlier in this volume, transposition is a parasexual mechanism. The likelihood that it still operates in eukaryotes can be regarded as a heritage from the evolutionary past. The frequency of interorganellar transposition and its long-term importance in evolution remain open questions for future investigation.

REFERENCES

Ellis J (1982): Promiscuous DNA—Chloroplast genes inside plant mitochondria, news and views. Nature 299:678–679.

Farrelly F, Butow RA (1983): Rearranged mitochondrial genes in the yeast nuclear genome. Nature 301:296–301.

Jacobs HT, Posakony JW, Grula JW, Roberts JW, Xin J-H, Britten RJ, Davidson EH (1983): Mitochondrial DNA sequences in the nuclear genome of Strongylocentrotus purpuratus. J Mol Biol 165:609–632.

Kemble, RJ, Mans RJ, Gaby-Laughnan S, Laughnan JR (1983): Sequences homologous to episomal mitochondrial DNAs in the maize nuclear genome. Nature 304:744–747.

Laughnan JR, Gabay SJ (1973): Mutations leading to nuclear restoration of fertility in S male-sterile cytoplasm in maize. Theor Appl Genet 43:109–116.

Lonsdale DM, Hodge TP, Howe CJ, Stern DB (1983): Maize mitochondrial DNA contains a sequence homologous to the ribulose-1,5-bisphosphate carboxylase large subunit gene of chloroplast DNA. Cell 34:1007–1014.

Palmer JD (1983): Chloroplast DNA exists in two orientations. Nature 301:92–93.

Stern DB, Lonsdale DM (1982): Mitochondrial and chloroplast genomes of maize have a 12-kilobase DNA sequence in common. Nature 299:698–702.

Stern DB, Palmer JD (1984): Extensive and widespread homologies between mitochondrial DNA and chloroplast DNA in plants. Proc Natl Acad Sci USA 81:1946–1950.

Timmis JN, Scott N (1983): Sequence homology between spinach nuclear and chlorplast genomes. Nature 305:65–67.

Wright RM, Commings DJ (1983): Integration of mitochondrial gene sequences within the nuclear genome during senescence in a fungus. Nature 302:86–88.

The Origin and Evolution of Sex, pages 97–111

Sex in Budding Yeast: How and Why

James B. Hicks, John M. Ivy, Jeffrey N. Strathern, and Amar J.S. Klar
Cold Spring Harbor Laboratory, Cold Spring Harbor, NY 11724

INTRODUCTION

Sex, in one form or another, is basic to nearly all forms of life. The sexual cycle, sometimes known as the "alternation of generations," provides the means for continuous reorganization of the gene pool and, coupled with mutation, generates the diversity required for evolutionary adaptation. Despite its basic importance, our knowledge of the mechanisms of sexual determination and differentiation is limited. At the physiological and cytogenetic levels we know that the mechanisms for sexual determination are many and varied. As described elsewhere in this volume, sex can be controlled by environment (in certain fish and reptiles), by haplo-diploidy (insects), by the sex chromosome-to-autosome ratio (insects), by male heterogamy (XY males and XX females, mammals), by female heterogamy (ZW females and ZZ males, birds), or by the presence of the male-determining plasmid factors (bacteria). Baker's yeast, *Saccharomyces cerevisiae,* is one of the few experimental organisms in which these processes are understood in molecular detail. In this paper we will summarize the molecular mechanisms of sex determination in yeast with the hope that understanding any system in detail will allow comparison with other systems and help shed light on the origin and the evolution of sex in general. For a more comprehensive treatment of this material, the reader is referred to several recent reviews [Herskowitz and Oshima, 1981; Haber, 1983; Nasmyth, 1983a, Klar et al., 1984b].

THE BUDDING YEAST HAS TWO SEXUAL TYPES

In heterothallic fungi the mating-type genes control sexual compatibility. The budding yeast, *S. cerevisiae* an ascomycetes fungus, exhibits a bipotential sexual system in which the mating type is determined by the alternate states (alleles) of a single locus, *MAT*, situated on the right arm of chromosome III. Such sexual types are designated a and α, respectively, controlled by the *MAT*a and *MAT*α alleles.

Mating occurs only between cells of opposite mating type. Mating involves cell cycle synchronization mediated by pheromones and cell fusion facilitated by agglutination factors [reviewed by Thorner, 1981]. Successful fusion generates the third "*MAT*a/*MAT*α" diploid cell type. The *MAT*a/*MAT*α cells do not mate and have the unique ability to undergo meiosis and sporulation. Completion of the sexual cycle generates an ascus with 2a and 2α haploid meiotic segregants.

YEAST SWITCHES SEX EFFICIENTLY

The *MAT* allele of almost all the commonly used laboratory strains behaves as a stable Mendelian trait. These strains are termed *heterothallic* and were selected for study because of their stability [Lindegren and Lindegren, 1943]. Almost all strains isolated from nature, however, contain the *HO* (for homothallism) gene on chromosome IV. Such strains defy conventional genetic rules of gene stability; the *MAT*a and *MAT*α alleles interchange nearly as often as every generation in a programmed fashion [Winge and Roberts, 1949; Hawthorne, 1963; Oshima and Takano, 1971; Hicks and Herskowitz, 1976; Strathern and Herskowitz, 1979]. This interchange, often called *homothallic switching*, generates a population of *MAT*a and *MAT*α cells that are capable of mating. Once mated, the *HO* gene is turned off and the *MAT* alleles remain stable in the diploid [Jensen et al., 1983]. It has been determined that heterothallic laboratory strains maintain a defective copy of the *HO* gene [Jensen et al., 1983], indicating that they were derived from a normal homothallic stock.

Soon after the discovery of two mating types in yeast it was realized that haploid spore progenies can regenerate diploid *MAT*a/*MAT*α cultures, a process requiring the *HO* gene function. Conventional genetic crosses led to the discovery of two other loci, *HM*a and *HM*α. Naturally occurring variants, *hm*a and *hm*α, control the direction of switching [Takano and Oshima, 1970; Naumov and Tolstorukov, 1973; Klar and Fogel, 1977]. For simplicity they have been renamed *HML* (*L* for left) and *HMR* (*R* for right). *HML* is located on the left arm about 100 kb away from *MAT*; *HMR* is located on the right arm of chromosome III about 120 kb away from *MAT* [Harashima and Oshima, 1976; Klar et al., 1980a; Strathern et al., 1979b].

To define the role of the *HM* loci, Oshima and colleagues proposed that the *HM* loci code for controlling elements that attach to *MAT* and thereby differentiate it into *MAT*a or *MAT*α [Oshima and Takano, 1971; Takano et al., 1973]. A specific proposal of this kind, the cassette model, proposed that the *HML* and the *HMR* loci contain unexpressed mating type information, a copy of which is transposed to *MAT* [Hicks et al., 1977]. The new information replaces the old *MAT* allele, resulting in a switch. This proposal was suggested to explain the unexpected "curing" by homothallism of mutations mapping within the *MAT* locus [Takano et al., 1973; Hicks and Herskowitz, 1977; Klar et al., 1979b; Strathern et al., 1979a]. Definitive genetic experiments showed that mutations originally isolated in the *HM* loci can be faithfully copied and transferred to *MAT* where the mutant allele is expressed [Klar and Fogel, 1979; Kushner et al., 1979; Klar, 1980]. Confirmatory evidence came from characterization of the DNA clones containing *MAT, HML,* and *HMR* [Hicks et al., 1979; Nasmyth and Tatchell, 1980; Strathern et al., 1980; Astell et al., 1981]. Each cassette showed extensive DNA sequence homology with the others (Fig. 1), and the *MAT*a and *MAT*α alleles differed only by the allele-specific Y region. Most strains contain an α cassette at *HML* and an a cassette at *HMR*.

Sex determination in yeast is highly programmed and regulated with respect to 1) which cassette switches, 2) which cell in a cell lineage switches, 3) which of the *HM* loci is used as the donor, and 4) which cassette is expressed.

In wild-type homothallic strains, only the cassette at *MAT* switches. Recent studies have shown that *MAT* switching is initiated by a double-stranded DNA break at *MAT* [Strathern et al., 1982; Klar et al., 1984a; see also Malone and Esposito, 1980; Weiffenbach and Haber, 1981]. The reaction is catalyzed in vivo by the *HO*-encoded sequence-specific endonuclease. This activity cleaves the *MAT* locus 3 bases into the Z1 region from the Y/Z boundary (Fig. 1) and generates a 4-base sticky end [Kostriken et al., 1983]. The cut persists transiently for only about 5% of the cell cycle and probably is made early in DNA synthesis [Strathern et al., 1982]. It has been proposed that the broken ends at *MAT* invade the intact homologous sequences present at the *HM* loci. DNA repair synthesis using the *HM* loci as the template would produce a *MAT* switch. Such a DNA transposition mechanism [Klar et al., 1980b; Haber et al., 1980; Strathern et al., 1982] is analogous to the process of gene conversion so well established in *S. cerevisiae*.

Because yeast multiplies by budding, mother and daughter cells can easily be distinguished. It is quite interesting that there is an invariant program for which cell in a lineage switches. Figure 2 illustrates three rules of switching [Hicks and Herskowitz, 1976; Strathern and Herskowitz, 1979]. First, only the mother (older) cells switch, and daughter (new bud) cells never switch. Second, switches always occur in pairs of cells; both progenies of a mother cell switch efficiently. Third, the mother cells switch in about 86% of the cell divisions. Such a high percentage is much greater than the predicted maximum 50% should the choice of *HM* locus used be random. Normal

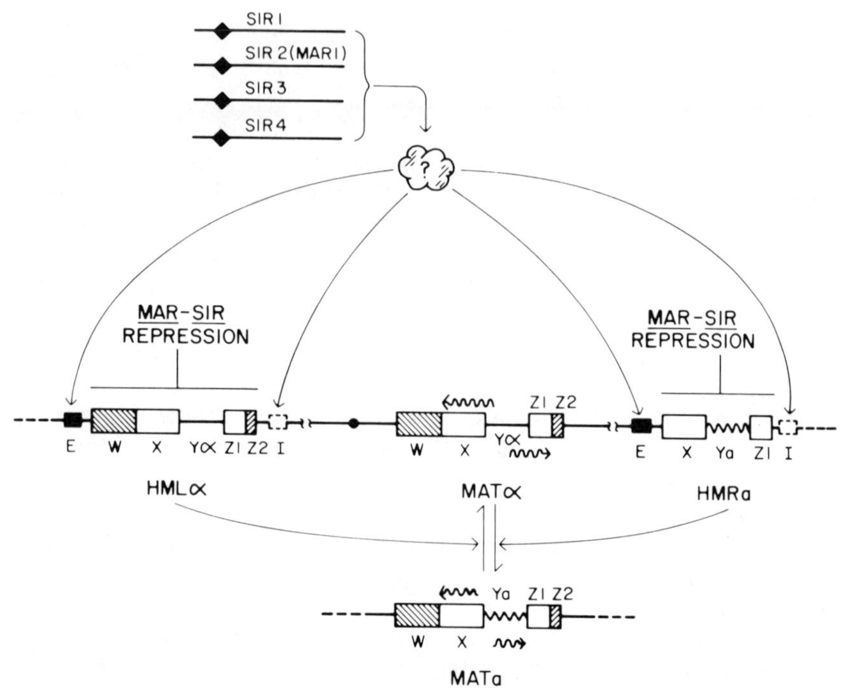

Fig. 1. Arrangement of *HML*, *MAT*, and *HMR* loci on chromosome III and the process of mating-type switching. Boxes represent regions of homology. The *MAT*a and the *MAT*α alleles differ by an allele-specific DNA substitution: Ya is a unique 642-bp and Yα is a unque 747-bp segment. The W, X, Z1, and Z2 regions are, respectively, 723, 707, 239, and 88 bp long. *MAT* switching occurs by substituting information residing at *MAT* with the information copied from either *HML* or *HMR* by a unidirectional transposition process. The DNA sequence replaced at *MAT* is apparently lost. *MAT*α and a code for two RNA transcripts (wavy lines) that are divergently transcribed [Klar et al., 1981a; Nasmyth et al., 1981]. *HML* and *HMR* loci are kept unexpressed by the concerted action of *trans*-acting *MAR/SIR* gene products and the *cis*-acting E and I regions.

homothallic cells, thus, must have a means to choose the *HM* locus containing the opposite cassette type in order to maximize switching efficiency.

Several of these features have now been explained. It has been shown that the *HO* gene expresses in mother cells and not in the newest buds (daughter cells) [Nasmyth, 1983b; R. Jensen and I. Herskowitz, personal communication]. The pair switching rule is probably explained by the precise timing of switching in the cell cycle such that both chromatids of the mother cell always switch together, possibly at the time of DNA replication of the *MAT* locus. Which cassette is inserted into *MAT* is dictated both by the expressed cell type and the contents of the *HM* loci. It has been demonstrated that a cells preferentially choose the *HML* locus as the donor, regardless of its

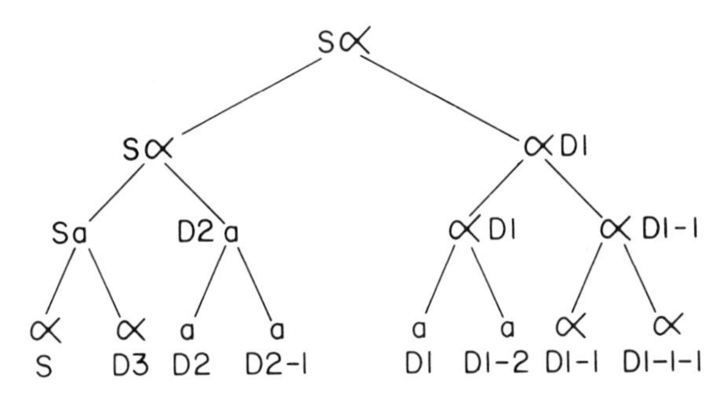

Fig. 2. The pedigree of mating-type switching in the descendants of a single homothallic spore (S) and its daughters (D1, D2, and so forth; D1–1 is the first daughter of D1, and so forth). The spore cell in the first division never switches. After producing D1 it is considered to be a mother cell and can switch efficiently. D1 will become competent to switch after it has produced its own unswitched offspring (D1–1). This asymmetric switching pattern continues indefinitely. Normally, however, if the cells are unperturbed, mating occurs between opposite types as early as at the four-cell stage to establish a stable *MAT*a/*MAT*α diploid. Switching in those diploids is shut off.

genetic content. Likewise, α cells choose the *HMR* locus preferentially. The preference seems to be mediated through competition between the *HM* loci; deletion of the efficient donor in a given cell type allows efficient use of the remaining normally inefficient donor [Klar et al., 1982]. In molecular terms, however, the basis for *HO* gene expression in mother cells only and the competition between the *HM* loci is not yet understood.

As is shown in Figure 1, the structure of *MAT* and the *HM* loci is quite similar at the DNA sequence level. Yet *MAT* behaves quite differently from the *HM* loci in two important aspects. The *MAT* locus switches and also is transcriptionally active. In contrast, the *HM* loci act only as donors of genetic information and remain unexpressed. Such behavior is even more interesting because the promotor(s) and the downstream sequences (over 1,000 bp) in the *HM* loci are identical to those found at the corresponding alleles of *MAT*. The task of keeping the *HM* loci unexpressed and unswitched is assigned to both *trans*- and *cis*-acting elements.

The *trans*-acting elements are four unlinked genetic loci, variously known as *MAR* (mating type regulator [Klar et al., 1979a]) or *SIR* (silent information regulator [Rine et al., 1979]) or CMT (change of mating type [Haber and George, 1979]) (Fig. 1). Mutations in any one of these unlinked genes allows expression of both the *HM* loci.

The negative *MAR/SIR* transcriptional control is presumably exerted through interactions with the specific *cis*-acting sites called *E* (essential) and I (important). Such sequences were defined by in vitro deletion analysis. These mutations allow expression of the adjacent *HM* cassette. The E site is situated on the left side over 1,500 bp away from the promoter; the I site is located over 1,000 bp to the right of the promoter region [Abraham et al., 1984; Feldman et al., 1984]. It is likely that the *MAR/SIR* control is mediated through differential organization of the chromatin at the silent loci [Nasmyth, 1982]. The *MAT* gene lacks the corresponding E and I sites, and therefore it is constitutively expressed.

Why does *MAT* switch and the *HM* loci do not? As was discussed above, the double-stranded break is found only at *MAT*. A few years ago it was shown that the *HM* loci also switch in a *mar*− background [Klar et al., 1981b; Haber et al., 1982]. It has been shown that in such strains the double-stranded break is also found at the *HM* loci [Klar, Strathern, and Abraham, 1984]. Therefore, the *MAR/SIR* control not only keeps the *HM* loci unexpressed but also keeps them from switching, presumably by not allowing the *HM* loci to get cleaved in vivo by the *HO*-encoded endonuclease.

In summary, then, budding yeast is a bipotential sexual organism. The two sexual types efficiently interconvert by transposing either a or α information to *MAT*, the sex-determining locus. Starting with a single homothallic spore, cells of the opposite mating type are generated, and efficient mating between the progeny ensues to generate a *MAT*a/*MAT*α meiotically competent cell type. The resulting heterozygous constitution of *MAT* turns off further switching by transcriptionally repressing the *HO* gene. Haploid segregants generated by meiosis will again repeat the switching and the mating cycle.

MAT IS THE MASTER REGULATORY GENE FOR DETERMINING SEX

MacKay and Manney [1974] originally proposed that *MAT* acts as the *master regulatory locus* maintaining the sexually regulated state, a function analogous to that served by the *sex lethal (Sxl)* locus of *Drosophila* (see chapter by Cline, this book). Recent studies have shown that α cells express a number of α-specific genes, whereas the a-specific genes are unexpressed. Likewise, the a-specific genes are on in a cells but off in α cells. The *MAT*α allele codes for two transcripts and specific roles of the *MAT* gene products have been assigned in the so-called α1-α2 hypothesis (Fig. 3) [Strathern et al., 1981]. The *MAT*α1 gene product acts as a positive regulator of the α-specific genes, whereas the *MAT*α2 gene product acts as a negative regulator of the bank of genes specific to the a cell type. The *MAT*a allele also codes for two transcripts, but no function has been assigned to them in a cells. In this scheme, the a-specific functions are proposed to be expressed in

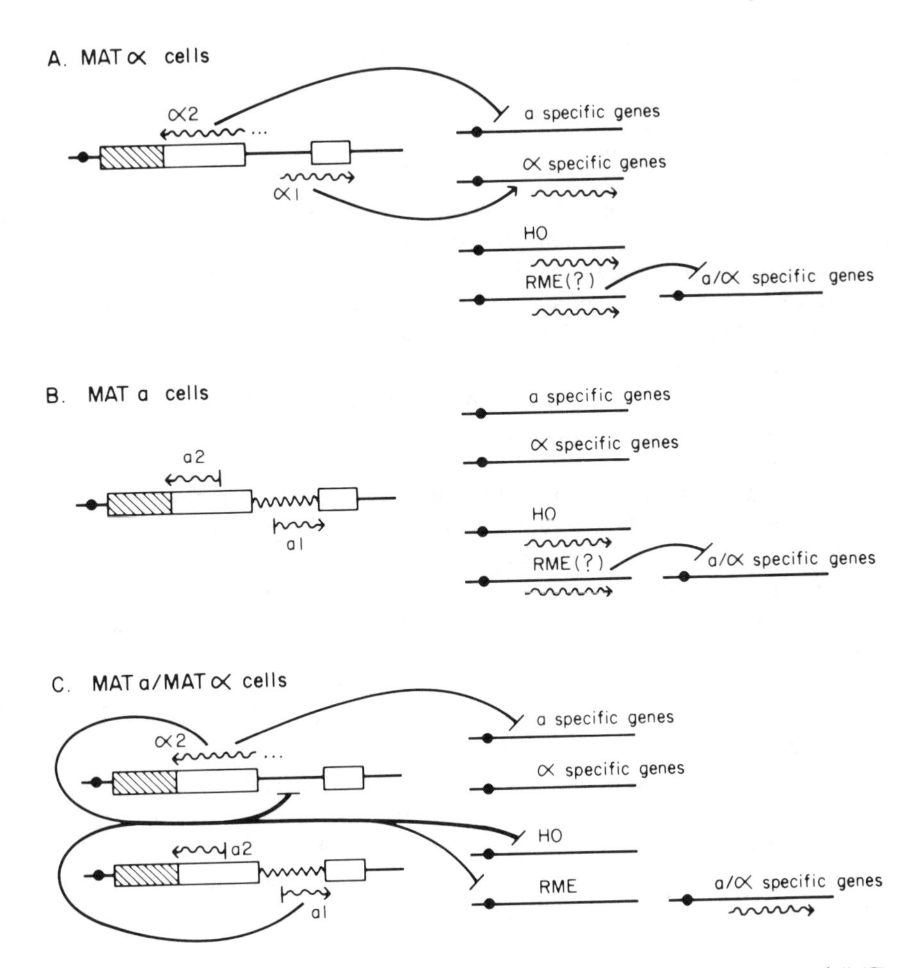

Fig. 3. The α1-α2 hypothesis: Role of the *MAT* transcripts in α (A), a̰ (B), and "a/α" (C) cell types. Arrows denote positive and flat lines indicate negative control over the unlinked cell type-specific genes. The *MATα1* gene is off in a̰/α cells.

*MAT*a̰ cells simply because of the lack of the negative regulatory *MATα2* function. Both *MAT*a̰ and *MATα* functions are, however, required in *MAT*a̰/*MATα* cells to turn off a̰- and α-specific functions and to turn on "a̰/α-specific" functions.

Recent molecular studies have shown several genes, those that were originally defined by genetic analysis as essential for mating, to be under the transcriptional control of *MAT* [Sprague et al., 1981, 1983], thus confirming the central elements of the α1-α2 hypothesis. Therefore, by interconverting alleles at one locus, *MAT*, budding yeast changes the state of expression of many unlinked genes.

EVOLUTIONARY ASPECTS

As in most organisms, the sexual cycle of yeast involves two distinct processes: mating or cell fusion between haploid gametes to form a diploid and meiosis or reductional division of diploid cells to form new haploid gametes, each containing a potentially novel complement of genetic material. This continual reassortment of the genetic material generates diversity, which is assumed to benefit survival of the individual and evolution of the species [Fisher, 1958]. Also, it is clear in higher plant and animal systems that maximum heterozygosity is usually beneficial, often resulting in so-called "hybrid vigor."

At first glance, however, the homothallic life cycle of yeast seems futile in both these regards. What is the advantage of allowing sibling progeny of a single haploid cell to mate and form a virtually homozygous diploid? Such questions are always difficult to answer in the laboratory far away from the subtle evolutionary pressures that exist in the natural environment. Nonetheless, they are ripe for speculation.

First, it is interesting to compare the *Saccharomyces* life cycle to that of other organisms and to note that it is by no means unique. Many lower plants are also homothallic and, like *Saccharomyces,* have free-living haploid and diploid phases. Ferns, mosses, liverworts, and hornworts, among others, fall into this category. Sexual reproduction in most common ferns, for example, begins when leaf cells in the familiar diploid form of the plant, the sporophyte, undergo meiosis, producing haploid spores. These spores are dispersed and, under appropriate conditions, germinate on the ground to give rise to simple, yet independent, haploid plants or gametophytes. As the name implies, this form of the plant is the source of the differentiated mating cells or gametes, in this case eggs and motile sperm. Although the haploid form of the fern is more complex than a colony of haploid yeast cells derived from a homothallic spore, its genetic lineage is exactly identical. Each gametophyte represents a clone of haploid cells that, within 11 rounds of cell division, undergoes several simple differentiation events resulting in mating cells. Homothallic yeast spores simply take a more direct path to arrive at the same point, i.e., one-cellular differentiation involving interconversion of *MAT* alleles. In both cases mating or sexual fertilization can occur between sibling gametes from the same clone resulting in a homozygous diploid, or between gametes from neighboring clones to form hybrid diploids. Thus this form of the sexual cycle is extremely versatile. Under conditions in which many spores germinate together, cross-fertilization will almost certainly occur, encouraging whatever advantage heterozygosity can provide. Alternatively, where spores germinate in isolation, homothallic differentiation and self-fertilization can reestablish the diploid phase of the sexual cycle.

It is interesting to speculate on whether the formation of homozygous diploids is due simply to the lack of opportunity to outbreed or there is some evolutionary advantage to that pathway. To that end, we note that the plants we have mentioned, unlike animals, do not set aside germ cells early in development. In fact, many or all diploid somatic cells can undergo meiosis and thus contribute to the reproductive gene pool. Perhaps much of the genetic variation available to these organisms is constantly generated in the diploid phase through movement of transposable elements or somatic recombination in addition to nucleotide-pair alterations. Thus even an originally homozygous diploid resulting from self-mating might contribute a significant level of new genetic combinations to the haploid meiotic progeny.

The combined effect of self- and cross-fertilization would be to test any genetic variant for fitness in both heterozygous and homozygous forms in the generation immediately following its occurrence. Coupled with the ability of each individual fern or yeast colony to produce millions of haploid spores (in which loss of even most of the spores to haplo-lethal events would not significantly affect reproductive efficiency [Haldane's cost of natural selection (Haldane, 1957)]), this arrangement of the sexual cycle provides a particularly powerful and responsive means of dealing with environmental stress both in the short-term and long-term evolutionary time frames. Because the rate of evolutionary change is directly proportional to the genetic variability available in the population (Fisher's fundamental theorem), the variety of possibilities for fertilization in homothallic organisms would generate additional genotypes and hence provide increased genetic variance.

Among organisms there is a rough correlation between morphological complexity and dominance of the diploid phase. In so-called higher plants and animals the haploid phase is reduced to gametes incapable of independent mitotic cell division. The lower plants described above, along with budding yeasts, represent several intermediate levels of codominance of the two forms. The familiar form of a fern is the diploid, the haploid being overlooked by most of us, whereas the dominant green form of most mosses is haploid. At the other extreme are prokaryotes and certain lower eukaryotes, such as the fission yeast *Schizosaccharomyces pombe* and some species of the unicellular green alga *Chlamydomonas,* which grow vegetatively as haploids but exhibit a homothallic sexual cycle.

The relative evolutionary advantage of haploid and diploid phases in microorganisms are not clear, but it is noteworthy that *S. pombe* and *S. cerevisiae,* occupying the same ecological niche, have dominant haploid and diploid phases, respectively. In fact, *S. pombe* [Beach, 1983; Beach and Klar,

1984] exhibits a switching system that closely parallels that described above for *S. cerevisiae*. Efficient alternation of mating-type genes occurs once again by a transposition-substitution reaction. The DNA segments used in the transposition event are derived from two closely linked storage loci. Despite the mechanistic parallels, the mating-type regulatory genes of the two organisms have no significant DNA sequence homology. In fact, no homology has been found between the *MAT* genes of *S. cerevisiae* and the genomes of several different homothallic yeast species. This could mean that convergent evolution has resulted in similar mechanisms of independent origin for switching mating-type genes.

Differentiation into sexual gametes in *S. pombe* and *Chlamydomonas* is a response to stress or starvation and the diploid zygotes thus formed undergo meiosis immediately, regenerating the haploid phase. The particular evolutionary advantages of this sexual cycle are more difficult to rationalize than even that of budding yeast, although it should be noted that the diploid zygospore formed after fertilization in *Chlamydomonas* is the stress-resistant spore form rather than the haploid meiotic product in the yeasts. In both varieties of yeast, however, the spore contains the meiotic product. Thus, as diploidy is a prerequisite for spore formation, rapid diploidization and meiosis through the homothallic life cycle might have an advantage under unfavorable conditions, allowing survival until a chance for outbreeding presents itself.

EVOLUTION OF THE CASSETTE MECHANISM

In the course of our studies on sex determination in *S. cerevisiae,* we have been struck by the order and complexity of the control systems involved. Basically, the haploid cell needs alternately to activate two alleles of *MAT*, a and α, at a frequency high enough that after two or three divisions cells of both types will be present in the progeny of a single spore and mating will ensue. Once mating is accomplished, switching is no longer required and is, in fact, turned off. Speaking in mechanical terms only, the simplest way to achieve that goal would be to have the two sets of genes located on opposite sides of an invertible promoter sequence. Inversion of the promoter would alternately activate one set and then the other. The classic example of this mechanism has been documented to account for another metastable switch, the alternate activation of flagellar antigen genes in *Salmonella typhimurium* [Zieg et al., 1977].

Baker's yeast, on the other hand, maintains extra copies of the *MAT* genes at all times, sequentially replacing the copy at *MAT* by unidirectional gene conversion. Furthermore, the cell-type gene expressed at *MAT* determines which cassette participates in the switch. No part of the switching process is left purely to chance.

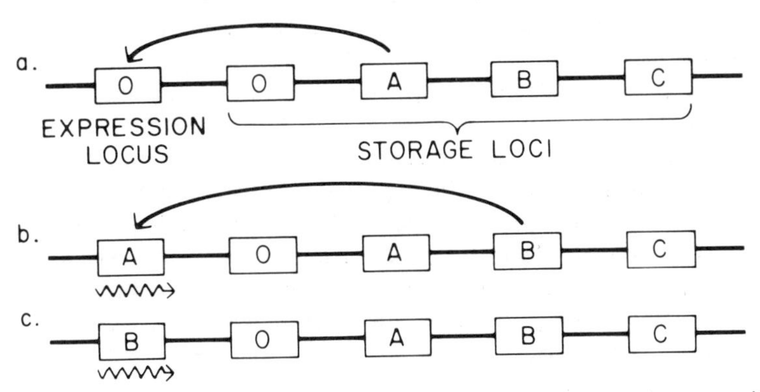

Fig. 4. Model for developmental cascade using components of the yeast mating-type switching mechanism. The cassette currently expressing determines which cassette next serves as source for the sequences transposed to the expression locus.

The existence of a mechanism capable of sequentially switching cassettes has led us to propose that such a mechanism, as it exists in yeast, is sufficient to drive an even more complex developmental pathway. The observed capacity of the expressed cassette to dictate which of several choices to activate next means that an indefinite series of cell-type switches could be programmed in this manner. The system is diagrammed in Figure 4 and consists of an indefinite number of silent cassettes (equivalent to the yeast *HML* and *HMR* loci) each containing the regulatory genes for a particular cell type in a multicellular organism. For example, in the haploid fern gametophyte described above, these might represent leaf, root hair, archegoniun, antheridium, egg, and sperm. Upon germination of the haploid spore, the cascade would be initiated by insertion of the first or leaf cassette. After two or three divisions a subset of the progeny leaf cells would activate the root cassette and differentiate into that cell type. The pedigree of this switching event might roughly parallel that shown in Figure 2 for yeast mating-type switching in which only a subset of cells can switch at any given time. At a later time, the sexual structures and gametes would be differentiated by sequential cassette insertion. After fertilization the expression locus for this cascade might be shut down until the next haploid generation.

An important feature of this model is that at the next generation, perhaps at meiosis, the whole cascade can be reactivated by reinsertion of the initial cassette, marked "O" in Figure 4. Because no information is lost during each developmental switching event, all cells remain developmentally totipotent at each step. Such a model could, therefore, account for the ability of most plant tissues to differentiate in culture and then reinitiate the complete developmental program of the whole plant. In fact, such an ordered, sequential transposition of "cassettes" to an expressed locus appears to be one mechanism able to account for the surface antigen variation of *Trypanosoma brucei* [Borst et al., 1980].

In summary, the sex-determination system in yeast, although complex, is probably the best understood of any system. It provides for the efficient interconversion of the two mating types, which allows the organism to diploidize if it fails to outcross. Although homothallism is not unique among lower eukaryotes, there appears to be no DNA sequence homology among the elements involved where it has been examined in *S. cerevisiae* and *S. pombe*. Finally, extension of such a cassette system provides a heuristic model of one kind of mechanism that should be able to generate diverse cell types in multicellular organisms.

ACKNOWLEDGMENTS

We thank all our associates and colleagues in the mating-type field for making exciting contributions to the story summarized here. We thank Russell Malmberg for his suggestions on the manuscript and Regina Schwarz for typing it. The work was funded by grants from NIH (A. K., J. H.) and by an NCI contract to Litton Industries (J. S.).

REFERENCES

Abraham J, Nasmyth KA, Strathern JN, Klar AJS, Hicks JB (1984): Regulation of mating-type information in yeast: Negative control requiring sequences both 5′ and 3′ to the regulated region. J Mol Biol 176:307–331.

Astell C, Ahlstrom-Jonasson L, Smith M, Tatchell K, Nasmyth K, Hall BD (1981): The sequence of the DNAs coding for the mating-type loci of *Saccharomyces cerevisiae*. Cell 27:15–23.

Beach DH (1983): Cell type switching by DNA transposition in fission yeast. Nature 305:682–687.

Beach DH, Klar AJS (1984): Rearrangement of the transposable mating type cassettes of fission yeast. EMBO 3:603–610.

Borst P, Frasch ACC, Bernards A, Van der Ploeg LHT, Hoeijmakers JHJ, Arnberg AC, Cross GAM (1980): DNA rearrangements involving the genes for variant antigens in *Trypanosoma bucei*. Cold Spring Harbor Symp Quant Biol 45:935–943.

Feldman JB, Hicks JB, Broach JR (1984): Identification of sites required for repression of a silent mating type locus in yeast. J Mol Biol 178:815.

Fisher RA (1958): "The Genetical Theory of Natural Selection." New York: Dover Publications, Inc., pp 22–51.

Haber JE (1983): Mating-type genes of *Saccharomyces cerevisiae*. In "Mobile Genetic Elements." Shapiro J (ed): New York: Academic Press, Inc., p 559.

Haber J, Comeau A, Livi P, Rogers D, Stewart S, Resnick M, Weiffenbach B (1982): Mechanism of homothallic switching of yeast mating type genes. In Berkeley Workshop on Recent Advances in Yeast Molecular Biology: Recombinant DNA. Vol. 1. Berkeley: Lawrence Berkeley Laboratory, University of California, p 332.

Haber JE, George JP (1979): A mutation that permits the expression of normally silent copies of mating-type information in *Saccharomyces cerevisiae*. Genetics 93:13–35.

Haber JE, Rogers DT, McKusker JH (1980): Homothallic conversions of yeast mating type genes occur by intrachromosomal recombination. Cell 22:277–289.

Haldane JBS (1957): The cost of natural selection. J Genet 55:511–524.

Harashima S, Oshima Y (1976): Mapping of the homothallic genes, *HMα* and *HMa*, in *Saccharomyces* yeasts. Genetics 84:437–451.

Hawthorne DC (1963): Directed mutation of the mating type alleles as an explanation of homothallism in yeast (abstract). Proc 11th Int Cong Genet 1:133.

Herskowitz I, Oshima Y (1981): Control of cell type in *Saccharomyces cerevisiae:* Mating type and mating-type interconversion. In Strathern JN, Jones EW, Broach JR (eds): "The Molecular Biology of the Yeast *Saccharomyces:* Life Cycle and Inheritance." Cold Spring Harbor, New York: Cold Spring Harbor Laboratory, p 181.

Hicks JB, Herskowitz I (1976): Interconversion of yeast mating types. I. Direct observations of the action of the homothallism (*HO*) gene. Genetics 83:245–258.

Hicks JB, Herskowitz I (1977): Interconversion of yeast mating types. II. Restoration of mating ability to sterile mutants in homothallic and heterothallic strains. Genetics 85:373–393.

Hicks JB, Strathern JN, Herskowitz I (1977): The cassette model of mating-type interconversion. In Bukhari AI, et al. (eds): "DNA Insertion Elements, Plasmids, and Episomes." Cold Spring Harbor, New York: Cold Spring Harbor Laboratory, pp 457–462.

Hicks JB, Strathern JN, Klar AJS (1979): Transposable mating type genes in *Saccharomyces cerevisiae*. Nature 282:478–483.

Jensen R, Sprague GF Jr, Herskowitz I (1983): Regulation of yeast mating-type interconversion: Feedback control of *HO* expression by the mating-type locus. Proc Natl Acad Sci USA 80:3035–3039.

Klar AJS (1980): Interconversion of yeast cell types by transposable genes. Genetics 95:631–648.

Klar AJS, Fogel S (1977): The action of homothallism genes in *Saccharomyces* diploids during vegetative growth and the equivalence of *hma* and *HMα* loci functions. Genetics 85:407–416.

Klar AJS, Fogel S (1979): Activation of mating type genes by transposition in *Saccharomyces cerevisiae*. Proc Natl Acad Sci USA 76:4539–4543.

Klar AJS, Fogel S, MacLeod K (1979a): *MAR1*—A regulator of *HMa* and *HMα* loci in *Saccharomyces cerevisiae*. Genetics 93:37–50.

Klar AJS, Fogel S, Radin DN (1979b): Switching of a mating-type a mutant allele in budding yeast *Saccharomyces cerevisiae*. Genetics 92:759–776.

Klar AJS, Hicks JB, Strathern JN (1982): Directionality of yeast mating-type interconversion. Cell 28:551–561.

Klar AJS, McIndoo J, Hicks JB, Strathern JN (1980a): Precise mapping of the homothallism genes, *HML* and *HMR*, in yeast *Saccharomyces cerevisiae*. Genetics 96:315–320.

Klar AJS, McIndoo J, Strathern JN, Hicks JB (1980b): Evidence for a physical interaction between the transposed and the substituted sequences during mating type gene transposition in yeast. Cell 22:291–298.

Klar AJS, Strathern JN, Broach JR, Hicks JB (1981a): Regulation of transcription in expressed and unexpressed mating type cassettes of yeast. Nature 289:239–244.

Klar AJS, Strathern JN, Hicks JB (1981b). A position-effect control for gene transposition: State of expression of yeast mating-type genes affects their ability to switch. Cell 25:517–524.

Klar AJS, Strathern JN, Abraham JA (1984a): The involvement of double-strand chromosomal breaks for mating-type switching in *Saccharomyces cerevisiae*. Cold Spring Harbor Symp Quant Biol 49:77–88.

Klar AJS, Strathern JN, Hicks JB (1984b): Developmental pathways in yeast. In Losick R, Shapiro L (eds): "Microbial Development." Cold Spring Harbor, NY: Cold Spring Harbor Laboratory, pp 151–195.

Kostriken R, Strathern JN, Klar AJS, Hicks JB, Heffron F (1983): A site specific endonuclease essential for mating-type switching in *Saccharomyces cerevisiae*. Cell 35:167–174.

Kushner PJ, Blair LC, Herskowitz I (1979): Control of yeast cell types by mobile genes—A test. Proc Natl Acad Sci USA 76:5264–5268.

Lindegren DC, Lindegren G (1943): A new method for hybridizing yeast. Proc Natl Acad Sci USA 29:306–308.

MacKay VL, Manney TR (1974): Mutations affecting sexual conjugation and related processes in *Saccharomyces cerevisiae*. II. Genetic analysis of nonmating mutants. Genetics 76:273–278.

Malone RE, Esposito RE (1980): The *RAD52* gene is required for homothallic interconversion of mating types and spontaneous mitotic recombination in yeast. Proc Natl Acad Sci USA 77:503–507.

Nasmyth K, (1982): Regulation of yeast mating-type chromatin structure by *SIR:* An action at a distance affecting both transcription and transposition. Cell 30:567–578.

Nasmyth K (1983a): Molecular genetics of yeast mating type. Ann Rev Genet 16:439–500.

Nasmyth K (1983b). Molecular analysis of cell lineage. Nature 302:670–676.

Nasmyth KA, Tatchell K (1980): The structure of transposable yeast mating type loci. Cell 19:753–764.

Nasmyth KA, Tatchell K, Hall BD, Astell C, Smith M (1981): A position effect in the control of transcription at yeast mating type loci. Nature 289:244–250.

Naumov GI, Tolstorukov II (1973): Comparative genetics of yeast. X. Reidentification of mutators of mating types in *Saccharomyces*. Genetika 9:82–91.

Oshima Y, Takano I (1971): Mating types in *Saccharomyces:* Their convertibility and homothallism. Genetics 67:327–335.

Rine JD, Strathern JN, Hicks JB, Herskowitz I (1979). A suppressor of mating type locus mutations in *Saccharomyces cerevisiae:* Evidence for and identification of cryptic mating type loci. Genetics 93:877–901.

Sprague GF Jr, Jensen R, Herskowitz I (1983): Control of yeast cell type by the mating type locus: Positive regulation of the α-specific *STE3* gene by the *MATα1* product. Cell 32:409–415.

Sprague GF Jr, Rine J, Herskowitz I (1981): Control of yeast cell type by the mating type locus. II. Genetic interactions between *MATα* and unlinked α-specific *STE* genes. J Mol Biol 153:323–335.

Strathern JN, Blair L, Herskowitz I (1979a): Healing of *mat* mutations and control of mating type interconversion by the mating type locus in *Saccharomyces cerevisiae*. Proc Natl Acad Sci USA 76:3425–3429.

Strathern JN, Newlon CS, Herskowitz I, Hicks JB (1979b): Isolation of a circular derivative of yeast chromosome III: Implications for the mechanism of mating type interconversion. Cell 18:309–319.

Strathern JN, Herskowitz I (1979): Asymmetry and directionality in production of new cell types during clonal growth: The switching pattern of homothallic yeast. Cell 17:371–381.

Strathern JN, Hicks JB, Herskowitz I (1981): Control of cell type by the mating type locus: The α1-α2 hypothesis. J Mol Biol 147:357–372.

Strathern JN, Klar AJS, Hicks JB, Abraham JA, Ivy JM, Nasmyth KA, McGill C (1982): Homothallic switching of yeast mating type cassettes is initiated by a double-stranded cut in the *MAT* locus. Cell 31:183–192.

Strathern JN, Spatola E, McGill C, Hicks JB (1980): The structure and organization of the transposable mating type cassettes in *Saccharomyces*. Proc Natl Acad Sci USA 77:2839–2843.

Takano I, Kusumi T, Oshima Y (1973): An α mating-type allele insensitive to the mutagenic action of the homothallic gene system in *Saccharomyces diastaticus*. Mol Gen Genet 126:19–28.

Takano I, Oshima Y (1970): Allelism tests among various homothallism-controlling genes and gene systems in *Saccharomyces*. Genetics 64:229.

Thorner J, (1981): Pheremonal regulation of development in *Saccharomyces cerevisiae*. In Strathern JN, Jones EW, Broach JR (eds): "The Molecular Biology of the Yeast *Saccharomyces:* Life cycle and inheritance." Cold Spring Harbor, NY: Cold Spring Harbor Laboratory, pp 143–180.

Weiffenbach B, Haber JE (1981): Homothallic mating type switching generates lethal chromosome breaks in *rad52* strains of *Saccharomyces cerevisiae*. Mol Cell Biol 1:522–534.

Winge O, Roberts C (1949): A gene for diploidization in yeast. C R Trav Lab Carlsberg Ser Physiol 24:341–346.

Zeig J, Silverman M, Simon M (1977): Recombination switch for gene expression. Science 196:170.

The Origin and Evolution of Sex, pages 113–121

Sex in *Chlamydomonas:* Sex and the Single Chloroplast

Ruth Sager and Constance Grabowy

Division of Cancer Genetics, Dana-Farber Cancer Institute, and Department of Microbiology and Molecular Genetics, Harvard Medical School, Boston, Massachusetts 02115

INTRODUCTION

A fundamental property of chloroplast inheritance in the sexual algae and in higher plants is maternal inheritance of chloroplast DNA [Kirk et al., 1967]. Like most fundamental properties in biology, this one has its exceptions, but they too appear to be regulated by the same general mechanism. In the green alga *Chlamydomonas,* the sexual transmission of chloroplast DNA is controlled by the mating type locus, a region on chromosome VI in the nucleus with numerous functions in the sexual life cycle. The focus of interest in this essay is the molecular mechanism that links mating type with chloroplast inheritance.

Figure 1 illustrates the haploid life cycle of *Chlamydomonas* and distinguishes the patterns of inheritance of nuclear and chloroplast genes. The zygote is the only diploid stage in this life cycle: Meiosis follows zygote formation, and the vegetative cells are haploid. As a consequence, nuclear genes segregate 2:2 among the four meiotic products. The first reported exception to this rule, resistance to the antibiotic streptomycin, exhibited a distinctive 4:0 pattern [Sager, 1954] as is shown in Figure 1. Subsequently, numerous genes affecting photosynthesis as well as protein synthesis in the chloroplast were shown to express the same 4:0 pattern: Those genes carried by the mt^+ parent were transmitted to all progeny of sexual crosses, and those from the mt^- parent were transmitted to none. Rarely, individual zygotes transmitted genes from both parents.

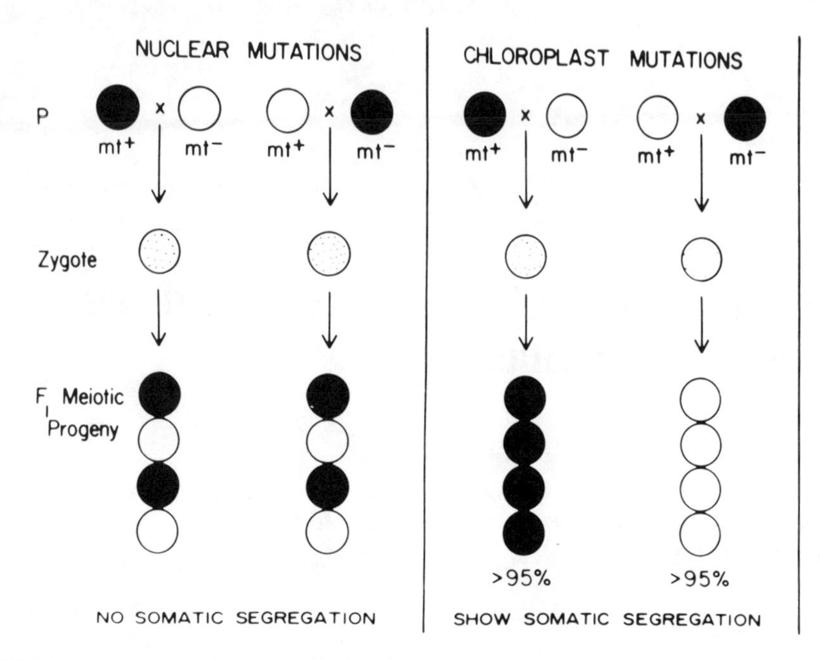

Fig. 1. Tetrad analysis of transmission patterns of nuclear and chloroplast mutations in *Chlamydomonas*. Black and white circles represent alternative alleles. From Gillham [1978], with permission.

Genes showing this 4:0 pattern were found to undergo recombination [Sager et al., 1963, 1965], providing a basis for genetic mapping. Several mapping procedures were then devised [summarized in Sager, 1977; Gillham, 1978], and the genetic map was shown to be circular [Singer et al., 1976]. The 4:0 pattern was generalized as maternal inheritance by analogy with higher plants in which chloroplast genes are transmitted from the female parent entirely or preferentially, depending on the species [Kirk et al., 1967]. Continuing the analogy, in *Chlamydomonas* the mt^+ parent is referred to as female and the mt^- parent as male.

MOLECULAR BASIS OF MATERNAL INHERITANCE

In higher plants, several mechanisms have been suggested for maternal inheritance: exclusion of plastids from male gametes, loss of plastids from the sperm, exclusion during fertilization, and degradation of plastid DNA after fertilization. In *Chlamydomonas reinhardi,* in which the isogamous gametes fuse completely during fertilization, the mechanism of maternal inheritance must involve loss of the male chloroplast or at least loss of its chlDNA after gametic fusion. Recently, this destruction has been made visible using DAPI staining to identify both nuclear and chloroplast DNAs [Kuroiwa et al., 1982; Tsubo et al., 1984]. Nucleoids are seen in both chloroplasts immediately after zygote formation but have disappeared within 2 hr from the chloroplast of mt^- origin.

Although this observational evidence is recent, the first demonstration of the loss in zygotes of chlDNA from mt^- cells was published in 1972 [Sager et al., 1972]. The same publication also provided the first evidence that the chlDNA from mt^+ cells is preserved in the zygote. These findings were the opening wedge for an experimental attack on the underlying question, what is the molecular basis of the differential fate of chlDNA from mt^+ and mt^- parents after zygote formation? How can homologous DNAs present in the same cell at the same time undergo different fates—one preserved and the other destroyed? The hypothesis initially proposed [Sager et al., 1973] to account for these findings is shown diagrammatically in Figure 2.

The hypothesis is based on the methylation-restriction (M-R) process responsible for the destruction of foreign DNA in bacteria elucidated initially by Arber and Linn [1969] and Meselson et al. [1972] (for a current review see Yuan [1981]). These authors showed that phage DNAs entering a bacterial cell may or may not be degraded, depending on whether or not particular endonucleolytic recognition sites were methylated. If methylated, the sites were protected from cleavage.

Applying this mechanism to *Chamydomonas,* it was postulated that a methylating enzyme is produced in mt^+ cells and a restriction enzyme is produced in mt^- cells and that both are localized in the chloroplast and are inactive until gametic fusion to form zygotes. After mating, about 5 hr elapses before the chloroplasts fuse; it was postulated that during this period the chloroplast DNA of mt^+ origin becomes methylated in its chloroplast compartment, whereas in the chloroplast compartment of mt^- origin the chlDNA is degraded. After chloroplast fusion, the mt^+ chlDNA is protected from enzymatic degradation by its prior methylation.

Direct supporting evidence came from the discovery that chlDNA of mt^+ origin is present in methylated form in zygotes whereas as that of mt^- origin remains unmethylated and is degraded and lost [reviewed in Sager, 1977]. The time course of chloroplast fusion, originally determined with unstained cells as occurring approximately 5–6 hr after zygote formation, was confirmed by electron microscopy [Cavalier-Smith, 1970]. The loss of DAPI staining by chlDNA of mt^- origin has been shown to occur 2–3 hr before chloroplast fusion, further supporting the time course required by the methylation-restriction hypothesis.

Further studies from this laboratory that have confirmed and extended our understanding of the methylation and restriction events that control chloroplast

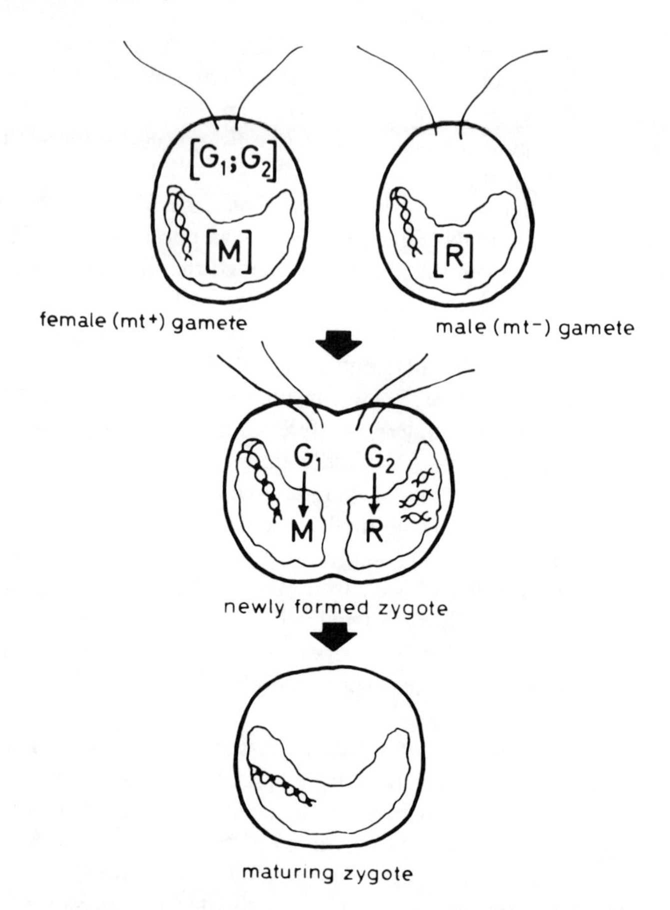

Fig. 2. Postulated control of maternal inheritance of chloroplast DNA in *Chlamydomonas* by a methylation-restriction mechanism. Female (mt^+) gamete contains inactive modification enzyme M in chloroplast and two regulatory substances, G_1 and G_2, in the cytosol. The male (mt^-) gamete contains inactive restriction enzyme R in its chloroplast. Before fusion of chloroplasts, the methylase is activated by G_1 to modify chloroplast DNA in the female chloroplast, and the restriction enzyme is activated by G_2 to degrade chloroplast DNA in the male chloroplast. The two chloroplasts then fuse, and only the chloroplast DNA from the female parent is available for replication. (Nuclei are not shown for sake of clarity.) From Sager et al. [1973].

inheritance have been summarized recently [Sager et al., 1984] and are listed in Table I. A modification of the original model is based on the finding that methylation of the mt^+ chlDNA begins during gametogenesis, so that at the time of gametic fusion the mt^+ DNA is already considerably methylated.

The finding of a mutant (*me-1*) in which vegetative chlDNAs of both mating types are partially methylated [Bolen et al., 1982] led to further

TABLE I. Differential Methylation of Chloroplast DNA in *mt*$^+$ Gametes of *Chlamydomonas*

1. Indirect identification of 5mC in *mt*$^+$ chlDNA by bouyant density shift in CsCl gradients from 1.695 (vegetative cells) to 1.690 (gametes) [Sager et al., 1972]
2. Direct identification of 5mC in *mt*$^+$ chlDNA by radioisotope labeling [Burton et al., 1979]
3. Differential methylation in gamete chlDNAs detected in restriction fragment patterns altered by methylation [Royer et al., 1979]
4. Absence of methylation in chlDNA of *mt*$^-$ gametes and presence in chlDNA of *mt*$^+$ gametes shown with anti-5mC antibody [Sano et al., 1980]
5. Differential activity of methyl transferase in *mt*$^+$ and *mt*$^-$ gametes [Sano et al., 1981]
6. Overmethylation of chlDNA in *mt*$^+$ but not *mt*$^-$ gametes of *me-l* mutants [Sager et al., 1983]

clarification of the hypothesis. In bacteria, the endonuclease involved in this process is highly site specific, and only a few sites are present per DNA molecule. Both the methylation and restriction enzymes recognize the same site, one methylating it and the other cleaving it only if it is unmethylated. The two enzymes share subunits that are coded by the same gene, and the endonuclease contains an additional unique subunit. The recognition sites are quite long and complex, as would be anticipated for sites present so few times in a DNA of some $6 \times 10^-$ nucleotides in length. A recognition site of similar complexity, present in only a few copies per genome, would also be anticipated in chlDNA if the methylation-restriction mechanism is similar to the bacterial one.

Support for this concept comes from studies of the *me-1* mutant [Sager et al., 1983]. As reported by Bolen et al. [1982], chloroplast genes show maternal inheritance despite the substantial methylation of chlDNA in vegetative cells of both mating types. If methylation protects against degradation, one might expect biparental rather than maternal inheritance of chlDNA in the *me-1* mutant. However, during gametogenesis, chlDNA in *mt*$^+$ cells undergoes further methylation, whereas the homologous chlDNA in *mt*$^-$ cells does not. Clearly then, the specific sites responsible for the methylation-restriction process must lie in the sequences differentially methylated in the gametes. Studies to identify those specific sites are in progress in this laboratory.

DEGRADATION OF *mt*$^-$ chlDNA IS REGULATED BY *mt*$^+$ CELLS

Because maternal inheritance of chloroplast genes obviates the possibility of recombination between chloroplast genomes of *mt*$^+$ and *mt*$^-$ cells, it has been important to devise means to induce biparental inheritance to carry out genetic mapping. Studies of inhibitors of DNA, RNA, or protein synthesis led to the discovery that of all treatments tested the most successful was UV irradiation. Surprisingly, it was UV irradiation of the mt^+ parent that led to protection of the chlDNA of the mt^- parent [Sager et al., 1967]. The next most effective treatment was with ethidium bromide; here, too, treatment of the mt^+ gametes induced biparental inheritance, whereas treatment of the mt^- gametes had little or no effect [Sager et al., 1973].

Subsequently, Wurtz et al. [1977] found that pretreatment of mt^+ cells with FdUrd induced biparental and paternal transmission of chloroplast genes, and here, too, a similar treatment of mt^- cells had little or no effect. Most recently, we found that 5-azacytidine (azaC) is effective in reducing the frequency of maternal inheritance when mt^+ are treated during gametogenesis with 8 mM azaC or more [Sager and Grabowy, in preparation]. There is a sharp concentration-dependent cut-off in the biological effectiveness of azaC; 4 mM azaC has virtually no effect on the frequency of maternal inheritance. In a recent paper, Feng and Chiang [1984] report that azaC does not alter maternal inheritance, but the highest concentration used was 3 mM, which is below the minimal concentration that we found to be effective.

In all these studies, only treatment of the mt^+ parent, not of the mt^- parent, was effective in decreasing the frequency of maternal inheritance. It has been argued (see Gillham [1978] for a summary) that these results are not in agreement with the methylation-restriction hypothesis; treatment of the mt^+ parent damages chlDNA or decreased its amount but does not affect the mt^- parent. However, this argument does not consider the possibility, first suggested in 1967 [Sager et al., 1967] that the mt^+ parent regulates not only methylation of mt^+ chlDNA but also the degradation of the mt^- chlDNA that occurs in the chloroplast compartment of mt^- origin in the zygote.

When the frequency of maternal inheritance is decreased, two classes of zygotes can be formed: biparental, with chloroplast genes from both parents, and paternal, containing only the mt^- chloroplast genome. The relative frequencies of these classes depends not only on protection of the mt^- chlDNA but also on the fate of the mt^+ chlDNA. Studies with UV irradiation have shown that, as the UV dose is raised, the fraction of paternal zygotes increases relative to the maternal and biparental classes. Photoreactivation reverses this shift, partially restoring maternal and biparental zygotes as well as increasing zygote viability. Elevated concentrations of FdUrd lead to the recovery only of paternal zygotes as the replication of mt^+ chlDNA is increasingly inhibited.

The results with azaC are particularly relevant here. The drug interferes with DNA methylation in at least two ways: 1) It is incorporated into DNA in place of cytidine, but, because it cannot be methylated, it inhibits potential methylation at each site of incorporation [Jones et al., 1980]. 2) When present in DNA, methyltransferase binds very tightly to azaC [Santi et al., 1984]. Because the enzyme is present in a very low amount, the azaC-DNA-enzyme complexes essentially deplete the cell of methyltransferase activity. The effects can be seen within a few replication cycles, because azaC is rapidly incorporated into replicating DNA. Thus azaC may decrease maternal inheritance by making some of the mt^+ chlDNA molecules subject to degradation as a result of hypomethylation. The decreased transmission of mt^+ chlDNA after treatment of mt^+ gametes with UV, ethidium bromide, FdUrd, or azaC is readily explained by direct effects on the mt^+ chlDNA itself or on its replication.

The survival of chlDNA from the mt^- parent is a separate question. Because total loss of chlDNA from both parents is lethal, if the treatments of mt^+ gametes that lead to destruction of mt^+ chlDNA had no effect on mt^- chlDNA, zygotic lethality would parallel loss of maternal inheritance. In the treatments that have been carefully studied, UV irradiation, FdUrd pretreatment, and azaC pretreatment, biparental and paternal zygotes are recovered demonstrating preservation of mt^- chlDNA. As is shown in Figure 2, the M-R hypothesis proposes that a regulatory system (G_2) in the mt^+ parent controls the activation of the restriction endonuclease. If G_2 is not active, then the endonuclease does not become activated, and the mt^- chlDNA is not degraded.

Support for this interpretation comes from recent studies of the mode of action of azaC. DNA containing azaC has been shown to react not only with methyltransferase but also with other proteins in an unusual way. Normally, proteins in chromatin can be readily dissociated from DNA by high salt, but, after incorporation of azaC into the DNA, many proteins are very tightly bound [Christman et al., 1985]. Thus decreased methylation might not be the only biologically important effect of azaC. It could induce the binding and inactivation not only of the methyltransferase but also of the postulated endonuclease.

In summary, the existing evidence is consistent in showing that the degradation of mt^- chlDNA that occurs soon after zygote formation is regulated by events occurring in mt^+ cells prior to zygote formation. The most likely explanation is that each of the treatments that lead to biparental inheritance has the same consequence: namely, inhibition of the endonuclease that is responsible for restriction of mt^- chlDNA.

METHYLATION AND RESTRICTION OF CHLOROPLAST DNA IN VEGETATIVE DIPLOIDS AND FUSANTS

In addition to the usual haploid sexual life cycle shown in Figure 1, *Chlamydomonas* can be induced to form vegetative diploids that are mt^+/mt^- heterozygotes [Gillham, 1978]. Furthermore, by treatment with an agent such as polyethylene glycol that induces cell-cell fusion, mt^+/mt^+ and mt^-/mt^- homozygous diploid lines, called *fusants,* have been obtained [Matagne et al., 1979]. Chloroplast gene transmission has been investigated in the progeny of crosses among these cell lines, principally by Matagne and colleagues in Belgium [cf. Matagne et al., 1983] and by Tsubo and colleagues in Japan [cf. Tsubo et al., 1984].

The results of the two groups are in agreement in showing that mt^+/mt^- heterozygous diploids behave as mt^- in mating exclusively with mt^+ cells, either mt^+ haploids or mt^+/mt^+ diploids, but that their chloroplast genes are transmitted biparentally. Thus the mt^- allele is dominant in mating, but the mt^+ allele is dominant in controlling the methylation-restriction process.

Tsubo and Matsuda [1984] correlated the transmission of chloroplast genes in crosses with loss or retention of nucleoids in DAPI-stained zygotes. They showed that maternal inheritance occurred in crosses between homozygous diploids: $mt^+/mt^+ \times mt^-/mt^-$, just as in haploid crosses, and that nucleoids from the mt$^-$ parent were lost with the same time course in both crosses. However, in crosses of $mt^+/mt^+ \times mt^+/mt^-$ heterozygotes, biparental inheritance occurred, and nucleoids were retained from both parents.

This result confirms findings of this laboratory with the *mat*-1 mutant, *now known to be a* mt$^+$/mt$^-$ *heterozygote* [Coleman and Sager, in preparation]. We previously reported that chloroplast DNA of the *mat-1* gametes is methylated [Sager et al., 1981] and contains the active form of the methylating enzyme [Sano et al., 1981]. Thus the presence of the mt^+ gene in mt^+/mt^- has an effect similar to that of UV, FdUrd, or azaC treatment of mt$^+$ cells, namely, to inhibit the degradation in zygotes of chloroplast DNA of mt^- origin.

In summary, chloroplast inheritance is controlled by the mating type locus. Whereas mating type preference is determined by the mt^- allele, chloroplast DNA transmission is controlled by the mt^+ allele. The mechanism of this control involves methylation of the DNA that is transmitted and destruction of the DNA that is not methylated at specific sites recognized by the restriction endonuclease. Detailed analysis of the mechanism awaits isolation of the enzymes and identification of the recognition sites in chloroplast DNA that are utilized.

REFERENCES

Arber W, Linn S (1969): DNA modification and restriction. Rev Biochem 38:467.

Bolen PL, Grant DM, Swinton D, Boynton JE, Gillham NW (1982): Extensive methylation of chloroplast DNA by a nuclear gene mutation does not affect chloroplast gene transmission in Chlamydomonas. Cell 28:335–343.

Burton W, Grabowy C, Sager R (1979): The role of methylation in the modification and restriction of chloroplast DNA in *Chlamydomonas*. Proc Natl Acad Sci 76:1390–1394.

Cavalier-Smith J (1970): Electron microscopic evidence for chloroplast fusion in zygotes of *Chlamydomonas reinhardi*. Nature 228:33.

Christman JK, Schneiderman N, Acs G (1985): Formation of highly stable complexes between 5-azacytosine-substituted DNA and specific nonhistone nuclear proteins. J Biol Chem 260:4059–4068.

Feng T-Y, Chiang K-S (1984): The persistence of maternal inheritance in *Chlamydomonas* despite hypomethylation of chloroplast DNA induced by inhibitors. Proc Natl Acad Sci USA 81:3438–3442.

Gillham NW (1978): "Organelle Heredity." New York: Raven Press.

Kirk JTO, Tilney-Bassett RAE (1967): "The Plastids." London, W.H. Freeman and Company.

Jones PA, Taylor SM (1980): Cellular differentiation, cytidine analogs and DNA methylation. Cell 20:85–93.

Kuroiwa T, Kawano S, Nishibayashi S (1982): Epifluorescent microscopic evidence for maternal inheritance of chloroplast DNA. Nature 298:481–483.

Matagne RF, Mathieu D (1983): Transmission of chloroplast genes in triploid and tetraploid zygospores of *Chlamydomonas reinhardtii:* Roles of mating-type gene dosage and gametic chloroplast DNA content. Proc Natl Acad Sci USA 80:4780–4783.

Matagne RF, Deltour R, Ledoux L (1979): Somatic fusion between cell wall mutants of *Chlamydomonas reinhardi*. Nature 278:344–346.

Meselson M, Yuan R, Heywood J (1972): Restriction and modification of DNA. Ann Rev Biochem 41:447–466.

Royer H-D, Sager R (1979): Methylation of chloroplast DNA in the life cycle of Chlamydomonas. Proc Natl Acad Sci 76:5794–5798.

Sager R (1954): Mendelian and non-Mendelian inheritance of streptomycin resistance in *Chlamydomonas reinhardii*. Proc Natl Acad Sci USA 40:356–363.

Sager R (1977): Genetic analysis of chloroplast DNA in *Chlamydomonas*. Adv Genet 19:287–340.

Sager R, Grabowy C (1983): Regulation of maternal inheritance by differential methylation of chloroplast DNA in me-1 mutant of *Chlamydomonas*. Proc Natl Acad Sci 80:3025–3029.

Sager R, Grabowy C, Sano H (1981): The *mat-1* gene in *Chlamydomonas* regulates DNA methylation during gametogenesis. Cell 24:41–47.

Sager R, Lane D (1972): Molecular basis of maternal inheritance. Proc Natl Acad Sci USA 69:2410–2413.

Sager R, Ramanis Z (1963): The particulate nature of non-chromosomal genes in *Chlamydomonas*. Proc Natl Acad Sci USA 50:260–268.

Sager R, Ramanis Z (1965): Recombination of non-chromosomal genes in *Chlamydomonas*. Proc Natl Acad Sci USA 53:1053–1061.

Sager R, Ramanis Z (1967): Biparental inheritance of non-chromosomal genes induced by ultraviolet irradiation. Proc Natl Acad Sci USA 58:931–937.

Sager R, Ramanis Z (1973): The mechanism of maternal inheritance in *Chlamydomonas*. Biochemical and genetic studies. Theor Appl Genet 43:101–108.

Sager R, Sano H, Grabowy C (1984): Control of maternal inheritance by DNA methylation in *Chlamydomonas*. In Trautner TA (ed): "Current Topics in Microbiology and Immunology." Heidelberg: Springer-Verlag, pp 157–172.

Sano H, Grabowy C, Sager R (1981): Differential activity of DNA methyltransferase in the life cycle of *Chlamydomonas reinhardi*. Proc Natl Acad Sci 78:3118–3122.

Sano H, Royer H-D, Sager R (1980): Identification of 5-methylcytosine in DNA fragments immobilized on nitrocellulose paper. Proc Natl Acad Sci 77:3581–3585.

Santi DV, Norment A, Garrett CE (1984): Covalent bond formation between a DNA-cytosine methyltransferase and DNA containing 5-azacytosine. Proc Natl Acad Sci 81:6993–6997.

Singer B, Sager R, Ramanis Z (1976): Chloroplast genetics of *Chlamydomonas:* III. Closing the circle. Genetics 83:341–354.

Tsubo Y, Matsuda Y (1984): Transmission of chloroplast genes in crosses between *Chlamydomonas reinhardtii* diploids: Correlation with chloroplast nucleoid behavior in young zygotes. Curr Genet 8:223–229.

Wurtz EA, Boynton JE, Gillham NW (1977): Perturbation of chloroplast DNA amounts and chloroplast gene transmission in *Chlamydomonas reinhardtii* by 5-fluorodeoxyuridine. Proc Natl Acad Sci 74:4552–4556.

Yuan R (1981): Structure and mechanism of multifunctional restriction endonucleases. Ann Rev Biochem 50:285–315.

The Origin and Evolution of Sex, pages 123–140

An Essay on the Origins and Evolution of Eukaryotic Sex

Ursula W. Goodenough

Department of Biology, Washington University, St. Louis, Missouri 63130

Cell biologists and biochemists, trained as experimental scientists, are reluctant to engage in speculations about evolution unless they can study it directly. Thus, we are comfortable with computer simulations of hemoglobin evolution based on amino acid or nucleotide sequence data, but we shy away from formulating theories on a subject such as the origin and evolution of sex.

Since the present volume encourages such formulations, I have set forth here some of the thoughts on the origins of sex that I have accrued during the 15 years I have studied *Chlamydomonas* sexuality. Some of these ideas may well have been elegantly stated by previous or current generations of population geneticists (I confess to a major ignorance of their literature); but the population geneticists often prove to be equally ignorant of many of the experimental observations that have been made on mating systems, and I would propose that in these times of burgeoning literatures, we should not allow ignorance of all that has ever been said on a subject to prevent us from putting forth the ideas that our own thinking has generated.

The essay begins with some general thoughts on the "driving forces" that might have led to eukaryotic sex. I then speculate on how fertilization mechanisms may have come into being.

DEFINING SEX

Because prokaryotes presumably inhabited the earth before eukaryotes, and because modern prokaryotes are capable of exchanging and recombining their genomic DNA, it has been proposed that eukaryotic sex evolved from prokaryotic sex [see, for example, Bernstein et al., 1981]. The authors of these proposals appear to be defining sex in a different way than I would define it. There is no question that both prokaryotes and eukaryotes possess enzymes capable of pairing, breaking, and rejoining homologous DNA duplexes, enzymes that presumably evolved initially as a means to effect post-replication recombinational repair between daughter duplexes [Rupp et al., 1971; Bernstein, 1977]. These enzymes have been recruited by prokaryotes to effect gene recombination between host chromosomes and homologous DNA introduced via transformation, transduction, or conjunction, and they have been recruited as well by eukaryotes to effect crossing over during meiosis. To my mind, however, the term "sex" refers to *all* those features of an organism's phenotype that serve to bring two genomes together and distribute recombinant genomes to daughter cells. By this definition, the enzymes of DNA metabolism would certainly participate in the sexual process, but the crux of the matter is the means by which the genomes are first brought together and then redistributed. Here the prokaryotes and the eukaryotes have come up with fundamentally different forms of sexuality. Transduction is clearly a feature of the phage's phenotype and not of the host, and plasmid-mediated conjugation is designed to promote the transmission of plasmids and not the transfer of genomic DNA; indeed, the pili specified by plasmid DNA seem more analogous to phage coats than to eukaryotic fertilization proteins. Transformation, therefore, emerges as the closest thing to eukaryotic sex, and the infrequent nonspecific uptake of DNA from the medium and its occasional insertion into the host genome is so dissimilar to the recognition and fusion of gametes and the meiotic distribution of chromosomes that an evolutionary link seems very unlikely.

It is also important to stress at the offset that gametic compatibility and successful zygote formation—the focal topics of this essay—are not the only factors that serve to isolate modern eukaryotic species from one another. As speciation is occurring, it is common to find that gametes remain capable of fertilization but the resultant zygotes are either inviable or sterile [cf. Stebbins, 1966]; indeed, hybrid inviability has been shown to be operative in matings between sibling species of *Chlamydomonas* [Cain, 1979]. As the process of speciation continues, of course, gametic incompatibility is eventually established. My thesis here is that to understand the origins of sex, one must understand how the first sexual eukaryotes invented the means of bringing together compatible gametes that would fuse and exchange chromosomes with one another in an efficient and accurate fashion. The relative importance of these inventions in determining present-day patterns of speciation would appear to be an independent question.

A SCENARIO FOR THE EVOLUTION OF SEX

From the organelles and proteins that are ubiquitous to all modern eukaryotic cells, it is a straightforward matter to list the components that must have been present in their primitive forebearers: This would include an enveloped nucleus and endoplasmic reticulum, a mitotic type of chromosome segregation involving microtubules, mitochondria, actin, basal bodies and cilia, histones and clathrin. Some, if not all, also must have had plastids. Because all major classes of modern eukaryotic organisms (except the rotifers) have a sexual option to their life cycle, moreover, the protozoa that gave rise to the metazoa also must have evolved mechanisms of fertilization (cell fusion) and the meiotic segregation of chromosomes. Assuming that the basic blueprint of eukaryotic cell structure was established preceding the evolution of eukaryotic sexuality (although of course many features could have evolved concurrently), the question is how fertilization and meiosis might have come into being.

A major hallmark of eukaryotic cells is their possession of membrane systems that fuse with one another, allowing the development of intracellular vesicular traffic, endocytosis, and exocytosis. That emerging eukaryotic cells occasionally fused with one another, and thereby combined their genetic material, is a likely explanation for the increase in genome size and complexity which appears to have occurred during the dawn of eukaryotic life. Indeed, if one accepts the theory championed by Margulis [1971] that the proto-eukaryotes acquired their organelles by the engulfment and subsequent endosymbiosis of prokaryotes, then one must accept the correlative theory that proto-eukaryotes also must have undergone occasional fusions with one another.

The promiscuous fusions we are envisioning, however, quickly become hazardous to the individual cell, for once two nuclei have been created with different complements of genetic information, their co-residence in a single cell is usually lethal. Indeed, in modern experiments of this kind wherein interspecific cell hybrids are created *via* Sendai virus or polyethylene glycol, most of the cells fail to survive. An obvious solution to this problem is to prevent cell-to-cell fusions by, for example, surrounding the cell membrane with an extracellular coat. Most modern protozoa, in fact, possess such coats; and their presence normally precludes cell fusion. A second solution is to develop a system which allows fusions only between cells of homologous genotypes. Such a system is, of course, synonymous with most modern fertilization systems, wherein like-to-like fusions are ensured by some type of species-specific adhesion event between compatible gametes. Because most modern groups of protozoa engage in sex, we are led to conclude that the proto-eukaryotes adopted both "solutions": They surrounded themselves with extracellular coats which prevented fusion most of the time, and they came to display specific fertilization molecules which promoted fusions between compatible genotypes. Such fusions have clear short-term and long-term selective advantages: Diploidy is usually associated with faster growth rates; the resultant gene complementation can correct the effects of deleterious genes; and, with two copies of a genome per nucleus, one copy is free to undergo mutational experimentation.

Laboratory-induced fusions between cells of the same genotype often can yield stable polyploids if the ploidy is increased by a factor of two (e.g., diploid to tetraploid), but attempts to triple or quadruple the ploidy are almost invariably unsuccessful, presumably because the cells become imbalanced by such a sudden large increment in chromosome number. Similarly, a fertilization system which promotes like-to-like cell fusion will quickly become deleterious if left unchecked, as the ploidy will double with every fertilization event. The development of meiosis from mitosis solved this problem, and the linkage of meiosis to fertilization was clearly a critical evolutionary compromise.

We have thus far envisioned the origins of fertilization and meiosis as mechanisms that at once exploit the potential physiological and genetic advantages of diploidy and avert the potential disasters of unregulated cell fusion. The usual textbook explanation for the origin of sex is its ability to provide progeny with new gene combinations generated via the process of meiosis. In fact, this feature of meiosis is realized only when cells of different genetic makeup fuse with one another. If cells of the same mitotic clone fuse and form a zygote, its meiotic progeny are genetically identical to each other and to their parents (unless, of course, both parents have acquired mutations following their mitotic separation). Therefore, the recombinational potential of meiosis could be efficiently exploited only with the evolution of mating types, which ensure that mating occurs within the species (to avoid genetic disaster) but among cells of different clones (to generate new gene combinations).

Among the modern protozoa, the mating-type systems of *Saccharomyces* and *Chlamydomonas* [Goodenough and Thorner, 1983; Sprague et al., 1983] are perhaps the best understood. In each organism there are two mating types, termed *a* and α for yeast and *plus* and *minus* for *Chlamydomonas*. A single gene locus, denoted *MAT* or *mt*, controls mating type and exists in two allelic forms, *MATa*/*MATα* for yeast and mt^+/mt^- for *Chlamydomonas*. Each is a regulatory locus governing the expression of unlinked genes. Thus in yeast, the *MATa* locus directs the expression of surface recognition molecules that define the *a* mating type, whereas *MATα* directs the construction on an α mating phenotype.

The mating-type loci not only govern fertilization; they also exert a critical control over meiosis, but only in the heterozygous diploid state. Thus *MATa*/*MATα* cells are capable of meiosis, as are mt^+/mt^- cells, but when homozygous *MATa*/*MATa* or mt^+/mt^+ diploid lines are constructed in the laboratory, these are restricted to mitotic vegetative growth. It would appear, therefore, that the evolution of both fertilization and meiosis were intimately coupled with the evolution of the mating-type loci.

At this point consider an additional feature of protozoan sex that is undoubtedly critical to its evolution, namely, the fact that for most modern organisms, fertilization and meiosis are coupled with a highly adaptive stage in the organism's life cycle. Thus, in *Chlamydomonas,* the ability to fertilize is acquired only when cells are starved for nitrogen, and the resultant zygote is uniquely capable of forming a spore-like cell wall which resists dessication and permits indefinite survival in the face of adverse environmental conditions [Sager and Granick, 1954]. This coupling of gametogenesis with starvation is found in most of the algae, fungi, and ciliates, and the coupling of meiosis with spore formation is ubiquitous among the alga and fungi as well. The ciliates have instead coupled meiosis with the regeneration of a macronucleus [Nanney, 1977], while the metazoa have taken this notion to its logical conclusion, generating a mortal soma whose ultimate biological purpose is to effect the dissemination and fertilization of the immortal germ line.

It turns out that for both yeast and *Chlamydomonas,* the mating-type loci control not only fertilization and meiosis, but also spore formation. Thus, the unique coats that surround spore cells cannot be produced by homozygous diploids (*MATa/MATa* or mt^+/mt^+), but only by heterozygous diploids given the appropriate environmental stimuli [Haber and Halvorson, 1972; Matagne, 1981].

Not only do the mating-type loci in yeast and *Chlamydomonas* control the same constellation of genes; they also appear to control them in the same fashion: gene products of the mating type loci "turn on" unlinked gametogenesis genes in haploids and, in the heterozygous state, "turn on" spore wall and meiosis genes in diploids [Goodenough and Thorner, 1983]. The most straightforward way to imagine the evolution of such a system is to propose that "fertilization genes," "meiosis genes," and "spore genes" originated separately and that a master mating-type locus evolved to combine these three adaptive traits into an effective "supergene." It is not necessary to propose that mating type was the last to evolve; in fact, it is more attractive to speculate that, say, "spore genes" specifying an asexual sporulation program came to be regulated by a locus sensitive to nitrogen starvation and that, as other genes adaptive for the sexual process were acquired, they too came under the regulation of this same locus.

The phenomenon of homothallism should be considered in this context. Homothallic strains of yeast [Nasmyth, 1982] or *Chlamydomonas* [van-Winkle-Swift and Burruscano, 1983] also require two mating types for fertilization, spore formation, and meiosis. However, the two mating types are generated epigenetically within one clone, so that each cell can fertilize an identical sib. As a consequence, the recombinational potential of meiosis is lost. It seems likely that homothallism arose as a secondary derivative of heterothallism; that is, that heterothallism came first and was a driving force in the evolution of sexuality and that some species later devised the means to circumvent it. This proposition can be readily applied to the yeasts, who lack all the means normally utilized for the dispersion of gametes (ciliary motility, air-born gametes, hyphal growth) and whose opportunities for finding a partner of the opposite mating type are thereby severely restricted; in this case, homothallism represents an obvious solution, since at least the advantages of diploid growth rates and spore formation can be realized. The argument is also applicable to selfing in the ciliates, wherein the cell prepares for fertilization with the opposite type but, if no mate appears, proceeds to fertilize itself so that at least macronuclear regeneration can take place.

In summary, my central thesis is that cell fusion, a key "invention" of the proto-eukaryote, was rendered more adaptive by cell surface molecules which limited such fusions to genetically appropriate partners; that meiosis evolved from mitosis, preventing the logarithmic increase in genome size which would otherwise accompany cell fusion; that complementary mating types evolved to exploit the potential for diversification inherent in the independent assortment of chromosomes; and that spore formation made possible the survival of new meiotic combinations in adverse environments. As genes regulating these processes evolved, I propose that they came to be controlled by a central regulatory locus denoted mating type.

VARIATION IN SURFACE RECOGNITION MOLECULES

The scenario presented may imply a uniformity of sexual strategies which must at once be corrected. Whereas it is true that meiosis itself is a highly conserved process which is likely played out by a similar group of genes in all the eukaryotes [see article by P. Bruns in this volume], fertilization mechanisms in eukaryotes are characterized by a bewildering diversity. Enormous variation is found in the degree of gamete anisogamy [see article by G. Bell in this volume], in the modes of gamete dispersal, and, pertinent to this essay, in the molecules that engage in gamete-to-gamete recognition. Thus, the unicellular fungi recognize one another by glycoproteins displayed on their cell walls [Pierce and Ballou, 1983]; the ciliates and the sperm utilize proteins associated with their membranes [Hitwatashi, 1981; Wassarman, 1983]; the echinoderms have recruited long-chain carbohydrates that make up the jelly coats of their eggs [SeGall and Lennarz, 1979]; mammalian eggs utilize the extracellular proteins of the zona pellucida [Wassarman, 1983]; the higher plants display specific proteoglycan-type molecules, the arabinogalactans, on their stigma surfaces [Fincher et al., 1983] and so on. It follows that whereas we would anticipate, and are in fact beginning to identify, common "meiosis genes" throughout the eukaryotes, we in no way anticipate the existence of common "fertilization genes," even when both traits have come to be regulated by a common mating-type locus.

In considering this phenomenon, I have been led to another central thesis. If, as proposed above, proto-eukaryotes evolved fertilization molecules to promote appropriate cell fusions, they would still be locked into some very primitive stage of existence if these molecules had the evolutionary conservatism of, say, histone proteins, tubulins, or cytochrome c. The fact is that the eukaryotes have engaged in spectacular evolution, and because all the eukaryotes are sexual creatures, each evolutionary event has necessitated sexual speciation. For the lower eukaryotes and the sexual proto-eukaryotes, this means that as two sub-species diverged, they had to come up with two new sets of gametic recognition molecules, both of which differ from each other *and* from the set displayed by the parental species. It is very difficult to imagine how this feat could be accomplished via natural selection of random gene mutations. I therefore propose that another possibility be considered, namely, that fertilization molecules (and hence fertilization genes) are *designed* for variation.

Proteins designed for variation are, of course, familiar to us in the form of the immunoglobulins and the T-cell antigen receptors, and the genes specifying these proteins are programmed to undergo an elaborate series of recombinational and mutational events during the ontogany of an individual organism. Clearly, the fertilization molecules are far more stable. Nonetheless, it is attractive to consider the possibility that they possess the capacity to mutate, or undergo recombinational events, with sufficiently high frequency that new complementary dyads are occasionally created, and the option for sexual speciation is left perpetually open.

The lower eukaryotes provide possible examples of this phenomenon. In *Tetrahymena,* there exist not two, but multiple mating types, and mating is possible with all but one's own type [Nanney, 1977], a rule that is also followed by most of the higher fungi [Raper, 1966]. The 14 sibling species of the *Paramecium aurelia* complex are of particular interest. Each species has only two mating types, designated odd (O) and even (E) [Sonnenborn, 1978]. All the O types are genetically homologous to one another as are all the E types, yet only species-specific O/E mating mixtures can effect conjugation. Except for these well-defined differences in mating ability, the 14 species are so similar that until recently they were designated as syngens.

Chlamydomonas provides another example of possible interest. When I first began to study its mating reaction, I sought wild-type strains that would mate with one another with high efficiency. When the available strains proved to be only about 60 percent efficient, I cloned a mating-type *plus* and a mating-type *minus* strain and tested pairwise mixtures of individual clones. I found a number of cases wherein clone A (mt^+) would mate efficiently with clone B (*mt*⁻) but poorly with clone C (*mt*⁻), whereas clone D (mt^+) would mate well with clone C. Similar observations have been published by Chiang et al. [1970]. I also found that when I chose two clones that mated well and then kept them in culture, their mating efficiency began to deteriorate with time. Fortunately for our mating research projects, I came up with two clones that have lost this lability and have continued to mate with high efficiency for many years. These observations suggest, however, that in most strains, fertilization-specific molecules may be prone to undergo variation.

These observations on the protozoa may have relevance to the properties of metazoan germ lines. It is my impression from the literature and from conversations with students of metazoan fertilization that a rather high proportion of the gametes produced by an individual are nonfunctional for one reason or another, that is, gametogenesis is a rather sloppy affair compared, for example, with the precision displayed by the cells of the early embryo. Possibly this sloppiness, or at least some fraction of it, is the result of the "programmed variability" proposed herein, and the resultant fractional infertility is more than compensated by the large number of gametes produced by the individual.

Variation in Surface Molecules

The thesis that fertilization molecules have an inherent ability to change, and that sexual speciation is a consequence of that ability, carries an unacceptable level of altruism: It implies that proto-eukaryotes somehow "knew" that it was a good idea to speciate and designed their fertilization molecules accordingly. In this section a less troublesome sequence of events will be developed. I will propose that eukaryotes initially developed "morphogenesis genes" which specified the structure of their extracellular coats, that these acquired a propensity for change which was adaptive to the individual cell, and that fertilization molecules are simply special examples of such extracellular molecules.

My thinking on this subject has been greatly influenced by a stimulating article by Lamport [1980]. He proposes that the appearance of glycoproteins was a major evolutionary step in the transition from prokaryotes to eukaryotes, that glycoproteins probably first arose as constitutents of protective cell coats and that they soon came to be involved in recognition phenomena at the cell surface. "The great interest and significance of glycoproteins," Lamport writes, "lies in their being the morphogenetic substrate of multicellular organisms. Glycoproteins together with other macromolecules create the extracellular matrix which is the essential component of form. That is why I regard the prokaryotic-eukaryotic transition as a change from chemical machinery to morphogenetic machinery." Lamport goes on to review the evidence, summarized as well in many other volumes [e.g., Hay, 1981; Yamada, 1983;

Trelstad, 1984], that extracellular proteins, the majority of which are glycoproteins, represent the complementary cell surface macromolecules that mediate cell recognition during embryology, molecules whose existence was anticipated by such early biologists as E.B. Wilson and Paul Weiss.

If extracellular matrices mediate growth and form, then an overview of the eukaryotes that presently inhabit our planet quickly generates the conclusion that extracellular matrices must be prone to diversity: whereas the ribosomes, mitochondria, nucleosomes, microtubules, and other internal organelles of eukaryotic cells are very similar to one another, their investment of extracellular materials (and hence their morphologies) is extremely different. An instructive case in this regard is provided by the extracellular coats of modern-day *Chlamydomonas* species. All that have been studied are similar to one another in that they contain hydroxyproline-rich glycoproteins which form several discrete layers; moreover, one of these layers is invariably an intricate crystalline lattice [Roberts et al., 1982; Goodenough and Heuser, 1985a] (see Fig. 1). However, the coat of each species is unique, as judged by the ultrastructure of its crystalline lattice [Roberts, 1974], its antigenic determinants (unpublished observations from this laboratory), and its sensitivity to wall-lysing enzymes [Schlosser, 1976]. This uniqueness is not a sexual adaptation, as the walls play no apparent role in fertilization, and there is as yet no selective explanation for why so much variation exists. The same obviously can be said for the diatoms and dinoflagellates, whose spectacular walls would appear to be the consequence of an orgy of morphogenetic experimentation that has no obvious adaptive "purpose."

This leads to the third point made at the outset of this section, namely, the notion that fertilization molecules are capable of variation because they are special examples of extracellular matrix molecules. This concept first occurred to us when we found that the sexual agglutinin of *C. reinhardi* is a fibrous, hydroxyproline-rich glycoprotein [Adair, et al., 1983; Cooper et al., 1983], very similar in morphology to several of the polypeptides that form the crystalline cell wall of this species [Goodenough and Heuser, 1985] (Fig. 2). Moreover, most monoclonal antibodies that react with the agglutinin also react with the wall proteins [Adair et al., 1985]. Roberts and his colleagues have shown that the isolated *C. reinhardi* cell wall can be disassembled into its component proteins in the presence of chaotropic agents and will reassemble into crystalline walls when these agents are removed by dialysis [Catt et al., 1978]. We were thus led to propose that the sexual agglutinins may recognize each other by the same sorts of cues utilized by the wall proteins as they co-assemble [Cooper et al., 1983]. Supporting this notion is the finding that as the agglutinins interact during the mating reaction, they form an interconnecting meshwork of filaments that is very reminiscent of the meshworks found in the wall layers [Goodenough et al., 1985].

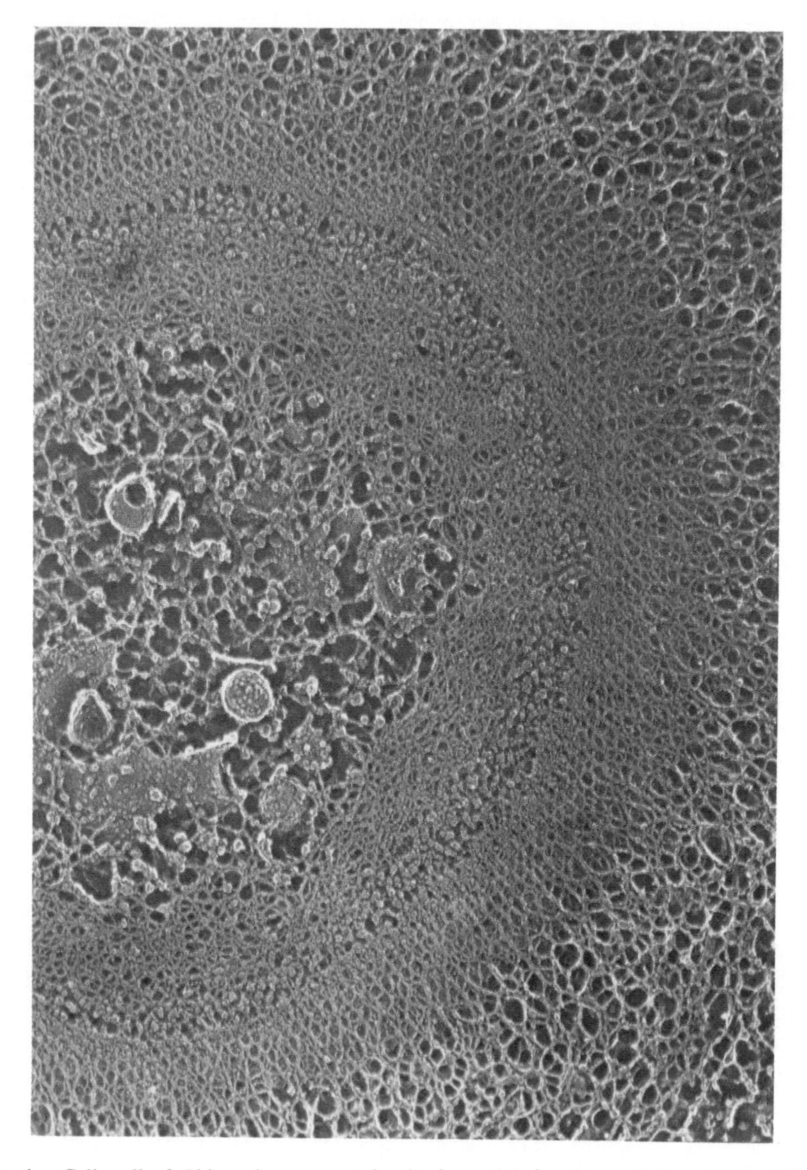

Fig. 1. Cell wall of *Chlamydomonas reinhardi* after quick freezing and deep etching. The cell was treated with detergent so that the plasma membrane in the center of the field appears as a series of vesicles. The wall has three discrete layers, the outermost displaying a crystalline regularity. Micrograph provided by J.E. Heuser. Magnification, ×100,000.

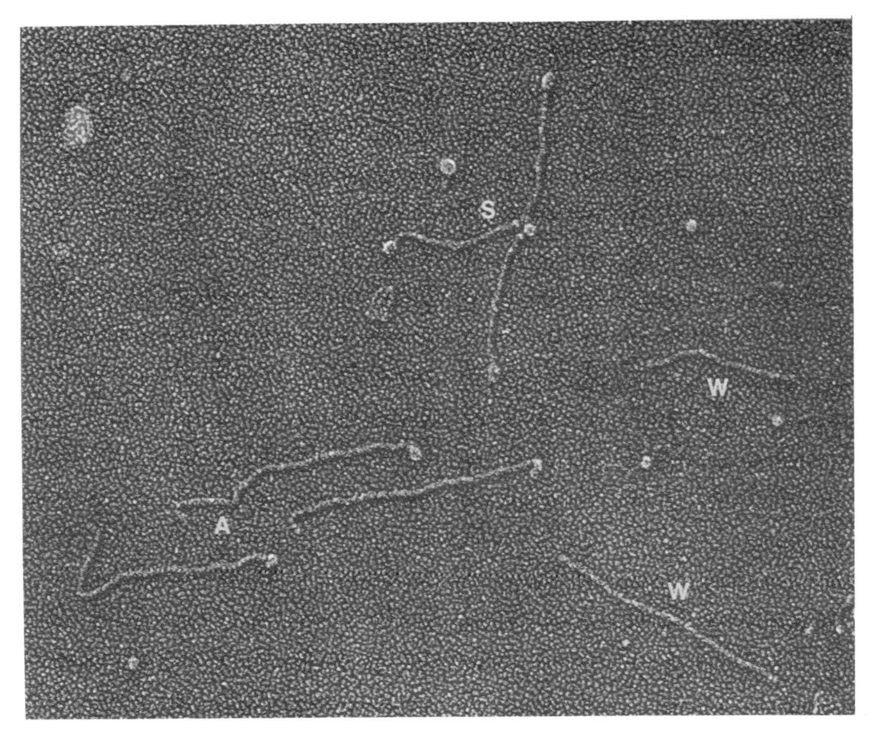

Fig. 2. *Chlamydomonas* cell surface proteins. Those labeled A are mt^+ sexual agglutinins; those labeled S are short flagellar proteins related to the agglutinins; and those labeled W derive from the cell wall. Molecules purified by W.S. Adair; micrograph provided by J.E. Heuser. Magnification, ×200,000.

There are two salient differences between the sexual agglutinins and the wall proteins of *Chlamydomonas*. The first is the sexual agglutinins insert into the plasma membrane, whereas most of the wall proteins do not. In fact, a survey of present-day eukaryotes reveals that some utilize membrane-associated recognition molecules (e.g., the green algae, the ciliates, and most metazoan sperm), whereas others utilize recognition molecules displayed solely in the extracellular matrix (e.g., the fungi, pollen, and most metazoan eggs); therefore, there seems to be no "rule" in this regard. The second difference, and the key difference, is that each sexual agglutinin is displayed by only one mating type and has an affinity only for its partner on the surface of the opposite mating type, whereas the wall proteins are expressed on all cells. To imagine the evolution of such a system, it is easiest to propose that the two ancestral molecules of the sexual agglutinins were components of an extracellular matrix and that they adhered to one another during the course of wall assembly. These then came under mating-type control such that they were expressed only under conditions of starvation and on only one cell type. The ancestral adhesion reaction may therefore have been wall-to-wall, as it continues to be in the fungi, and the flagellar location of the *Chlamydomonas* agglutinins may have been a secondary development designed to improve the chances of contact.

To my knowledge, the only other eukaryotic protists whose fertilization molecules have been biochemically characterized are those of the yeasts. In several different genera, one mating type displays a large, carbohydrate-rich, heat-stable agglutinin (e.g., type 5 of *Hansenula wingei* and type 16 of *Saccharomyces kluyveri* [Crandall and Brock, 1968; Pierce and Ballou, 1983] on its cell wall, whereas the other displays a smaller, less-glycosylated, and heat-labile agglutinin (e.g., type 21 of *H. wingei* and type 17 of *S. kluyveri*) on its cell wall. The fact that the two types of yeast agglutinins are very different biochemically, whereas the two types of *Chlamydomonas* agglutinins are very similar biochemically [Collin-Osdoby and Adair, 1985], should be borne in mind; the hypothesis proposed has no inherent requirement that the interacting dyad recruited by the mating type system be composed of proteins that are similar to one another.

MECHANISMS FOR MODIFYING CELL SURFACE PROTEINS

We therefore come to the heart of the matter for the experimental biologist: How have eukaryotes managed to come up with such an array of morphogenetic "ideas"? Do the properties of known cell-surface molecules suggest why they can assemble in such different combinations? If we examine the glycoproteins that are present in extracellular coats, many prove to be fibrous. This has been documented directly by electron microscopy for the collagens, laminins, and proteoglycans of the multicellular animals [Hay, 1981] and the extensin protein of higher plants [Van Holst and Varner, 1984]. In other cases it can be inferred by the high proportions of proline, hydroxyproline, and/or glycine residues in the coat proteins. Fibrous molecules, whether carbohydrate or protein, are clearly the materials of choice as the "building blocks" of extracellular coats, as they can display along their length binding sites for other fibrous molecules and thereby create the three-dimensional networks that are the basis of form.

A common feature of the fibrous glycoproteins examined to date is that they are internally repetitive: collagens are largely Gly XY, where X or Y is usually Pro or Hyp [Linsenmayer, 1981], and the plant extensins have repeating modules of Ser Hyp_4 [Lamport, 1980]. A well-documented property of repetitive DNA sequences is that they are prone to undergo addition and deletion events during meiosis and possibly even during mitotic replication. It is therefore not implausible to imagine that the genes specifying repetitive fibrous proteins might have a built-in propensity to change at a faster rate than the average gene.

A second major group of proteins in extracellular matrices are lectins [Barondes, 1983]. Most, although not all, lectins are glycoproteins, and their substrate is by definition a carbohydrate. One notion for the origin of lectins, first proposed by Roseman [1970], is that they are modified glycosyltransferases that have retained an affinity for a particular carbohydrate configuration but have lost their ability to add an additional monomer to the chain. By this model, a cell that could produce a particular extracellular carbohydrate would by definition possess a putative lectin for that carbohydrate, and if that intracellular lectin were given an extracellular "address," it would come to reside at the cell surface. As most lectins are at least divalent, they could cross-link carbohydrate species in the matrix and create a variety of three-dimensional patterns. If a lectin/carbohydrate dyad came to be displayed on heterogametes, then fertilization could be triggered by their interaction. There is considerable evidence that the echinoderms have exploited just such a system, creating ever more bizzare carbohydrates on their egg surfaces and complementary lectin-like proteins on their sperm surfaces [Glabe et al., 1982].

An interesting case is given by extracellular proteins that have both fibrous *and* lectin-like domains. Thus, many of the lectins found in the cell walls of the solonaceous plants (tomato, potato, etc.) are fibrous hydroxyproline-rich glycoproteins [Leach et al., 1982], and proteins such as fibronectin and laminin have lectin-like domains that interact with glycosaminoglycan moieties in the animal ECM [Yamada, 1983]. Gene fusions between lectin-like proteins and fibrous repetitive proteins might represent particularly "creative" morphogenetic genes.

Although gene rearrangements are a key mechanism for the production of new proteins, other mechanisms exist as well, and these have also been exploited by components of extracellular matrices. The most conspicuous is post-translational processing. It has been known for some time that many of the collagens found in interstitial tissues are secreted as larger proteins which are processed by extracellular proteases to produce smaller final products, those products being uniquely capable of assembling into the massive collagen fibrils of the higher animals [Bornstein and Sage, 1980]. More recently, it has become clear that one type of collagen, known as type IV, is not processed by such enzymes and instead interacts to form a delicate branching network in the extracellular matrix that surrounds individual cells [Yurchenco and Furthmayr, 1984]. To my knowledge, there is as yet no example of an adhesive "fertilization molecule" created from an ancestral matrix molecule by an endoprotease, although such examples may well emerge with further study. There is, however, an interesting case in this regard. The yeast *S. cerevisiae* initiates its mating reaction via complementary hormones, denoted *a*-factor and α-factor, which, although not adhesins per se, stimulate the display of adhesins in the walls of opposite-type cells and can thereby be considered as adhesin accessories. The α-factor proves to be synthesized and secreted as a large prepro-α-factor protein, and the active hormone is clipped out of this precursor by several endoproteases [Julius et al., 1984]. One of these proteases proves to be dictated by the STE13 locus, whose expression is under direct control of the MATα mating-type locus [Julius et al., 1983].

Post-translational processing is usually thought of as a proteolytic event, but a recent article by Politz and Edgar [1984] indicates that extracellular proteins may also be created in a reciprocal fashion, namely, by the covalent joining of small polypeptides to produce a larger final product. Their analysis involves the cuticle of the nematode *C. elegans*. It has been documented [Cox et al., 1981] that the worm produces four distinct types of cuticle as it proceeds through its various larval stages and that each type of cuticle contains a unique array of collagens that are held together by disulfide bonds. When the disulfides are reduced, the individual collagen polypeptides range in molecular weight from 60 to >210 kd by SDS/PAGE. When, however, the mRNAs for these collagens are isolated from the larvae, they prove to be small (1.2 kd) species which dictate the synthesis of comparably small collagenous polypeptides, most <50 kd. Politz and Edgar suggest, therefore, that the cuticle collagens are synthesized as small species that are subsequently joined together covalently to form the larger species extracted from the cuticle. The enzyme(s) that effect such crosslinking could conceivably come up with different combinations at different larval stages, thereby accounting, at least in part, for the morphological and biochemical differences between the four cuticle types.

Cross-linking of extracellular matrix proteins in fact often occurs, examples being the dityrosine bridges that form within the sea urchin egg cortex after fertilization [Foerder and Shapiro, 1977], the iso-di-tyrosine linkages between the extensin proteins of the higher plant cell wall [Fry, 1982; Cooper and Varner, 1983] and the lysine-derived crosslinks in elastin [Franzblau and Faris, 1981]. All of these crosslinks, however, usually are considered as secondary events which reinforce the stability of coats that have already assembled; indeed, they are analogous to the disulfide bridges that stabilize the nematode cuticles. Novel in the Politz and Edgar article is the notion that the building blocks themselves might be created by protein-to-protein crosslinks.

There are two examples in the limited literature on fertilization molecules that are reminiscent of the nematode strategy. As noted earlier, the four types of yeast whose agglutinins have been analyzed all produce the same dyad, a large carbohydrate-rich species and smaller heat-labile species [reviewed in Pierce and Ballou, 1983]. The large species proves to be >95% carbohydrate, the carbohydrate being long mannan polymers comparable to those utilized to form the bulk of the cell wall. To this mannoprotein is attached, via disulfide bonds, a small glycopeptide which, when separated by reducing agents, proves to carry all the recognition activity of the whole. Thus, a mannoprotein, whose ancestor was likely a bona-fide wall component, has been converted into a sexual molecule via its covalent attachment to an adhesin.

The second example comes from recent observations in our laboratory [Goodenough et al., 1985]. We find that the surface of the gametic flagellum carries not only agglutinin proteins but also proteins that are half as long as agglutinins and have the morphology and antigenic properties of half the full-length species (Fig. 2). When isolated, these half-proteins are inactive as agglutinins. Mutant strains that are devoid of active agglutinin, and whose mutant gene loci are under *mt* control, produce large quantities of these half-proteins but no full-length proteins. These findings suggest three alternate interpretations: the half-molecules are the degradation products of a surface protease which is hyperactive in the mutants; the half-molecules serve other, non-adhesive functions and their morphological and antigenic resemblance to agglutinins is fortuitous; or the half-molecules are spliced together at the flagellar surface to produce an adhesive, full-length protein, and the mutants lack this activity. Should this last explanation prove to be correct, then it is directly analogous to the nematode cuticle story. In the context of the propositions put forward in this paper, it is attractive to consider that mating molecules might in some cases be generated by splicing, as new molecules, available for speciation, could be generated by heritable alterations in the sites at which the splicings take place or the enzymes that effect the splices. The validity of this notion may be testable by a morphological study of the agglutinins produced by the various extant species of *Chlamydomonas*.

THE FORMATION OF SPORES

As noted earlier, many lower eukaryotes have interposed spore formation between zygote formation and meiosis and have placed the three activities under the control of a master mating-type locus. Sporulation is not, however, a eukaryotic invention, and it is instructive to summarize current information on spore formation in prokaryotes. The gram-positive bacteria and the cyanobacteria have evolved highly sophisticated means for producing spores, albeit asexual spores. The structural genes involved in spore production are, interestingly, controlled by master loci that are sensitive to nitrogen starvation [reviewed in Losick and Shapiro, 1984]. These loci specify polypeptides which associate with the vegetative RNA polymerase and restrict it to the transcription of sporulation-specific genes, very much in the way that some phages alter the specificity of the host polymerase so that it prefers phage gene promoters. Many of the sporulation-specific genes specify the formation of a cell coat that is very different from the vegetative coat and highly resistant to adverse environments. It is clearly tempting to speculate that such a mechanism was coopted directly by the proto-eukaryotes and that the nitrogen-sensitive mating-type loci derive directly from such prokaryotic master loci; DNA sequence data presumably will soon be available to test this notion.

Of interest to the question of the evolution of multicellularity is the fact that the fungi and the algae can produce one type of cell coat when they are haploid or homozygous for mating-type and acquire the ability to produce a second type of cell coat when they are heterozygous for mating type. The switch in genetic programs that is initiated at fertilization in the higher eukaryotes may be similarly dependent on heterozygosity at critical loci.

HOMOLOGY BETWEEN MATING TYPE AND HOMOEO GENES

A recent paper by Shepherd et al. [1984] documents a partial homology between the homoeotic genes of *Drosophila,* which specify the developmental fate of the larval and adult cells, and the *MAT*α gene of yeast. It also has been shown that these same "homoeo domains" are present in a *Xenopus* gene expressed early in development [Carrasco et al., 1984]. Taken together, it would appear that there is a good chance that mating-type loci represent early examples of genes designed to control cell differentiation. If, as is widely believed, metazoan development is ultimately dictated by the display of cell-surface proteins, then the mating-type control of fertilization molecules in the lower eukaryotes may well be an excellent paradigm for the control of embryonic patterns in the higher eukaryotes.

REFERENCES

Adair WS, Hwang CJ, Goodenough UW (1983): Identification and visualization of the sexual agglutinin from mating-type plus flagellar membranes of *Chlamydomonas*. Cell 33:183–193.

Adair WS, Long J, Mehard WB, Heuser JE, Goodenough UW (1985): Production and characterization of monoclonal antibodies to the mt^+ agglutinin of *Chlamydomonas* (submitted).

Barondes SH (1983): Developmentally regulated lectins. In Yamada K (ed): "Cell Interactions and Development: Molecular Mechanisms." New York: Wiley, pp 185–202.

Bernstein H (1977): Germ line recombination may be primarily a manifestation of DNA repair processes. J Theor Biol 69:371–380.

Bernstein H, Byers GS, Michod RE (1981): Evolution of sexual reproduction: Importance of DNA repair, complementation, and variation. Am Natur 117:537–549.

Bornstein P, Sage H (1980): Structurally distinct collagen types. Annu Rev Biochem 49:957–1003.

Cain JR (1979): Survival and mating behavior of progeny and germination of zygotes from intra- and interspecific crosses of *Chlamydomonas eugamatos* and *C. moewusii* (Chlorophyceae, Volvocales). Phycologia 18:24–29.

Carrasco AE, McGinnis W, Gehring WJ, DeRobertis EM (1984): Cloning of an X. laevis gene expressed during early embryogenesis that codes for a peptide region homologous to Drosophila homeotic genes. Cell 37:409–414.

Catt JW, Hills GJ, Roberts K (1978): Glycoproteins from *Chlamydomonas reinhardii*, and their self-assembly. Planta 131:165–171.

Crandall MA, Brock TD (1968): Molecular basis of mating in the yeast *Hansenula wingei*. Bacteriol Rev 32:139–163.

Chiang KS, Kates JR, Jones RF, Sueoka N (1970): On the formation of a homogeneous zygotic population in *Chlamydomonas reinhardi*. Dev Biol 22:655–669.

Collin-Osdoby P, Adair WS (1985): Characterization of the purified *Chlamydomonas minus* agglutinin. J Cell Biol (in press).

Cooper JB, Varner JE (1983): Insolubilization of hydroxyproline-rich cell wall glycoprotein in aerated carrot root slices. Biochem Biophys Res Commun 112:161–167.

Cooper JB, Adair WS, Mecham RP, Heuser JE, Goodenough UW (1983): *Chlamydomonas* agglutinin is a hydroxyproline-rich glycoprotein. Proc Natl Acad Sci USA 80:5898–5901.

Cox GN, Staprans S, Edgar RS (1981): The cuticle of *Caenorhabditis elegans* II. Stage-specific changes in ultrastructure and protein composition during postembryonic development. Dev Biol 86:456–470.

Fincher GB, Stone BA, Clarke AE (1983): Arabinogalactan-proteins: Structure, biosynthesis, and function. Ann Rev Plant Physiol 34:47–47.

Foerder CA, Shapiro BM (1977): Release of ovoperoxidase from sea urchin eggs hardens the fertilization membrane with tyrosine crosslinks. Proc Natl Acad Sci USA 74:4214–4218.

Franzblau C, Faris B (1981): Elastin. In Hay ED (ed): "Cell Biology of Extracellular Matrix" New York: Plenum Press, pp 65–93.

Fry SC (1982): Isodityrosine, a new cross-linking amino acid from plant cell-wall glycoprotein. Biochem J 204:449–455.

Glabe CG, Grabel LB, Vacquier VD, Rosen SD (1982): Carbohydrate specificity of sea urchin sperm bindin: a cell surface lectin mediating sperm-egg adhesion. J Cell Biol 94:123–128.

Goodenough UW, Heuser JE (1985): The *Chlamydomonas* cell wall and its constituent proteins analyzed by the quick-freeze deep-etch technique (submitted).

Goodenough UW, Adair WS, Collin-Osdoby P, Heuser JE (1985): Structure of the *Chlamydomonas* agglutinin and related flagellar surface proteins in vitro and in situ (submitted).

Goodenough UW, Thorner J (1983): Sexual differentiation and mating strategies in the yeast *Saccharomyces* and in the green alga *Chlamydomonas*. In Yamada KM (ed): "Cell Interactions and Development: Molecular Mechanisms." New York: Wiley, pp 29–75.

Haber JE, Halvorson HO (1972): Regulation of sporulation in yeast. Curr Top Dev Biol 7:61–82.

Hay ED (1981): "Cell Biology of Extracellular Matrix." New York: Plenum Press.

Hiwatashi K (1981): Sexual interactions of the cell surface in *Paramecium*. In O'Day DH, Horgen PA (eds): "Sexual Interactions in Eukaryotic Microbes." New York: Academic Press, pp 351–377.

Julius D, Blair L, Brake A, Sprague G, Thorner J (1983): Yeast α factor is processed from a larger precursor polypeptide: The essential role of a membrane-bound dipeptidyl aminopeptidase. Cell 32:839–852.

Julius D, Brake A, Blair L, Kunisawa R, Thorner J (1984): Isolation of the putative structural gene for the lysine-arginine-cleaving endopeptidase required for processing of yeast prepro-α-factor. Cell 37:1075–1089.

Lamport DTA (1980): Structure and function of plant glycoproteins. In: Preiss J (ed): "The

Biochemistry of Plants," New York: Academic Press, Vol. 3, pp 501–541.
Leach JE, Cantnell MA, Sequira L (1982): Hydroxyproline-rich bacterial agglutinin from potato. Plant Physiol 70:1353–1358.
Linsenmayer TF (1981): Collagen. In Hay ED (ed): " Cell Biology of Extracellular Matrix." New York: Plenum Press, pp 5–37.
Losick R, Shapiro L (1984): *Microbial Development*. Cold Spring Harbor, New York: Cold Spring Harbor Press.
Margulis L (1972): "Origin of Eukaryotic Cells." New Haven: Yale University Press.
Nanney DL (1977): Cell-cell interaction in ciliates: Evolutionary and genetic constraints. In JL Reissing (ed): "Microbial Interactions." London: Chapman and Hall, pp 351–397.
Nasmyth KA (1982): Molecular genetics of yeast mating type. Ann Rev Genetics 16:439–500.
Pierce M, Ballou CE (1983): Cell-cell recognition in yeast. J Biol Chem 258:3567–3582.
Politz JC, Edgar RS (1984): Overlapping stage-specific sets of numerous small collagenous polypeptides are translated in vitro from Caenorhabditis elegans RNA. Cell 37:853–860.
Raper JR (1966): *Genetics of Sexuality in Higher Fungi*. New York: Ronald Press.
Roberts K (1974): Crystalline glycoprotein cell walls of algae: their structure, composition, and assembly. Philos Trans R Soc London Ser 3, 268:129–146.
Roberts K, Hills GJ, Shaw PJ (1982): The structure of algal cell walls. In "Electron Microscopy of Proteins." London: Academic Press, Vol. 3, pp 1–40.
Roseman S (1970): The synthesis of complex carbohydrates by multi-glycosyl-transferase systems and their potential function in intercellular adhesion. Chem Phys Lipids 5:270–297.
Rupp WD, Wilde CE, Reno DL, Howard-Flanders P (1971): Exchanges between DNA strands in ultraviolet-irradiated *Escherichia coli*. J Mol Biol 61:25–44.
Sager R, Granick S (1954): Nutritional control of sexuality in *Chlamydomonas reinhardi*. J Gen Physiol 37:729–742.
Schlösser UG (1976): Entwicklungsstadien- und sippenspezifische Zwellwand-Autolysine bei der Freisetzung von Fortpflanzungzellen in der Gattung *Chlamydomonas*. Ber Deutsch Bot Ges 89:1–56.
SeGall GK, Lennarz WJ (1979): Chemical characterization of the component of the jelly coat from sea urchin eggs responsible for induction of the acrosome reaction. Dev Biol 71:33–48.
Shepherd JCW, McGinnis W, Carrasco AE, DeRobertis EM, Gehring WJ (1984): Fly and frog homoeo domains show homologies with yeast mating-type regulatory proteins. Nature 310:70–71.
Sonnenborn TM (1978): Genetics of cell-cell interactions in ciliates. In Lerner RA, Bergsma D (eds): "The Molecular Basis of Cell-Cell Interaction." New York: Alan Liss.
Sprague CF Jr, Blair LC, Thorner J (1983): Cell interactions and regulation of cell type in the yeast *Saccharomyces cerevisiae*. Annu Rev Microbiol 37:623–660.
Stebbins GL (1966): "Processes of Organic Evolution." Englewood Cliffs, NJ: Prentice Hall.
Trelstrad RL (1984): "Role of Extracellular Matrix in Development." New York: Alan Liss.
van Holst GJ, Varner JE (1984): Reinforced polyproline II conformation in a hydroxyproline rich cell wall glycoprotein from carrot root. Plant Physiol 74:247–251.
van Winkle-Swift KP, Burruscano CG (1983): Complementation and preliminary linkage analysis of zygote maturation mutants of the homothallic alga, *Chlamydomonas monoica*. Genetics 103:429–445.
Wassarman PM (1983): Fertilization. In Yamada KM (ed): "Cell Interactions and Development: Molecular Mechanisms." New York: Wiley, pp 1–27.
Yamada K (1983): Cell surface interactions with extracellular materials. Annu Rev Biochem 52:761–799.
Yurchenco PD, Furthmayr H (1984): Self-assembly of basement membrane collagen. Biochemistry 23:1839–1850.

The Origin and Evolution of Sex, pages 141–152

Part of a Meiosis Associated Gene Is Evolutionarily Conserved

Duane W. Martindale, Helen Martindale, and Peter J. Bruns

Section of Genetics and Development, Cornell University, Ithaca, New York 14853

Conjugation in the ciliated protozoan *Tetrahymena thermophila* involves a series of developmental steps, including meiosis, fertilization, and nuclear differentiation [see Nanney, 1980 for a review]. These cells contain two distinctly different nuclei, the germinal micronucleus and the somatic macronucleus [see Gorovsky, 1980 for a review]. During mating, a series of events leads to the formation of a new hybrid micro- and macronucleus. The events are summarized in Figure 1 and described below. The diploid micronucleus undergoes meiosis, with retention of one haploid product in each conjugant. This nucleus undergoes a mitotic doubling to form the stationary and the migratory pronuclei. Each mate transfers its migratory pronucleus to its partner, and the subsequent fertilization regenerates a diploid recombinant germinal zygote nucleus. This newly formed nucleus next divides twice in each conjugant to form four nuclei with identical genomes. The pairs separate and a period of development follows. Two of the progeny nuclei differentiate into new macronuclei, the third is retained to be the next generation's micronucleus, and the fourth is degraded.

The transition from vegetative growth to the sexual cycle is first triggered by starvation. It has been shown previously that cells are initiated only after 1 1/2 hr at 30°C in 10mM Tris before they will respond to similarly initiated cells of another mating type [Bruns and Brussard, 1974b; Wellnitz and Bruns, 1979; Finley and Bruns, 1980]. Cells will undergo all the nuclear events listed above, producing new macro- and micronuclei even after extensive starvation prior to mixing with another mating type. After 12 hr of starvation,

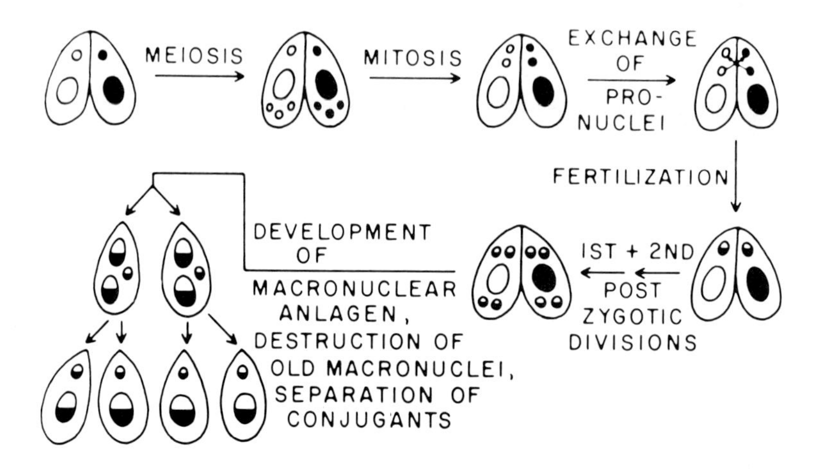

Fig. 1. Major events in conjugation [Bruns and Brussard, 1974b].

the cells have stabilized in a regimen of slow protein and ribosome turnover [Sutton and Hallberg, 1979]; they remain sexually competent for more than two weeks under these conditions. Thus, in the absence of any added nutrients, well-starved cells of two mating types simply can be mixed, and the events of conjugation commence. If no nutrients are added, the conjugants proceed to, but arrest at, the last stage mentioned above (pairs separated, each exconjugant with two new macronuclei, one new micronucleus). With the addition of food at any point during conjugation, the cells begin growth at this last stage by undergoing mitosis to double the new micronucleus and cytokinesis to yield two cells, each with one micronucleus and one macronucleus. Thus, the first part of this sexual system generates a new, recombinant germline; the second part delivers this new genome to the somatic nucleus. The focus of this discussion is on the early stages in the first part: gene activity during meiosis.

Technically, it is extremely useful that starvation is necessary but not sufficient for pair formation and that all subsequent events seem to be tied to pair formation. After 12 hr of starvation in 10 mM Tris, the level of polysomes is virtually undetectable; the rate of overall protein synthesis has dropped significantly. Two separate mating types thus may be starved separately and simply mixed to start the next steps on the way to pair formation and conjugation. Meiosis begins soon after pairs have formed. A series of stages can be recognized cytologically and used to monitor how many cells are in each stage, therefore providing a measure of the degree of synchrony. Figure 2 presents the stages that have been identified previously [Martindale et al., 1982] and Figure 3 provides an indication of the degree of synchrony that it is possible to achieve.

Protein synthetic activity was monitored during the process by administering 30-min ^{3}H leucine pulses at hourly intervals to a mating mixture and determining relative amount of TCA precipitable material. Figure 4 presents the results. The two mating types had been prestarved for 24 hr before the start of the experiment; activity was low at the beginning. There is a clear peak between hour 1 and 3, which is coincident with meiotic prophase.

In order to isolate genes that are expressed in this process, a cDNA library was created in pBR322 with poly A^{+} RNA isolated from mating mixtures at 2, 2.5, and 3 hr [Martindale and Bruns, 1983]. The resulting colonies were screened with labeled RNA from both mating and starving cultures. From 480 possible clones, 18 were chosen for further study because they showed a strong signal with RNA from conjugating cells or because they had large inserts. Cross-hybridization studies indicated that some of the clones were independent isolates of cDNA of the same RNA. Representative clones for each of these separate transcripts were used as probes against Northern blots of RNA taken from a variety of stages in the life cycle of *T. thermophila*.

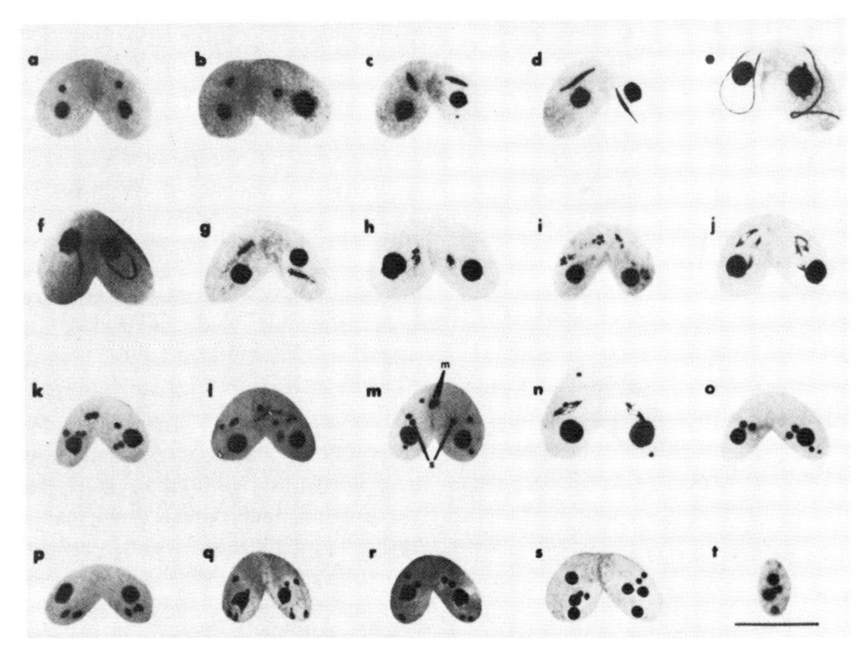

Fig. 2. The cytological stages of conjugation, visualized by Giemsa staining [Martindale et al., 1982]. a) Micronucleus no longer closely associated with macronucleus; b–g) various stages in meiotic prophase; h) meiotic metaphase; i) first prezygotic division; j,h) second prezygotic division; l,m) third prezygotic division (migratory and stationary pronuclei labelled m and s, respectively); n,o) first postzygotic division; p,q) second postzygotic division; r,s,t) various stages of macronuclear development.

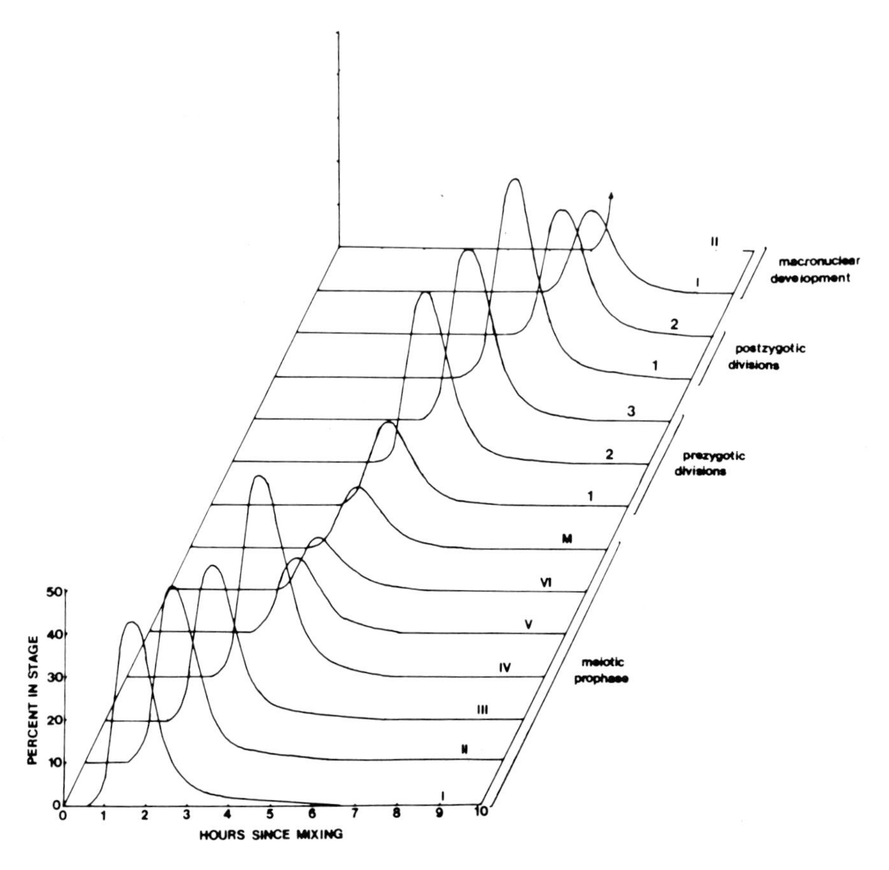

Fig. 3. Pairs in a particular stage versus time since mixing. Cytological stages were visualized with Giemsa stain as in Martindale et al. [1982].

Clones representing three different RNA molecules were found to give a positive signal solely with RNA from mating cells.

The presence of these sequences in RNA from conjugating cells was measured more carefully by preparing RNA from many time points from a mating mixture. The samples also were carefully monitored cytologically to determine the conjugation stage for each time point. Figure 5 presents the stages in conjugation for the culture (top panel) and the relative hybridization for each of the clones. All three probes show a peak of expression during meiotic prophase. An additional clone (pC3) also was assayed because, although present at other parts of the life cycle, it is strongly induced during conjugation. The association of transcription with meiosis led us to see if

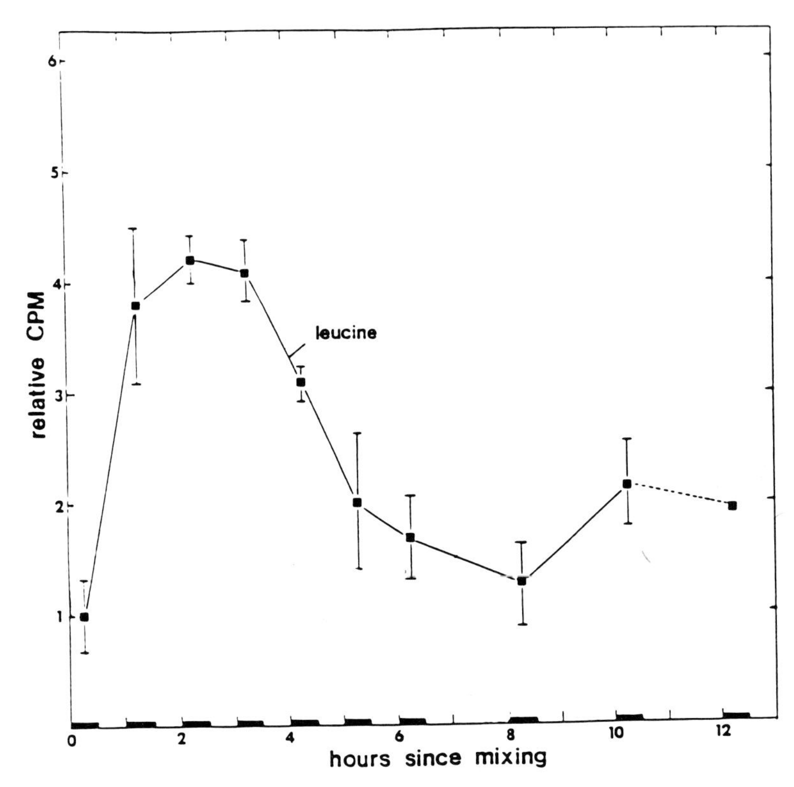

Fig. 4. Leucine incorporation during conjugation. Ordinate: incorporation level of ^{3}H-leucine into acid-precipitable material (protein). One unit equals 1.2–9.6 × 10^3 cpm. Abscissa: time zero is when two cultures of different mating type that had been starved for approximately 24 hr were mixed. The curve is constructed from the data of four separate experiments. Cultures were pulse labelled for 30 min at times indicated by bars on the abscissa.

any of these sequences are present in other genomes. Accordingly, the clones were used as probes against Southern blots of DNA from several different organisms. One of the conjugation-specific clones presented a clear signal. Figure 6 shows the result of such a blot against Tetrahymena, yeast, and Drosophila DNA. The hybridization was performed in fairly stringent conditions: 50% formamide, 5 × SSC, 42°C. It is interesting to note that there is a small family of sequences in the Drosophila blot. At lower stringencies, a similar family appears in the yeast blot (data shown below).

The Tetrahymena clone was used to screen a library of the Drosophila genome, inserted in bacteriophage λ (kindly supplied by Michael Goldberg, Cornell University). A Southern blot analysis of four phage showing ho-

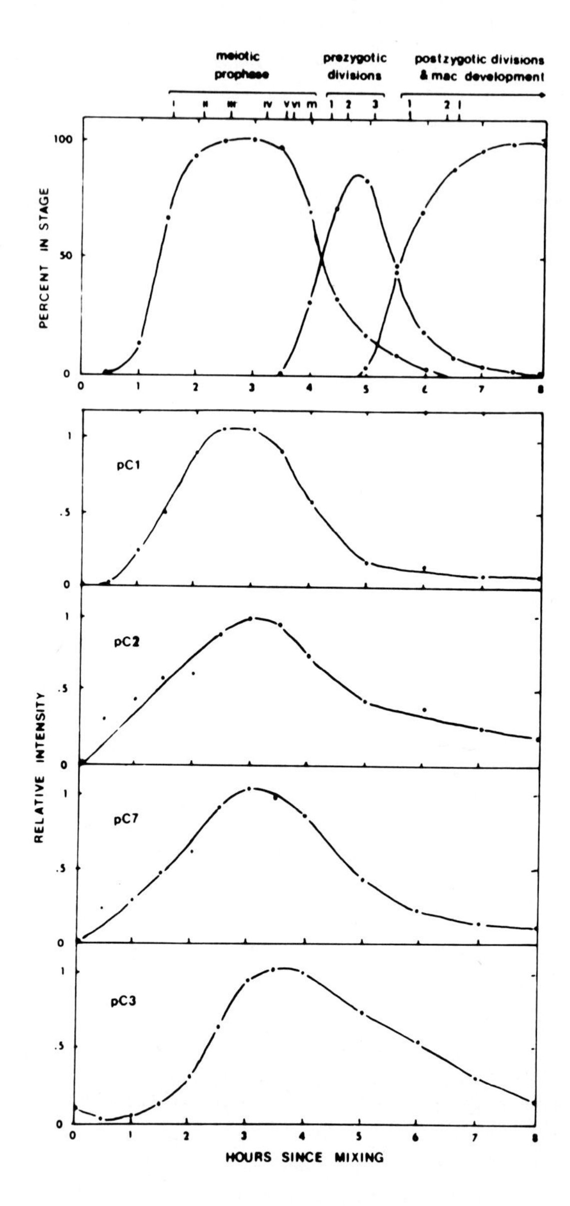

Fig. 5. Transcription patterns of conjugation-specific genes: A conjugating culture was monitored cytologically and at the same time had the mRNA levels of three conjugation-specific genes (pC1, pC2, and pC7) and a conjugation-induced gene (pC3) monitored. Ordinate in top panel: the percentage of pairs in a particular cytological stage. Ordinate in bottom panels: filters containing equal amounts of RNA from different times of conjugation were hybridized with a probe from each gene and exposed to x-ray film. The resulting autoradiographs were scanned with a densitometer, and the maximum value for each autoradiograph was assigned a relative intensity of one. Abscissa: hours since two starved cultures with different mating types were mixed.

mology to the Tetrahymena probe is shown in Figure 7. Although each phage yields a different EcoRI fragment pattern, each has a fragment bearing strong homology to the probe. The fragment labeled D6f in the figure was used in all subsequent analyses as the Drosophila probe.

This fragment was used for *in situ* hybridization to Drosophila polytene chromosomes initially by Ross MacIntyre and then by Michael Goldberg (Cornell). Interestingly, it mapped to position 3A3,4; and, by subsequent analysis by M. Goldberg, it has been localized just adjacent to the zeste locus. This is extremely intriguing because this locus is involved in transvection [Lewis, 1954] and may be involved somehow in chromosome pairing [Jack and Judd, 1979].

Mariana Wolfner (Cornell) prepared cDNA from total poly A^+ RNA extracted from third instar larvae, pupae, and adult male and female Drosophila. When this cDNA was used as a probe against blots of the phage DNA carrying the Drosophila sequences, a strong signal was seen in all conditions; it was therefore concluded that the conserved sequences are actively transcribed.

Both the Tetrahymena (pC7) and the Drosophila (D6f) cloned sequences were used as probes against Southern blots of DNA from both yeast and Drosophila at different concentrations of formamide (Fig. 8). Stringency is directly proportional to formamide concentration [McConaughy et al., 1969]; other conditions were as in Figure 6. The two probes reacted with the same fragments in each DNA preparation, although the degree of homology is somewhat different. Again, the presence of a small family of fragments is evident.

A λ charon 4A library of the yeast genome (kindly supplied by J. Woolford, Carnegie Mellon) was screened with both the Tetrahymena and the Drosophila clones, and 12 different phage clones were isolated. All contained a 2.2 Kb EcoRI fragment that hybridized with both the Tetrahymena (pC7) and Drosophila (D6f) probes. One has been further characterized. The extent of homology between the sequences cloned from Tetrahymena, Drosophila, and yeast was studied. Figure 9 presents restriction maps of the cloned sequences from the three organisms, indicating the fragment in each that contains the conserved sequences. The conserved regions are within a 320 bp Taq I-Pst I fragment in the Tetrahymena clone, a 520 bp HhaI-EcoRI fragment in the yeast clone, and a 380 bp PvuII-HhaI fragment in the Drosophila clone.

Finally, Y2f, a 1.2 Kb Pst I fragment from the yeast clone (see Fig. 9) was used to probe Northern blots of poly A^+ RNA (provided by J. Segall, University of Toronto) from diploid yeast which were growing either vegetatively or in sporulation conditions. Figure 10 demonstrates that the probe identifies two species of RNA, one of which is present only when the cells

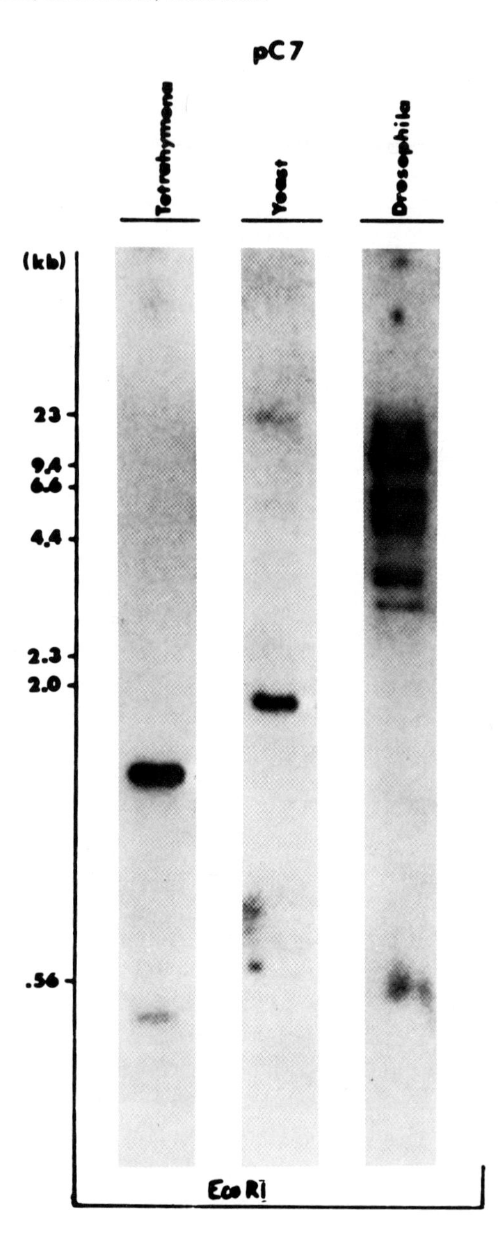

Fig. 6. Evidence for sequence conservation: Genomic DNA from *Tetrahymena*, yeast (*Saccharomyces cerevisiae*), and *Drosophila* were digested with the restriction endonuclease EcoRI, run on an agarose gel, transferred to a nitrocellulose filter, and hybridized with ^{32}P-labelled pC7 (a conjugation-specific cDNA clone). The hybridization was done under stringent conditions (50% formamide, 5 × SSC, 42°C). HindIII digested lambda DNA was used for size markers.

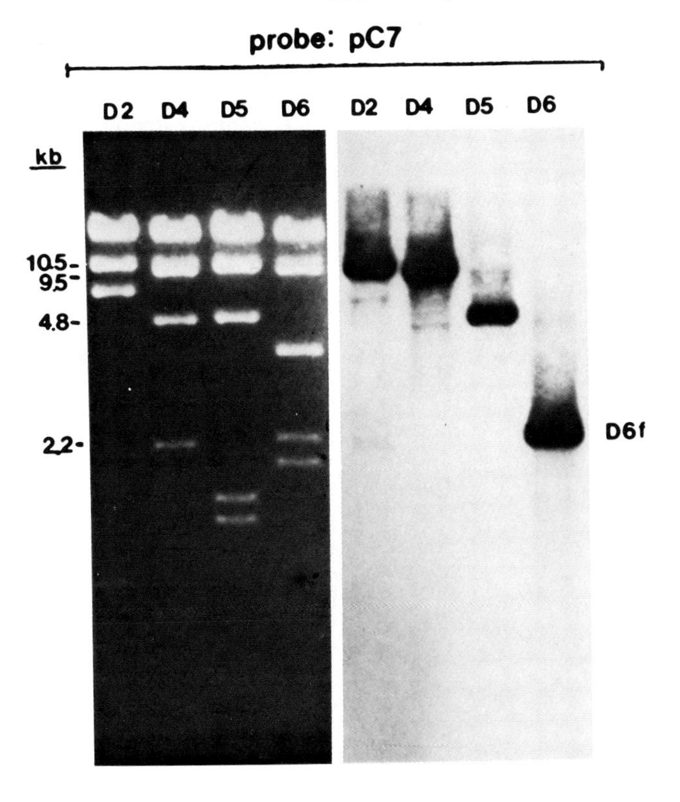

Fig. 7. Cloned Drosophila DNA showing homology to pC7. DNA was isolated from four recombinant phages selected by screening a Drosophila genomic DNA library with pC7. This DNA was cut with the restriction endonuclease EcoRI, separated on an agarose gel, transferred to nitrocellulose, and hybridized to ^{32}P-labeled pC7. Four different sizes of EcoRI fragments hybridized (10.5 kb, 9.5 Kb, 4.8 Kb, and 2.2 Kb) under conditions described in Figure 6.

are in sporulation conditions. It should be noted that the RNA from all conditions was derived from diploid yeast of the genotypes *aa, αα,* and *aα*. As can be seen, the sporulation condition transcript is present in all three genotypes when the cells are in the proper conditions for expression. Thus, the conserved sequence again appears to be transcribed in yeast and, as in Tetrahymena, may be subject to transcriptional regulation and expressed only in conditions when meiosis will occur. One complication is that these data suggest that in yeast the gene in question may not be under the control of the mating type locus, as a signal is seen not only in the *aα* cells, but also the *aa* and *αα* cells.

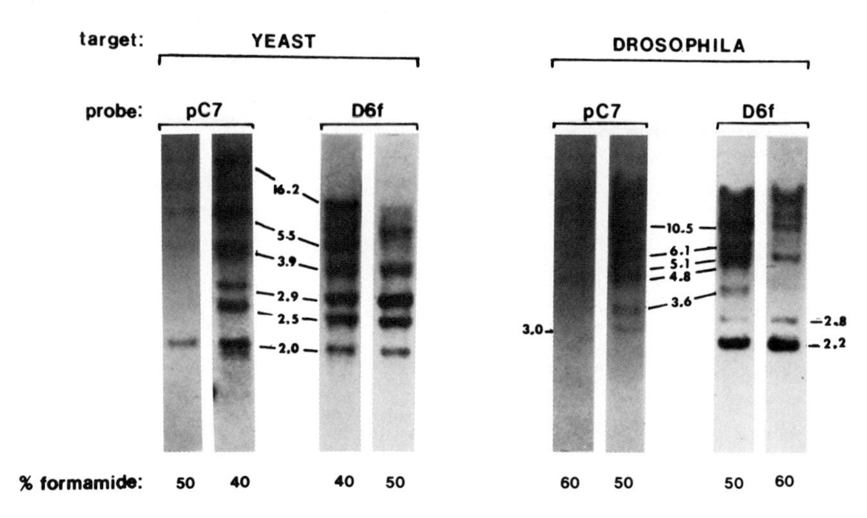

Fig. 8. Cross-hybridization under different stringencies. Genomic DNA from yeast and *Drosophila* was treated as in Figure 7 and then hybridized with ^{32}P-labeled pC7 or D6f under different stringencies. The stringency was varied by varying the formamide concentration from 40% (low stringency) to 60% (high stringency). The sizes are in kilobases.

In summary, the biology of Tetrahymena makes possible molecular and genetic studies on events in a sexual pathway with features common to many eukaryotes. Nuclear events, including meiosis, fertilization, and differentiation can be synchronized fairly well, and large quantities of material can be prepared. In this study, the focus was on genes expressed at the time when cells are undergoing meiosis. Sequences from a cDNA library were isolated that are expressed only at this point in the sexual cycle. It was found that one of the clones contains a short sequence (of approximately 300 bp) that shows strong homology with what may be a small family of genes in both Drosophila and yeast.

The first member of the Drosophila family studied maps to a region in which genes possibly involved with chromosome pairing reside [Jack and Judd, 1979]. At the very least, it is clear that the conserved sequences appear in an abundance in the RNA of Drosophila sufficient for their easy detection in cDNA.

Finally, the isolation of a sequence from yeast that bears the conserved region, and that may be active only in sporulation conditions, makes possible further studies to probe function of the gene. It will be disrupted by integrative recombination [Shortle et al., 1982], and loss of vegetative or sexual functions will be investigated.

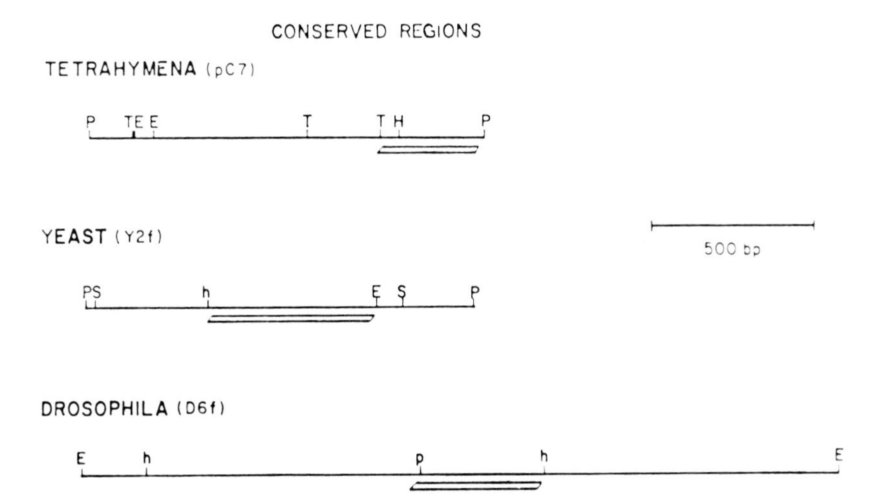

Fig. 9. Restriction maps of conserved regions: Shown are restriction enzyme maps for DNA containing the conserved regions in the three organisms: pC7 is a cDNA insert (PstI sites constructed during cloning); Y2f is a fragment of yeast DNA from a lambda recombinant clone, and D6f is the EcoRI fragment shown in Figure 7. E = EcoRI, H = HindIII, h = HhaI, P = PstI, p = PvuII, S = Sau3A, T = Taq I. The underlined fragments indicate the conserved regions (320 bp in *Tetrahymena*, 520 bp in yeast, and 380 bp in *Drosophila*).

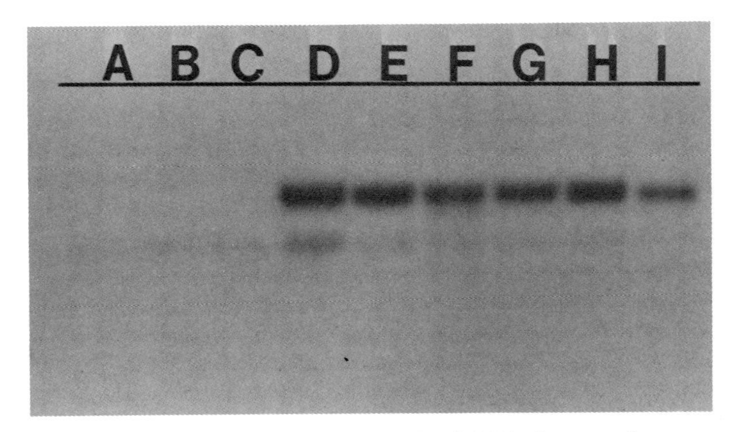

Fig. 10. Hybridization of Y2f to yeast RNA: $PolyA^+$ RNAs from growing yeast and yeast that had been in sporulation media for 7 hr and 10 hr were denatured by glyoxal treatment, separated on an agarose gel, transferred to nitrocellulose filter, and hybridized with ^{32}P-labeled Y2f. The yeast were diploid and one of three genotypes (aa, αα, and aα). The lanes in the filter are the following: A) vegetative cells, genotype *aa*; B) vegetative cells, genotype *αα*; C) vegetative cells, genotype *aα*; D) 7 hr sporulation cells, genotype *aa*; E) 7 hr sporulation cells, genotype *αα*; F) 7 hr sporulation cells, genotype *aα*; G) 10 hr sporulation cells, genotype *aa*; H) 10 hr sporulation cells, genotype *αα*; I) 10 hr sporulation cells, genotype *aα*.

The small, conserved sequence thus is found in the transcripts of genes that are expressed when cells undergo meiosis (Tetrahymena and probably yeast) or possibly are involved with chromosome pairing (Drosophila). It may be that they code for some common function in a highly conserved protein domain or reflect some aspect of regulation common to all and that this feature has not diverged much in these evolutionarily distant species. In any case the finding of a common sequence in a small set of genes may provide a first step in studies of some of the genes involved in a fundamental process common to most eukaryotic sex: meiosis.

ACKNOWLEDGMENTS

This research was supported by NIH Grant GM 27871.

REFERENCES

Bruns PJ, Brussard TB (1974a): Pair formation in *Tetrahymena pyriformis,* an inducible developmental system. J Exp Zool 188:337–344.

Bruns PJ, Brussard TB (1974b): Positive selection for mating with functional heterokaryons in *Tetrahymena pyriformis*. Genetics 78:831–841.

Finley MJ, Bruns PJ (1980): Costimulation in *Tetrahymena*. II. A nonspecific response to heterotypic cell-cell interactions. Dev Biol 79:81–94.

Gorovsky MA (1980): Genome organization and reorganization in Tetrahymena. Annu Rev Genet 14:203–239.

Jack JW, Judd BH (1979): Allelic pairing and gene regulation: A model for the zeste-white interaction in *Drosophila melanogaster*. PNAS 76:1368–1372.

Lewis EB (1954): The theory and application of a new method of detecting chromosomal rearrangements in *Drosophila melanogaster*. Am Naturalist 88:225–239.

Martindale DW, Allis D, Bruns PJ (1982): Conjugation in *Tetrahymena thermophila*: a temporal analysis of the cytological stages. Exp Cell Res 140:227–236.

Martindale DW, Bruns PJ (1983): Cloning of abundant mRNA species present during conjugation in *Tetrahymena*: Identification of mRNA species present exclusively during meiosis. Cell Mol Biol 3:1857–1865.

McConaughy BL, Laird CD, McCarthy BJ (1969): Nucleic acid reassociation in formamide. Biochemistry 8:3289–3295.

Nanney DL (1980): "Experimental Ciliatology. An Introduction to Genetic and Developmental Analysis in Ciliates." New York: John Wiley and Sons.

Shortle D, Haber JE, Botstein D (1982): Lethal disruption of the yeast actin gene by integrative DNA transformation. Science 217:371–373.

Sutton CA, Hallberg RL (1979): Regulation of ribosomal RNA degradation in growing and growth-arrested *Tetrahymena thermophila* cells. J Cell Physiol 101:349–358.

Wellnitz WR, Bruns PJ (1979): The pre-pairing events in *Tetrahymena thermophila*. Analysis of blocks imposed by high concentrations of Tris-Cl. Exp Cell Res 119:175–180.

The Origin and Evolution of Sex, pages 153–154

Sex in Unicellular Eukaryotes: Discussion

Leader: Peter J. Bruns

Section of Genetics and Development, Cornell University, Ithaca, New York, 14853

The discussion in this session centered principally on an attempt to identify general features of the sexual cycle in unicellular eukaryotes and to relate them to the events seen in prokaryotes. We hoped to identify some sort of evolutionary link. In general, such a link was not seen, but a number of useful points were certainly made.

Sex in these organisms is not associated with growth and reproduction; the cells grow vegetatively. Sexual encounters are aimed at genetic recombination and lead to the formation of new genotypes. It was pointed out that most of the organisms studied undergo the sexual cycle as a response to starvation, possibly reflecting a selective pressure favoring entry into the sexual cycle only when the parental genotype cannot cope with the environment. Simply stated, lower eukaryotes would rather eat than mate. In an organism such as the ciliate *Tetrahymena* this seems reasonable; the mating act yields four cells from every pair in the time two cells would generate over 2,000 cells (10 fissions) if they were in exponential growth.

It was pointed out that some sexual systems do not exchange complete genomes from both parents. In *Chlamydomonas,* the chloroplast genome of only one parent is transmitted to the progeny. Nuclear genomes that are derived from only one parent were also mentioned. Examples cited included the unfertilized eggs of Haemenoptera, which develop into males, and the elimination of male-derived chromosomes in the sexual cycle of Coccids. No simple selective advantage could be conceived for these strategies, although the uniparental transmission of plastid genomes should have reduced the rate of evolutionary change by eliminating chances for recombination.

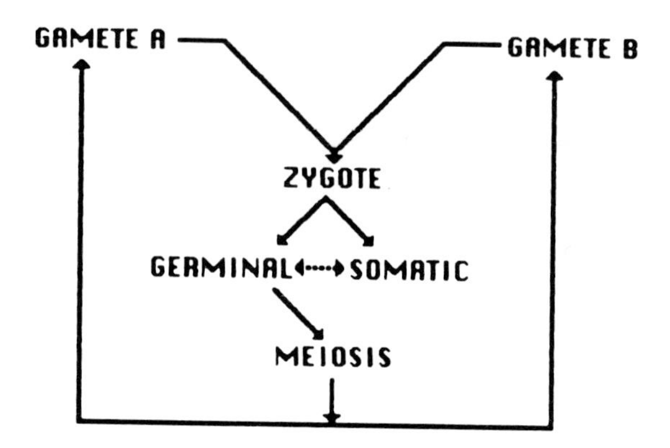

Fig. 1. General features of the sexual cycle in lower eukaryotes.

The group also discussed a unique problem associated with the origins of eukaryotic sex, namely the evolutionary origin of meiosis. Klar described several mutants in yeast and pointed out that studying appropriate mutants singly and in combination in systems such as yeast might shed light on features shared by meiosis and mitosis. Sager suggested that meiosis might well be an elaborate form of mitosis and might therefore have evolved from it.

A general consensus on several points was reached. The nutritional control of sexuality is probably worth further study. In addition, studies aimed at comparing specific elements of meiosis and mitosis should be pursued. Conceptually, an important aspect of these chromosomal maneuvers is the fact that eukaryotes have a nuclear membrane. Recent work was cited in which DNA was introduced into the cytoplasm of eukaryotic cells, either by injection or by being lost from the spindle apparatus; in either case a nuclear membrane formed around the DNA. The ability to form such a membrane around chromosomes, as well as its very presence, yields an extra layer of complexity for chromosomal maneuvers in the eukaryotic as opposed to the prokaryotic sexual cycle.

Finally, the evolutionary relationships between the origin of eukaryotes and the origin of eukaryotic sexual systems was discussed. At question was whether or not a sexual cycle was an obligate part of early eukaryotic life. Although lower eukaryotes having no sexual cycle are known, it was suggested that these might be degenerate forms of the original types.

Figure 1 summarizes many of the features associated with the nuclei in the sexual systems discussed in the session and was generally agreed upon as being a reasonable summary. Different systems vary in the point or points at which vegetative growth can occur and whether or not there are separate germinal and somatic nuclei.

EVOLUTIONARY PATTERNS IN SEGREGATION OF GERM CELLS

The Origin and Evolution of Sex, pages 157–162

Evolutionary Patterns in Germ Cell Segregation

Alberto Monroy

Zoological Station, 80121 Naples, Italy

This paper will discuss multicellular eukaryotes. A minimim requirement for any multicellular eukaryote is to be composed of two major cell lines, a somatic and a germ cell line. At some point in evolution, the segregation of a cell line uniquely specialized for the propagation of the species must have occurred. How was this dichotomy achieved? Did it occur in a single step or in a series of intermediate steps? In the latter case, are there clues with which to reconstruct the evolutionary pathway that led to the segregation of the two major cell lines? This is, however, one half of the problem, the other half is how, within the hypothetical ancestral germ cell line, the differentiation of male and female gametes occurred.

The origin of sexes has preoccupied philosophers and religious thinkers throughout the ages. The most common answer was a kind of splitting of a hermaphroditic creature into male and female. In this connection, I would like to show you the beautiful XIII century mosaic in the Cathedral of Monreale, near Palermo, Sicily, in which you can see Eve emerging from one of Adam's ribs (Fig. 1). However, I am told that the Hebrew word "zela" also means part; hence, the original biblical meaning may have been that the two sexes were the product of the "splitting" of a hermaphrodite creature. This would be more in line with ideas of the Hindu religion. When I look at that mosaic, I am struck by its similarity to the asexual reproduction by budding in Hydra (Fig. 2).

Coelentherates, and Hydra in particular, are of interest in connection with the problem of the origin of the germ cell line, and they will be discussed

Fig. 1. The birth of Eve from Adam's rib. Black and white photograph of a mosaic in the XIII century cathedral of Monreale, Italy.

in this session. Indeed, in Hydra there are cells, the so-called interstitial cells (IC), that can be made to differentiate either into the highly sophisticated nematocytes or into germ cells. Certain dense bodies that are present in the IC are lost when the cells differentiate into nematocytes. These same dense bodies are retained and considerably amplified when the IC differentiate into germ cells. This system may offer an excellent opportunity to ask questions as to whether and how this material exerts an action on the genome. Indeed, this material may be visualized as containing gene regulatory factors that prevent the restriction of the differentiative potencies in the cells of the somatic lines. However, it is interesting that in the Nematode (*Caernorhabditis elegans*) certain cytoplasmic granules which are uniformly distributed in the egg, are selectively segregated in the germ line precursor cells and finally localized exclusively in the Z_2 and Z_3 germ line stem cells [Strome and Wood, 1982], (Fig. 3). In nematodes the segregation of the germ from the somatic cell line begins at the first cleavage of the egg.

This was what Theodor Boveri discovered in 1899 in the development of the nematode *Ascaris*, a discovery that marks a turning point in the studies on the origin of germ cells. The most important part of Boveri's observations was that the segregation of the two lines was accompanied by the elimination of parts of the chromosomes of the blastomeres of the germ line (chromosome diminution) (Fig. 4). This discovery was the first step in understanding the

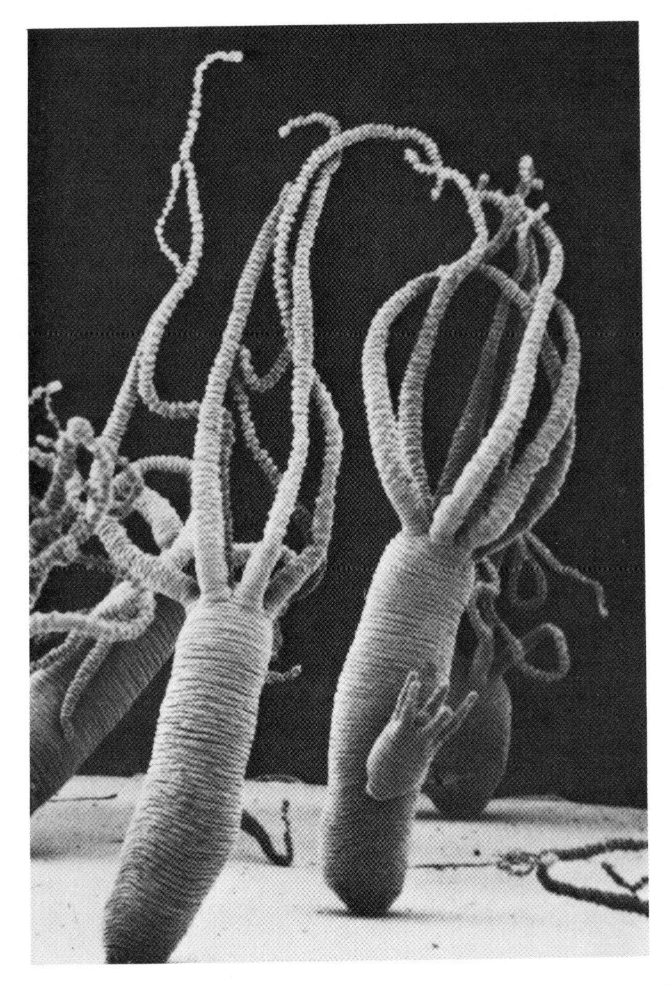

Fig. 2. Budding (arrow) in Hydra (Courtesy of Professor P. Tardent, Zürich).

restriction of the developmental potencies that occurs in the cells of the somatic line as opposed to the totipotency of the germ line cells. The analysis of this process at the molecular level has shown that in the somatic cells a massive elimination of repetitive sequences occurs [Streeck et al., 1982]. Hence the fact that these sequences are retained in the germ line suggests that they may be involved in the maintenance of the "genetic status" of the germ cells [Moritz and Roth, 1976].

The early segregation of the somatic and germ line, such as occurs in Nematodes, is certainly of great interest in connection with the molecular

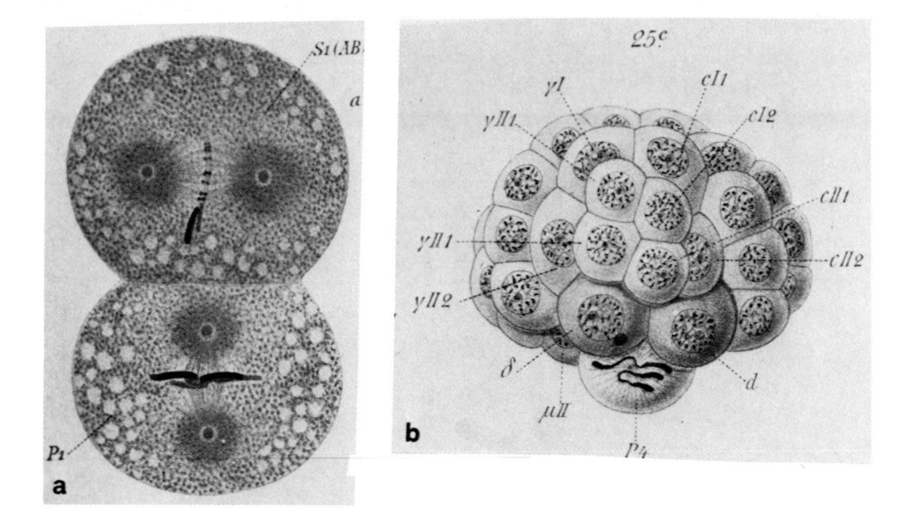

Fig. 3. a) Two cell stage of an embryo of *Ascaris*. In the upper blastomere, designated S_1 (AB), which is the precursor of the somatic line, the two chromosomes of the zygote have fragmented in a number of small chromosomes and their tips have been eliminated; in the lower blastomere, OP_1, which is the precursor of the germ line, the chromosomes have kept their shape intact. b) A later cleavage stage: the blastomere from which the germ line will originate has still its chromosomal complement intact. These are black and white photographs of the original color drawings illustrating Boveri's 1899 monograph [Boveri, 1899].

mechanisms of the restriction of the developmental potencies of the somatic cell lines. However, for the problem of the evolutionary history of the segregation of the germ cells, animals such as the Desmosponges and the *Hydra,* in which the differentiation of some cells may be switched one way or another, may be a more attractive and promising material.

The most spectacular examples of massive elimination of parts of the genome during segregation of the somatic cell line are those of insects. However, work carried out in recent years in a number of laboratories has shown that elimination of part of the genome occurs in the differentiation of the macro from the micronucleus in ciliates, which, most interestingly, is accompanied by extensive reorganization of the genome [see, for example, Klobutcher et al., 1984].

However, in the vast majority of animals, no such spectacular alterations have been observed, although minor rearrangements of the genome may have escaped detection. Hence, the restriction of the developmental potencies that occurs in the cells of the somatic cells must depend on more subtle and possible reversible processes such as methylation. From this point of view, the study of the reversible inactivation of the X chromosome in mammals may be very promising.

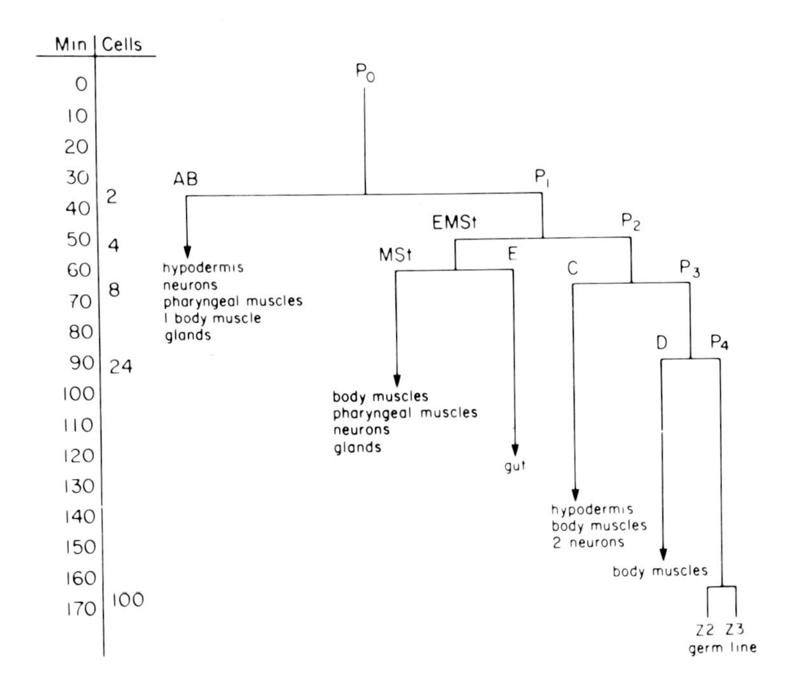

Fig. 4. Cell lineage in *Caenorhabditis elegans* showing the segregation of the germ line (P_0–P_4) through the formation of the germ line stem cells, z_2 and z_3 [From Strome and Wood, 1982].

A new and powerful tool to approach the problem of sex differentiation stems from the work of K. Jones on the sex linked Bkm sequences in snakes [see Singh et al., 1981; Jones and Singh, 1981; Singh et al., 1984]. These sequences appeared very early and have been conserved throughout invertebrate and vertebrate evolution. It will be important to find out how far back in evolution these or similar sequences can be traced. Indeed, they may be a valuable marker to approach at the molecular level not only the problem of the male-female dichotomy, but also that of the segregation of the germ cells. Simpler organisms, such as Volvox and the Sponges, may prove to be revealing in this sense. The most specific distinguishing character of the germ cells is their ability to carry out meiosis. If meiosis is part of the differentiative program of the germ cells, this program must be silenced in the somatic cells [Monroy et al., 1983].

One may then ask, when in the course of their differentiative pathway do presumptive germ cells acquire the ability to carry out meiosis, and when do they become committed to do so? In other words, can presumptive germ cells be switched to the somatic differentiative pathway? Of course, in animals in which the segregation of the somatic cells is accompanied by drastic rearrangements of their genome, there can be no question of a somatic cell becoming a germ cell. There is some evidence that in Planarian regeneration, spermatogonia and early spermatocytes are able to participate in the formation of the regeneration blastema [Gremigni and Puccinelli, 1977; Gremigni et al., 1980]. Hence, just prior to the onset of meiosis, the differentiative pathway of these cells can be reprogrammed in the somatic direction. This behavior implies the silencing of the meiosis-controlling genetic system. Differentiation of germ cells results eventually in development of the gametes. In order for fertilization to occur, gametes must recognize each other not only with respect to species specificity, but also in such a way that only gametes belonging to different genetic sex can meet and fuse. This property is reminiscent of the surface exclusion in prokaryotes [Monroy and Rosati, 1979]. It has been suggested that the gamete self, non-self recognition system may have been the ancestor of the histocompatibility system of higher animals. The genetic system controlling histocompatibility *and* self-fertilization in the hermaphroditic colonial Protochordates has provided important clues [see Scofield et al., 1982].

REFERENCES

Boveri T (1899): Die Entwicklung von *Ascaris megalocephala* mit besonderer Rücksicht auf die Kernvehältnisse. Festsch. F.C. von Kupfer, Jena 383–430.

Gremigni V, Miceli G, Puccinelli I (1980): On the role of germ cells in Planarian regeneration. 1. A karyological investigation. J Embryol Exp Morphol 56:53–63.

Gremigni V, Puccinelli I (1977): A contribution to the problem of blastema cells in Planarians: A karyological and ultrastructural investigation. J Exp Zool 199:57–72.

Jones KW, Singh L (1981): Conserved repeated DNA sequences in vertebrate sex chromosomes. Human Genet 58:46–53.

Klobutcher LA, Jahn CL, Prescott DM (1984): Internal sequences are eliminated from genes during macronuclear development in the ciliated protozoan *Oxytrichia nova*. Cell 36:1045–1055.

Monroy A, Parisi E, Rosati F (1983): On the segregation of the germ and somatic cell lines in the embryo. Differentiation 23:179–183.

Monroy A, Rosati F (1979): The evolution of the cell-cell recognition system. Nature 278:165–166.

Moritz KB, Roth GE (1976): Complexity of germ line and somatic DNA in *Ascaris*. Nature 259:55–57.

Scofield VL, Schlumberger JW, Weyt LAS, Weisman JL (1982): Protochordate allorecognition is controlled by a MHC like gene system. Nature 295:499–502.

Singh L, Philips C, Jones KW (1984): The conserved nucleotide sequences of Bkm, which define Sxr in the mouse, are transcribed. Cell 36:111–120.

Singh L, Purdom IF, Jones KW (1981): Conserved sex chromosomes associated nucleotide sequences in eukaryotes. Cold Spring Harbor Symp Quant Biol 45:805–813.

Streeck R, Moritz KB, Beer K (1982): Chromatin diminution in Ascaris suum: nucleotide sequence in the eliminated satellite DNA. Nucleic Acids Res 10:3495–3502.

Strome S, Wood WB (1982): Immunofluorescence visualization of germ-line specific cytoplasmic granules in embryos, larvae and adults of *Caenorhabditis elegans*. Proc Natl Acad Sci USA 79:1558–1562.

The Origin and Evolution of Sex, pages 163–197

The Differentiation of Germ Cells in *Cnidaria*

Pierre Tardent

Zoological Institute, University of Zürich-Irchel, Winterthurerstrasse 190, CH 8057 Zürich, Switzerland

INTRODUCTION

The most primitive of the recent *Metazoa* include the following four groups: 1) the recently rediscovered *Placozoa* [Grell and Benwitz, 1971, 1974, 1978], which document in a most convincing way what the first *Metazoa* may have looked like; 2) the *Mesozoa*, the simple cellular architecture which could be of secondary nature, as all members of this group are endoparasites [Lapan and Morowitz, 1972]; 3) the more conspicuous and better known *Porifera* and; finally 4) the *Coelenterata* composed of two phyla, the *Ctenophora* (*Acnidaria*) and the *Cnidaria*, with which this paper will be concerned.

Sexuality in one way or another, has been found in all the members of these groups thus far examined. However, as in protozoans, asexual multiplication by budding, fission and other means still plays a most important role. These two different modes of reproduction, sexual and asexual, which may be practiced by the same individual or by alternating generations (Fig. 3a) seem to complement each other in a most perfect manner: Sexual reproduction, through its various means of genetic recombination, guarantees an appropriate degree of variability. Asexual multiplication, on the other hand, stabilizes successful genotypes and, under favourable ecological circumstances, it promotes their propagation while avoiding a continuous and possibly contraproductive reshuffling of the genome [Tardent, 1984]. The switch from one type of reproduction to the other is mostly appropriately controlled by environmental signals such as temperature changes [Werner, 1958], sexual cycle of the host [Lapan and Morowitz, 1972], and others. In some cases one encounters puzzling similarities between sexual and asexual reproduction.

For example, in the *Mesozoa*, asexually initiated morphogenesis derives from a single diploid and previously somatic cell (axoblast), which must therefore be totipotent like the zygote [Lapan and Morowitz, 1972]. Similar situations although not as far reaching exist also in the *Cnidaria*.

In such cases, as in *Protozoa*, could the same cell assume either reproductive roles, that is, of a gamete or of a somatic cell, which through proliferation and differentiation could generate a new organism or at least part of it? This is, it seems, a central problem with which we are continuously confronted at this low level of metazoan organization, where the structural and functional diversification of cells and organs has barely begun. Is there already a true germ line, or is any somatic cell capable of becoming a germ cell? This question will be one of the main topics of the present paper.

At this point it is of interest to remember that at the turn of the century the great theoretician August Weismann [1834–1914], who created the germ line concept in his book "Das Keimplasma" [1892], emphasized how the Coelenterates, in particular, the *Cnidaria*, demonstrated in a most exemplary manner the existence of a germ line [Weismann, 1880a, 1880b, 1883, 1884]. Yet, we will show that in this phylum the presence or absence of a germ line is probably more difficult to demonstrate than in any other group.

SEXUALITY AND DEVELOPMENTAL CYCLES IN *CNIDARIA*

The *Cnidaria* are characterized by their extremely complicated, but highly efficient, stinging cells, the nematocytes or cnidocytes [Weill, 1930; Mariscal, 1974; Tardent and Holstein, 1982; Holstein and Tardent, 1984] feature two fundamentally different morphs, the sessile polyp (Figs. 1a, 2) and the free swimming medusae (Fig. 1b).

The polyp is usually a tube or a hollow cylinder consisting of only two epithelial layers, the ectoderm and the endoderm, separated from each other by an acellular mesoglea (Figs. 1a, 5, 8e). A spacious gastric cavity opens to the outside by a single opening, the mouth, itself surrounded by a varying number of mobile tentacles.

The mobile medusae are of a more elaborate structure (Fig. 1b, 3b, 7) which is also based on a diploblastematic architecture. They are composed of two main parts: the centrally located manubrium, with its mouth leading into the spacious gastric cavity, and the medusa's locomotory organ, the contractile bell or umbrella. Pulsations of the latter are brought about by rapid contractions of circularily arranged cross-striated muscle cells [Chapman et al., 1962]. These are antagonized by a bulky mesoglea situated between the ex- and subumbrellar ectoderm and the endoderm (Fig. 1b). The food digested within the manubrium is distributed throughout the umbrella by a system of endodermal radial and ring canals (Fig. 3b).

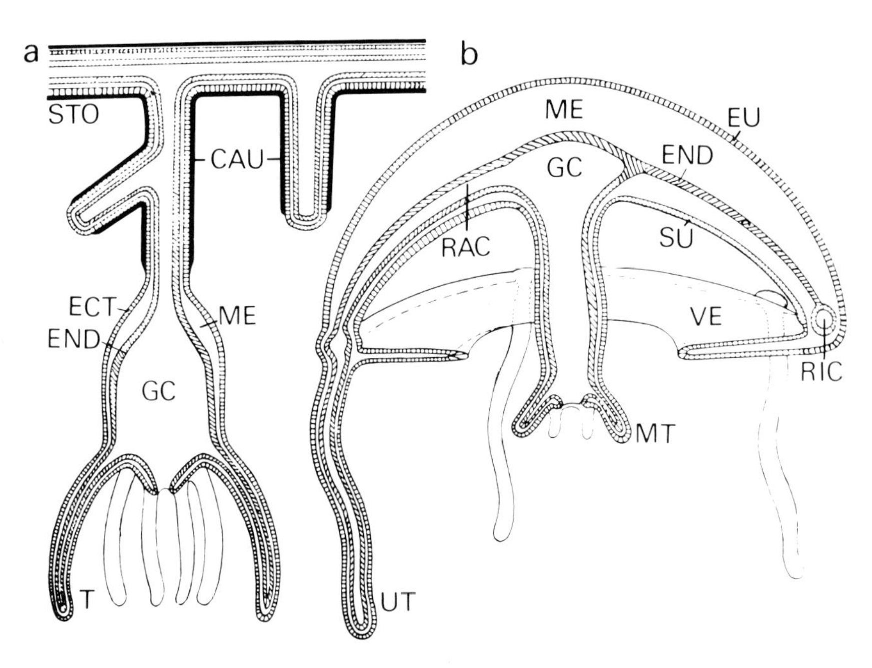

Fig. 1. Blueprint of the diploblastematic, epithelial architecture of a colonial polypoid and medusoid cnidarian, both members of the Hydrozoa. a) longitudinal section through a hydroid polyp with its stalk (caulus) and root-like stolon b)radial section through a hydromedusa (CAU = caulus; ECT = ectoderm; END = endoderm; EU = exumbrella; GC = gastric cavity; ME = mesoglea; MT = manubrial tentacles; RAC = radial canal; RIC = ring canal; STO = stolo; SU = subumbrella; T = tentacles; VE = velum.) [modified from Tardent, 1978].

The following sections characterize the four classes of the *Cnidaria* in an effort to understand sexual reproduction, gametogenesis, and the origin of germ cells [see also Tardent, 1978].

Anthozoa (Approximately 4800 Species: Actinians, Corals, Sea Pens)

This class features solitary or colonial polypoid morphs only (Fig. 2) that neither produce a medusa generation nor reveal any indication that such an additional morph has ever existed in this class. Asexual multiplication (budding, laceration, fission) and sexual reproduction are, therefore, both practiced by the predominantly gonochoristic, in some cases, hermaphroditic polypoid individuals.

The germ cells, which are believed to derive from the endodermal epithelium that covers the concentrically arranged gastric septa (Fig. 2), ripen within the mesoglea near their point of origin. From there, ripe gametes break

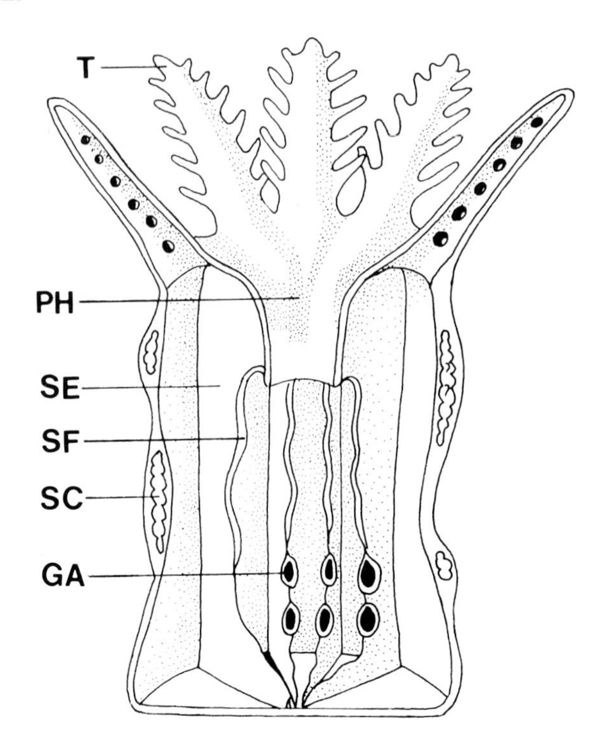

Fig. 2. Longitudinal section through an anthozoan polyp (Octocorallia) (GA = gametes; PH = pharynx; SC = sclerites; SE = endodermal septum; SF = septal filament; T = tentacle.)

through the endodermal epithelium and pass into the gastric cavity. The eggs are fertilized either in the gastrenteron or externally after expulsion through the mouth or through special openings in the body wall [Nyholm, 1943] or at the tip of the tentacles (*Ceriantharia*).

Hydrozoa (Approximately 2600 Species: Hydroid Polyps, Hydromedusae)

The recent *Hydrozoa* are "a priori" dimorphic (Figs. 1, 3a). They illustrate various examples of the classical alternation between asexually proliferating

---

Fig. 3. a. A typical metagenetic cycle of a hydrozoan (Limnomedusae: *Craspedacusta sowerbyi*). 1 = zygote; 2 = planula-larva; 3 = sessile founder polyp producing a frustules (bud); 4 = frustule; 5 = autozoid colony; 6 = gonozoid with a developing medusa-bud; 7 = young medusa; 8 = adult medusa (GA = gametes). [modified from Tardent, 1978] b–f. Step-wise reduction of hydromedusae. b. the normal, free-swimming and feeding anthomedusa of *Coryne tubulosa*. c–e. free-swimming, but non-feeding and short-lived medusae (c = *Orthopyxis integra;* d = *Clathrozoon wilsoni*, e = *Millepora* sp.). f. sessile eumedusoid of *Tubularia crocea*. (GA = gametes; MA = manubrium; NC = nematocytes; OC = ocellus; RAC = radial canal; RIC = ring-canal.)

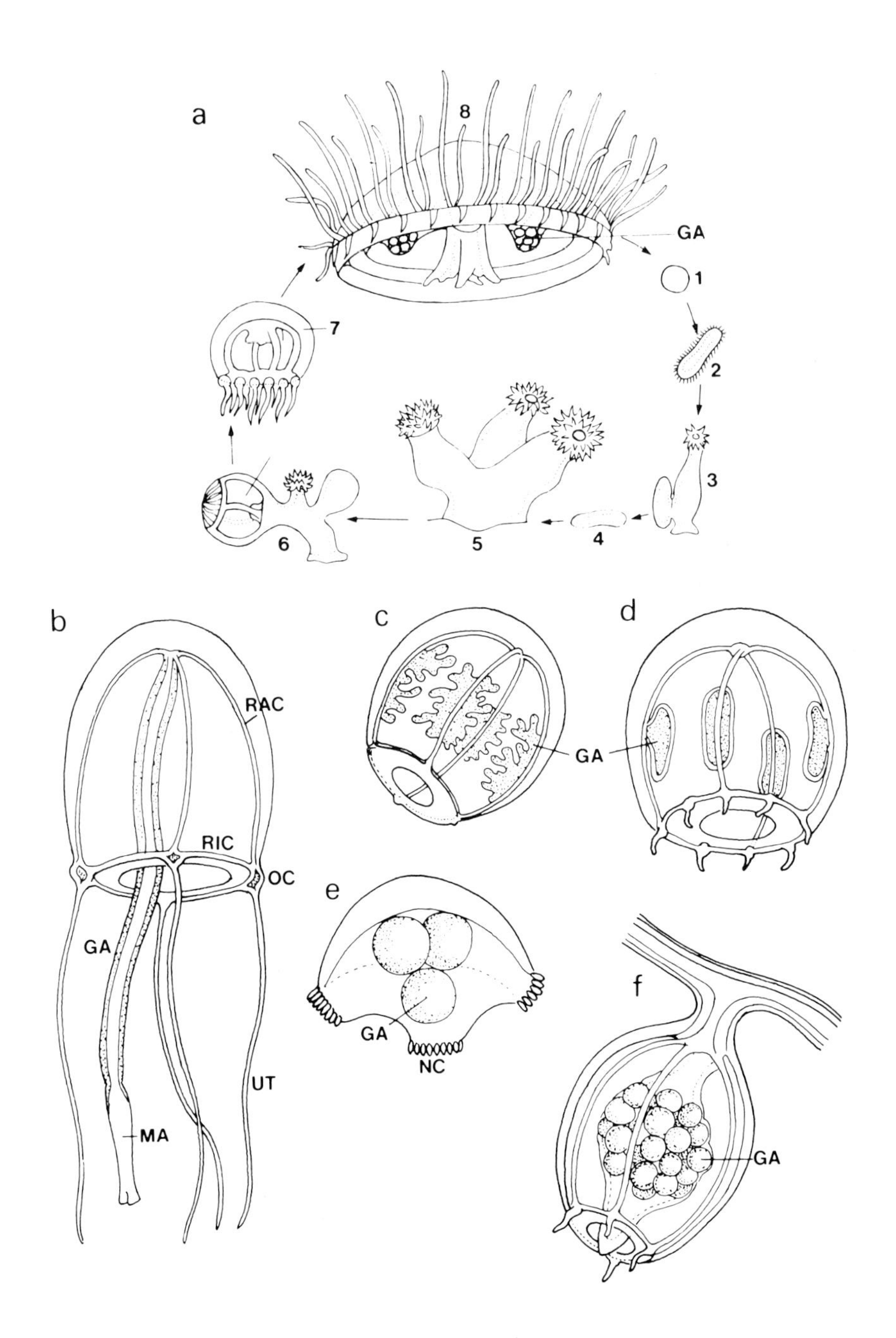
a
8
GA
1
2
3
4
5
6
7
b
RAC
RIC
OC
GA
MA
UT
c
GA
d
e
GA
NC
f
GA

polyp and the pelagic, sexually competent hydromedusa, which by a blastogenetic budding process is generated by the polyp (Fig. 3a). This metagenetic cycle has undergone various modifications. In rare cases it is such that not only the polyp but also the medusa multiplies asexually before it enters the terminal gametogenetic period of its existence [Werner, 1958]. More often, however, this cycle h as been modified so that either the polypoid or medusoid stage is partially (Fig. 3c-f) or totally (Fig. 5) suppressed.

In the order *Trachylina*, the zygotes produced by the medusae develop directly into another medusoid generation, thus bypassing the polyp stage completely. On the other hand, there are numerous examples of a progressive reduction of the medusa (Figs. 3, 4). In some *Hydroida* (*Orthopyxis integra,* Fig. 3c; *Clathrozoon wilsoni* Fig. 3d; *Millepora,* Fig. 3e), the gamete producing medusae are still capable of swimming but are extremely short-lived due the lack of digestive organs. In these cases the medusa is reduced to what could be called a "swimming gonad." In another step of reduction (as observed, for example, in *Tubularia crocea,* Fig. 3f) the medusa not only lacks a proper digestive system, but also functional umbrellar muscle cells. These eumedusoids consequently remain attached to the polyp from which they derive nutrition. In yet more dramatic steps in the rudimentation of the medusa (Fig. 4), the latter is essentially reduced to a sessile gonangium in which the umbrella encloses a brood chamber and the former manubrium, now called spadix, bears and nourishes the gametes. Finally, in the freshwater hydra (Fig. 5) all traces of a medusa stage have vanished. The eggs and spermatozoa now ripen within the ectodermal epithelium of the polyps body column.

The questions of how and why the original metagenetic cycles of the *Hydrozoa* evolved and which came first, the polyp or the medusa, remain open to speculation. All the various modifications of the dimorphic cycle as described above for some recent members of this group (Figs. 3c–f,4) are, we believe, of secondary nature and do not themselves represent the evolutionary steps on their way to the completion of the polyp-medusa cycle. Nevertheless, they may be used in an attempt to reconstruct the original phylogenetic events.

There are, in our opinion, a number of observations suggesting that, at least in the *Hydrozoa*, the original morph was the sedentary polyp (Fig. 6b).

Fig. 4. Various degrees of hydromedusan reduction. a) A normal free swimming anthomedusa previous to its detachment from the polyp. b) sessile eumedusoid (Anthomedusae) gonophore (cfr. Fig. 3f). c) sessile eumedusoid (Leptomedusae) without a manubrium. d–e) sessile cryptomedusoids with reduced and absent subumbrellar cavities. f) sessile heteromedusoid. g) sessile styloid (GAM = gametes; MA = manubrium; RAC = radial canal; SUC = subumbrellar cavity; UT = umbrellar tentacles; VE = velum). [modified from Kühn, 1914]

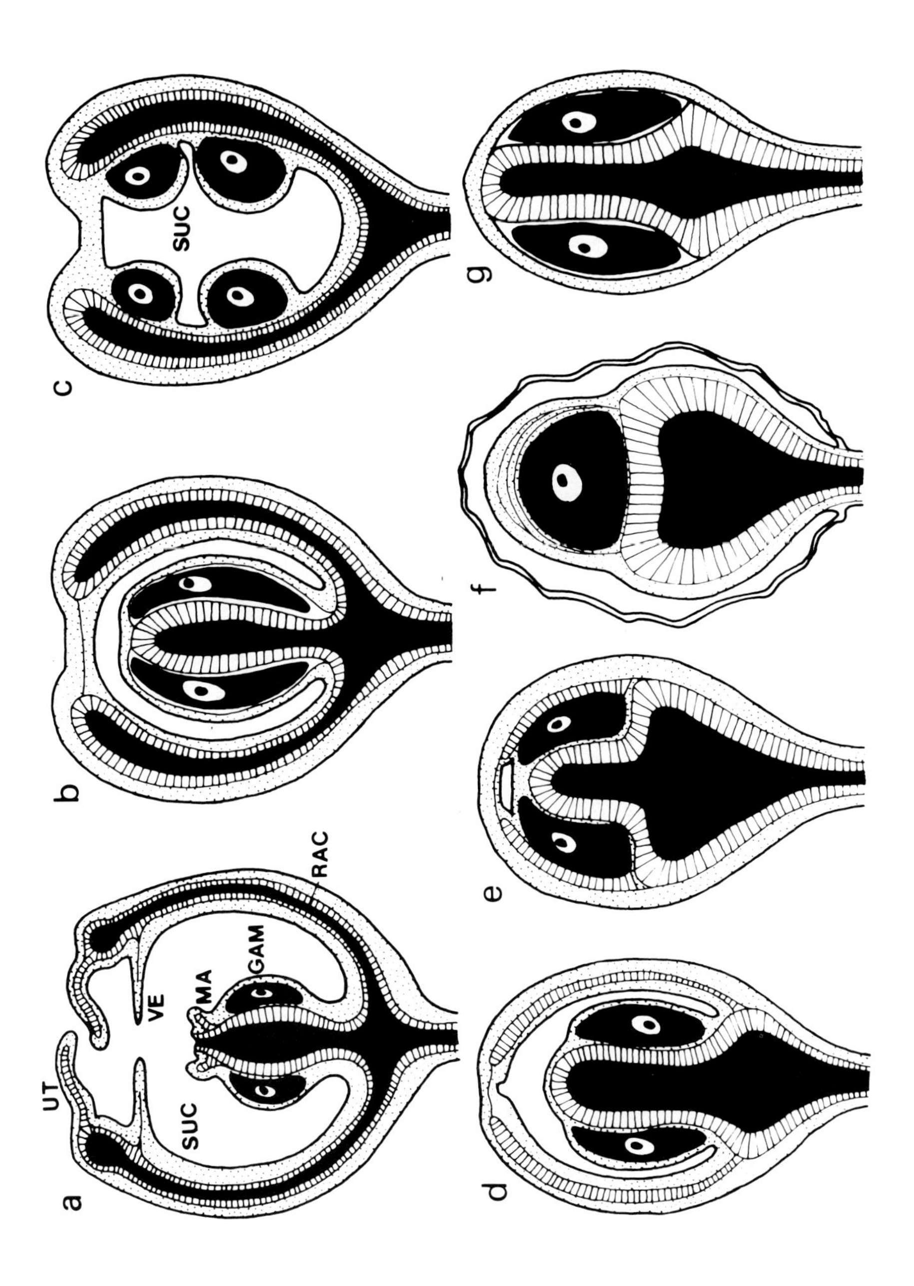
a
UT
SUC
VE
MA
GAM
RAC
b
c
SUC
d
e
f
g

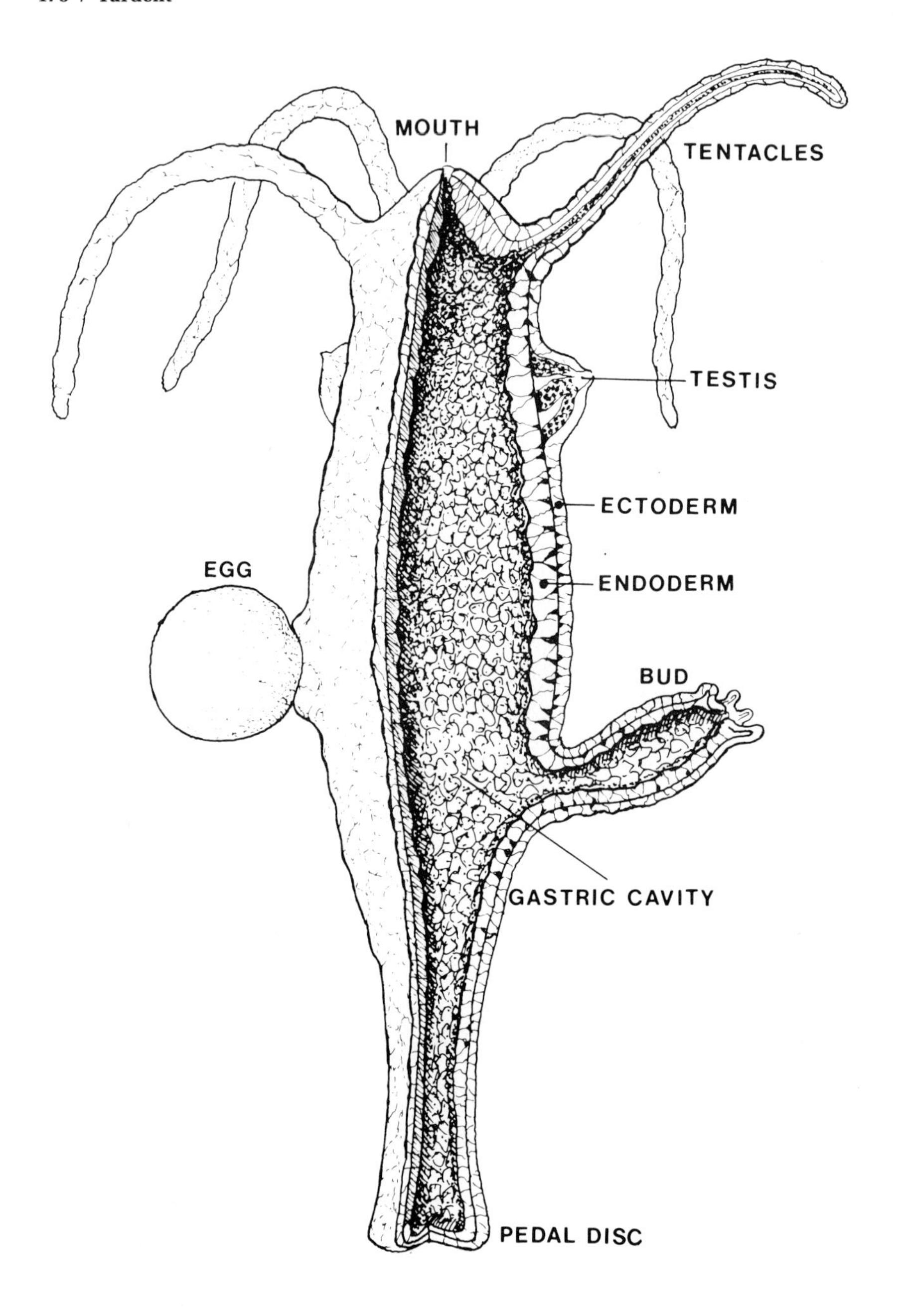

Fig. 5. Cut-away of a hermaphroditic fresh water *Hydra*.

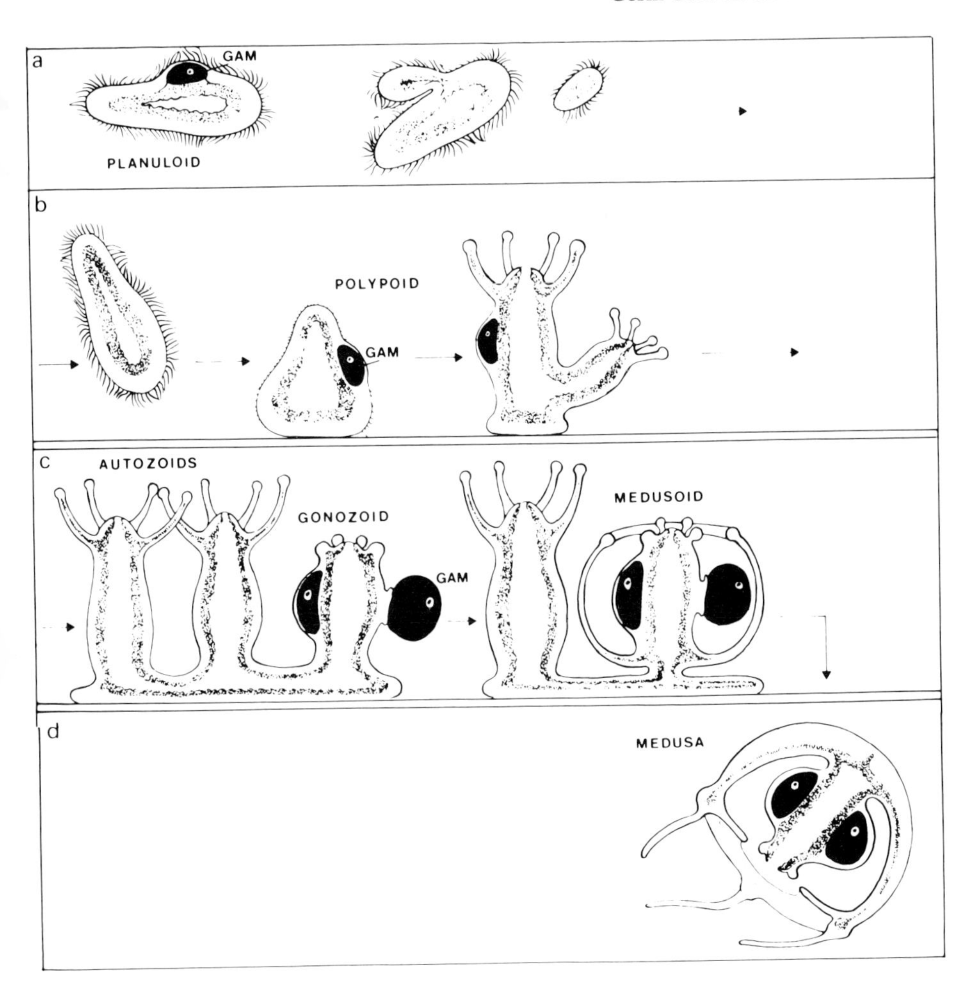

Fig. 6. Reconstruction of the hypothetical evolutionary steps, which may have led to the completion of the metagenetic alternation of polypoid and medusoid generations in Hydrozoa. It is assumed that the ancestor of the sessile polyp (b) was a pelagic, planuloid organism (a). Colony formation in polyps (c) could have led to the differentiation of autozoid- and gonozoid individuals (c). The latter then developed a bell-shaped brooding chamber for protection of the gametes and larvae. This new contractile structure, the umbrella, eventually led to the emancipation of the free swimming medusa (d). (GAM = gametes.)

Colony formation at this level allowed the development of functionally motivated polymorphisms similar to what is found in numerous recent forms. Here a division of labor has led, for example, to feeding autozoids and to gonozoids which dedicate their activities to sexual reproduction (Fig. 6c). The most primitive of the sexual polyps may have produced their gametes at the surface of their tubular body, as is seen in the modern hydra (Fig. 5). Such a situation may have called for protection for the exposed gametes.

This goal could have been achieved by the elaboration of a bell-shaped envelope, the future medusoid umbrella (Fig. 6c). In a later step, this new structure may have acquired the faculty to pulsate, thereby ventilating the brood chamber. Increasingly efficient umbrellar musculature may have eventually led to the detachment of a now fully evolved, free-swimming hydromedusa (Fig. 6d). This new morph not only contributed efficiently to the spacial propagation of the gametes and zygotes, favouring panmixy, but also allowed the species to draw from the resources of another ecosystem such as to render gametogenesis independent of the polyps trophic requirements. Therefore, according to this not uncontested scheme [see Tardent, 1978], the primary reason for the evolution of the hydromedusa is its relationship with sexuality and gametogenesis rather than with the necessity for exploiting additional ecological niches for the species.

Cubozoa (Some 16 Species: Cuboid Medusae)

This small group, which recently has been [Werner, 1973, 1984] separated from the *Scyphozoa*, features, like the latter and the *Hydrozoa* a typical metagenetic polyp-medusa cycle. The pecularity of this group is, however, that their free-swimming sexually functional medusae (Fig. 7a) are not the result of a budding (*Hydrozoa*) or a strobilating (*Scyphozoa*) process by which only parts of the polyp are invested into a medusa. Instead, they arise by a peculiar metamorphosis during which the entire sessile polyp turns into a swimming cubomedusa [Werner, 1973].

Scyphozoa (Approximately 200 Species: Common Jelly Fishes)

With some exceptions (e.g., *Pelagia noctiluca*) and the semisessile *Stauromedusae* the cycles of the *Scyphozoa* also include both a sessile, asexually multiplying polyp and a mobile gamete-producing medusa (Fig. 7b). However, the medusa develops by a process known as strobilation. In this process juvenile medusae (ephyra) originate from disk-like axial fragments of the scyphopolyp. It is of interest that in some species, e.g., *Aurelia aurita,* these transverse fragments can lead either directly to new polyps or alternatively to the medusoid ephyra [Tardent, 1978].

The hypothesis we have proposed above for explaining the evolution of the hydromedusae is applicable neither to the *Scyphozoa* nor to the *Cubozoa* because the morphogenetic pathways leading to the medusae of these two groups are fundamentally different. We therefore believe that the cnidarian medusae have polyphyletic origins.

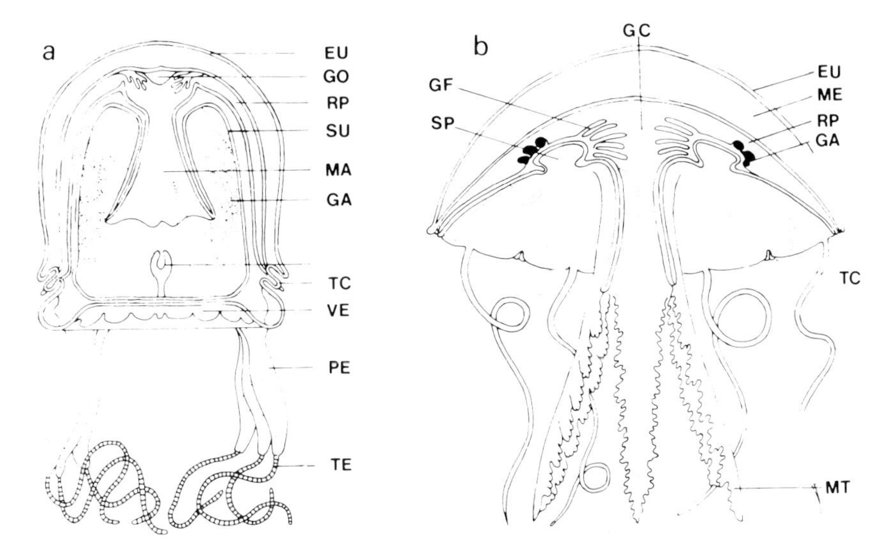

Fig. 7. Slightly schematized radial sections through representatives of the Cubozoa (a) and Scyphozoa (b). (EU = exumbrella; GA = gametes; GC = gastric cavity; GF = gastric filaments; GO = gastric ostium; MA = manubrium; ME = mesoglea; MT = mouth tentacles; PE = pedalia; RP = radial pocket; SP = subgenital pocket; SU = subumbrella; TC = tentaculocyst; TE = tentacles; VE = velum.)

GAMETOGENESIS

The Sites of Gametogenesis

With a few exceptions found in *Anthozoa* and *Scyphozoa* [Tardent, 1978], the *Cnidaria* lack true gonads, i.e., structures composed of germ cells and an envelope of closely associated follicle cells. There are no specialized gonadal epithelia for protecting and nourishing the gametes. Gametogenesis takes place instead in either intraepithelial spaces (*Hydrozoa*, *Cubozoa*, *Scyphozoa*, Figs. 8a, e) or within the extracellular matrix (mesoglea) which separates the ectoderm from the endoderm (*Anthozoa, Scyphozoa*). Any somatic cell which happens to be adjacent to the developing oocytes seem to be able to supply them with the precursors required for yolk synthesis. For this reason, oogenesis and spermatogenesis usually take place in close association with the gastrovascular cavity (Figs. 2, 3b) or with the tubular canals extending from it (Figs. 3c, d, 8a).

Furthermore, there are no other accessory gonadal structures, such as ducts, for the storage and/or release of the gametes to the outside. Usually, after rupturing through the epithelia, the ripe eggs and spermatozoa pass directly into the surrounding water or into the gastrovascular cavity. From the latter,

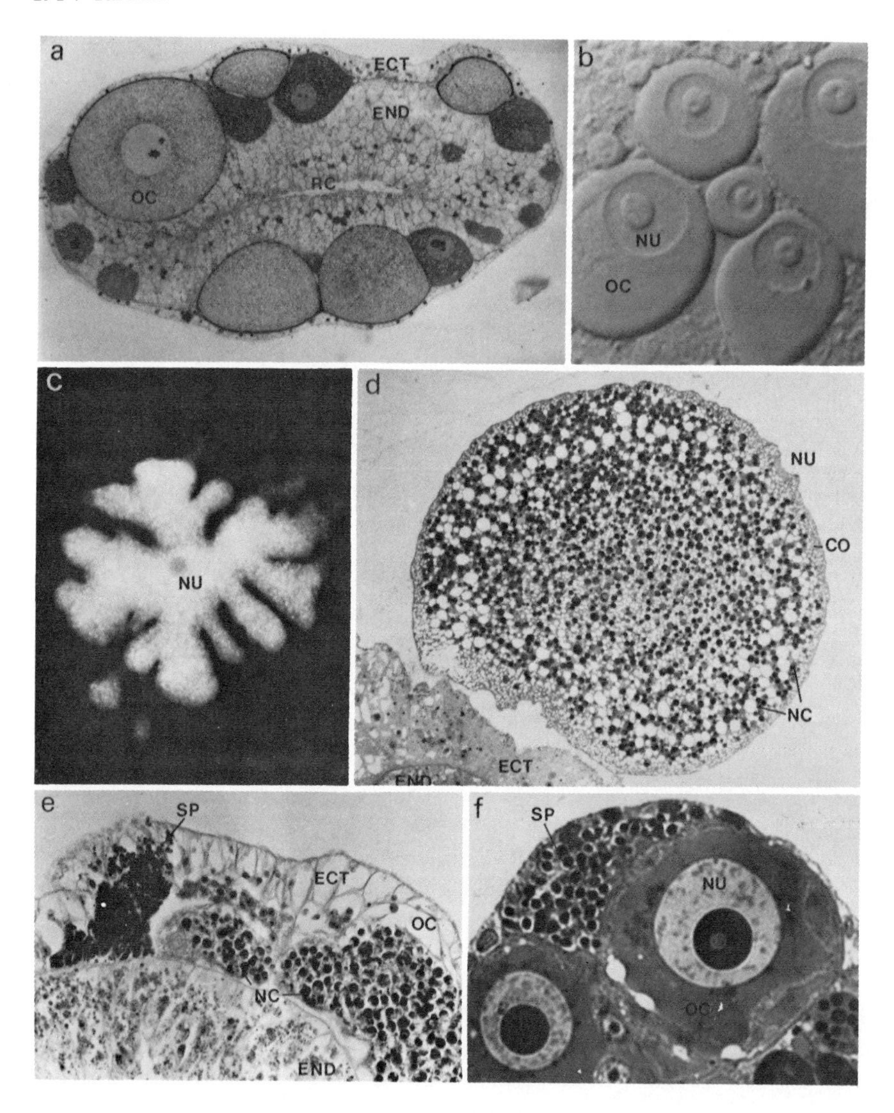

they may leave the polyp or medusa through the mouth or through pores opening either in the body wall (e.g., *Sagartia troglodytes, Anthozoa*) [Nyholm, 1943] or at the tip of the hollow tentacles (*Ceriantharia, Anthozoa;* personal observation). In a considerable number of species, the fertilization of the eggs occurs either within the polyp (*Anthozoa*) or medusa and medusoid respectively (*Scyphozoa*, *Hydrozoa*). Brooding of the offsprings up to variable stages of early development is also practiced by members of the mentioned three classes.

In spite of the absence of anatomically defined gonads and gonadal ducts, the tissues in which gametogenesis occurs are specific for the various taxonomic groups. In *Anthozoa* (Fig. 2), *Scyphozoa* (Fig. 7b) and *Cubozoa* (Fig. 7a) gametogenesis takes place within the endoderm, whereas in *Hydrozoa* it is the ectoderm that harbours the developing spermatozoa and eggs (Figs. 8a, e). This, however, does not necessarily mean that these are the sites where the dormant, not yet active germ cells are localized. The problem concerning the location of the primordial germ cells will be discussed later.

Oogenesis

In *Cnidaria*, as in *Echinodermata* [Czihak, 1975], fertilization takes place after the oocyte has fully completed its second maturation division. So far, no exception to this rule has been found. The sizes of the mature ova vary considerably [Tardent, 1978]. Oocyte diameters may range from 30 μm (*Haliclystus octoradiatus, Scyphozoa)* to 1300 μm (*Periphylla periphylla, Scyphozoa*).

How a developing cnidarian oocyte acquires nutrients is highly variable and often unusual. A functional association of the oocyte with specialized follicular cells is reported in only a few anthozoans [Nyholm, 1943; Chia and Crawford, 1973] and scyphozoans [Widersten, 1965]. In *Cerianthus lloydii* (*Anthozoa*), the oocytes take up their food directly from the polyps gastric cavity by means of thread-like cytoplasmatic processes (trophonema) that protrude into the cavity [Nyholm, 1943].

The oocytes of most *Hydrozoa*, however, seem to be able to do without the assistance of such mediating devices (Fig. 8a). They gather the soluble precursors for yolk synthesis either from unspecialized neighbouring cells or directly from the circulating content of the gastrovascular system. In many

---

Fig. 8. Gametogenesis in Hydrozoa. a) Section (7μm) through the pocket of the radial canal of the leptomedusa *Phialidium hemisphaericum* (*Leptomedusae*) containing various stages of oocytes. magnification, ×65. b) Interference phase-contrast picture of isolated young oocytes of *Phialidium hemisphaericum*. magnification, ×70. c)Early stage of an oocyte of *Hydra carnea* after its operational isolation from the polyps ectoderm (cfr. Fig. 9) [from Zürrer, 1983]. magnification, ×100. d) Cross-section (7μm) through an unfertilized egg of *Hydra carnea* after it has broken through the ectodermal cover. (photograph by Th. Honegger). magnification, ×210. e) Cross-section (7μm) through the gastric column of a rare hermaphroditic *Hydra attenuata* featuring intraepithelial pockets containing side by side spermatogonia, spermatids (SP) and a growing oocyte (OC) with nurse-cells (NC). magnification, ×268. f) Cross-section (7μm) through the ectoderm of the manubrium of a hermaphroditic *Podocoryne carnea* medusa (*Hydrozoa, Anthomedusae*). This specimen was produced by a polyp colony which had regenerated from a heterosexual reaggregate obtained by mixing cells of a male and a female clone. [From Bührer, 1981.]

hydrozoans, developing oocytes have been found to migrate considerable distances within either the solitary polyp or the polyp colony [Weismann, 1883; Mergner, 1957b]. In doing so, they also may move from one epithelium to another [Glätzer, 1971], picking up food from various sources. Investigations about the biochemical and ultrastructural events that accompany oocyte maturation and yolk synthesis are relatively scarce [e.g., Glätzer, 1971]. However, one very unusual manner of food gathering, practiced by the oocytes of *Hydra* [Kleinenberg, 1872; Zihler, 1972; Zürrer, 1983] and some other athecate hydroidpolyps such as *Tubularia* [Benoît, 1925; Nagao, 1965; Boelsterli, 1975], *Pennaria* [Smallwood, 1899], and *Millepora* [Mangan, 1909], has been described. Oocytes of these organisms actually "feed" on adjacent cells by phagocytizing them in large numbers (Fig. 9). The condensed bodies of these cells and their still Feulgen-positive nuclei which Kleinenberg [1872] called "pseudocells" and Zihler [1972] characterized as "shrunken cells" constitute the bulk of the egg cytoplasm (Fig. 8d), persist through embryonic development. During cleavage they are transferred to the cytoplasm of the presumptive endodermal cells where they are digested.

On the basis of their morphology, all these phagocytosed cells were considered as potential oocytes [Zihler, 1972]. It was believed these would gradually fuse to produce a single cell in which only one nucleus would become a germinal vesicle [Zihler, 1972]. The remainder of the cell mass would simply provide a nutritional supply for the developing oocyte. A more recent study [Zürrer, 1983] has, however, provided a new interpretation. Within an initial large cluster of interstitial cells (see chapter 5c), only one becomes determined as an oocyte (Fig. 9). The others begin an intense synthesis of yolk precursors before they are first partially then totally phagocytosed. Consequently they play a more active role in oocyte nutrition than being just accidental casualities of the oocyte [Zihler, 1972]. Experimental evidence at hand does not, however, allow us to decide if each of these nurse cells (formerly interstitial cells) is equally potent in becoming an oocyte or whether all but one must be considered as somatic components which are trophically associated with the oocyte.

Spermatogenesis

The known spermatozoa of the *Cnidaria* are of conventional sizes and shapes [Franzèn, 1956; Hanisch, 1970; Moore and Dixon, 1972; Campbell, 1974; Schmidt and Zissler, 1979]. They lack a distinct acrosomal apparatus but feature small apical and lateral vesicles in their cytoplasm which may function as primitive precursors of an acrosome [Weissman et al., 1969; Hanisch, 1970].

Spermatogenesis takes place in widened intercellular spaces of the epithelia (Fig. 5, 8e, f). Clusters of accumulated interstitial cells synchronously undergo

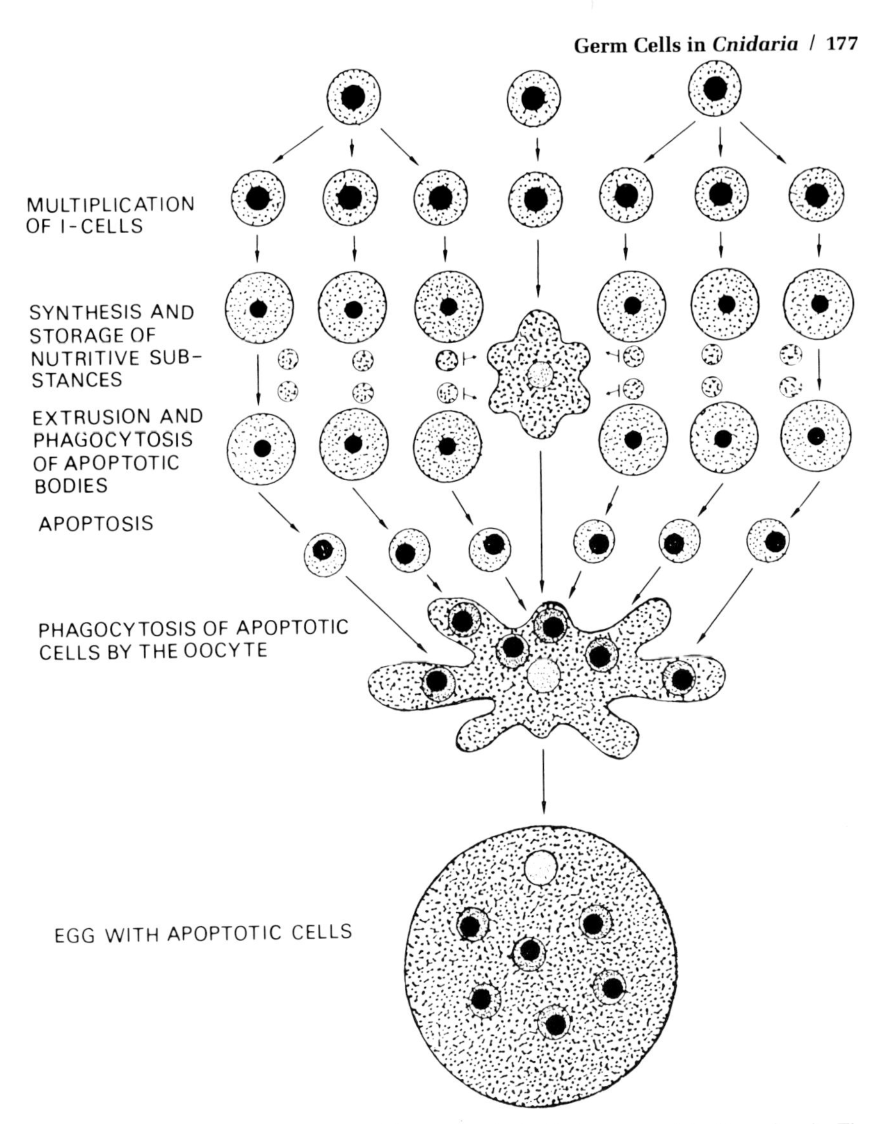

Fig. 9. Schematical representation of the oogenesis in *Hydra carnea* (cfr. Figs. 8c, d). The oocyte (center) phagocytoses cytoplasmatic fragments and whole nurse cells which derive from interstitial cells. [From Zürrer, 1983.]

the conventional meiotic divisions and the subsequent process of spermiogenesis [Szollosi, 1964; Hanisch, 1970; Zihler, 1972; Munck, 1983].

In *Hydra* (*Hydrozoa*), small batches of ripe spermatozoa, 5–100 of them at a time, break through the epithelium at intervals varying from 5 to 60 min [Zihler, 1972]. This spawning strategy provides for an uninterrupted supply of active spermatozoa whenever a population of *Hydra* engages in sexual activities.

In most cases, however, polyps and medusae release all available spermatozoa more or less synchronously with the shedding of the eggs, usually in response to photic stimuli [Ballard, 1942; Yoshida, 1959; Honegger et al., 1980].

So far, the only case of direct sperm transmission from male to female has been reported from the cubomedusa *Tripedalia cystophora* [Werner, 1973]. The male produces and transfers so called spermatozeugma with its umbrellar tentacles. Spermatozeugma are a mass of densely packed sperm heads held together by a mucous envelope, through which protrude the sperm tails. Thanks to their synchronous beating, the spermatozeugma are capable of moving autonomously without the aid of accessory structures.

GONOCHORISM, HERMAPHRODITISM, AND SEX DETERMINATION

Gonochorism

An overwhelming number of cnidarian species are dioecious and the respective state of functional determination appears to be extremely stable. This is also true for the metagenetic species which have two separate generations (Fig. 3a). Although the sex of the asexually multiplying polyps cannot be recognized morphologically, they invariably produce either male or female medusae. This means that the sexual determination of the potential germ cells remains stable from one generation to the next.

In order to test the degree of this stability, we have dissociated polyps of the athecate hydroid *Podocoryne carnea* (*Anthomedusae*) from known female and male clones [Bührer, 1981]. The heterosexual reaggregates obtained from mixed cell suspensions regenerated into stolons and polyps, which eventually generated mature medusae. These mosaic individuals simultaneously produced spermatozoa and eggs side by side within the manubrial ectoderm (Fig. 8f), thus confirming, in this particular case, the stability of the germ cells as to their sexual determination.

Few efforts have been made to observe the stability of sex determination over the long term in traditionally gonochoristic species such as *Podocoryne*. Neither is there any cytological evidence available as to the presence or absence of sex chromosomes. However, it seems reasonable to assume that in gonochorists this stability is a long-term phenomenon and has a genotypic basis.

Hermaphroditism

Although hermaphroditism is relatively rare in cnidarians, it has been extensively investigated in the genus *Hydra* (*Hydrozoa*), which features various degrees and expressions of this phenomenon. Species such as *Hydra viridis* [Brien and Reniers-Decoen, 1950] and *H. circumcincta* [Tardent et al., 1968] are classical hermaphrodites, that is, simultaneously producing both kinds of gametes within the ectoderm, although at different axial levels of the tubular body (Fig. 5). Other species (*H. attenuata, H. carnea, H. magnipapillata*) can be termed either unbalanced hermaphrodites or labile gonochorists [Bacci, 1950]. The sex of such individuals does not change for months, even years, encompassing many sexual periods [Tardent, 1968]. Thus, as a rule, the state of determination of these species is inherited by their asexual offsprings so that clones are fairly stable as male or females.

However, occasional "spontaneous" inversions of an individual from female to male or vice versa lead from one relatively stable state of determination to another. When such a transition takes place during the sexual period the polyp is temporarily hermaphroditic.

Hydra fusca [Brien and Reniers-Decoen, 1949] was, until recently, considered to be the only stable gonochoristic species of this genus, but we have witnessed rare cases of sex inversion in this species also.

The factors which in these cases trigger such a rare event in one specimen among many others are still not known. At least for *Hydra attenuata*, these inversions of the sexual status are not due to a programmed protandry or protogyny. Among freshly hatched specimens, both sexes are represented when they enter their first sexual period (personal observation). In addition, inversions have been found to occur in both directions. *Hydra carnea* has been described [Hyman, 1931; Ewer, 1948] as being a protandrous hermaphrodite, whereas Littlefield [1984a, b] describes this species as a protogynous hermaphrodite. In our view, sex reversal in the labile gonochoristic members of this genus is a randomly occurring phenomenon.

Sex Determination

It must be assumed that in *Hydra*, as with hermaphrodites of other animal groups, the germ cells must be able to develop into either spermatozoa or eggs. Questions such as how determination in one sense or another is brought about or to what extent environmental and/or somatically generated factors act as determinants are obvious.

It has been shown repeatedly that in *Hydra fusca* [Wiese, 1953a, b; Brien, 1962, 1963], *Hydra attenuata* [Tardent, 1966, 1968] and *Hydra carnea* [Littlefield, 1984a, b] sex inversion can be induced experimentally by parabiotically grafted male and female fragments (Fig. 10). In order to be successful, heterosexual parabiosis must last at least 72 hr in *H. attenuata* [Tardent, 1968] or 96 hr in *H. fusca* [Wiese, 1953a, b]. These experimental measures always resulted in a masculinization of the female partner even if the male fragment was considerably smaller than its complementary female counterpart (Fig. 10) [Tardent, 1968]. The masculinating effect has also been found to be successful in xenoplastic chimaeras (*H. attenuata* ♀/*H. carnea* ♂; Fig. 11) [Littlefield, 1984a].

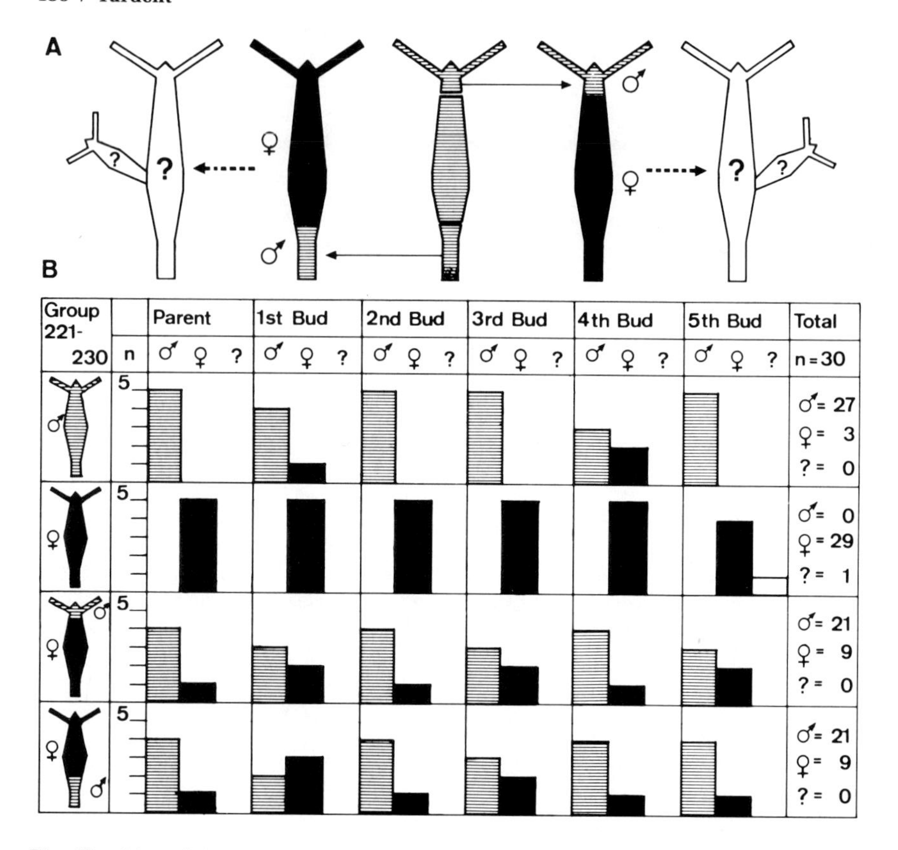

Fig. 10. Masculinization of female fragments of *Hydra attenuata* by grafted complementary male fragments. A) Experimental procedure; B) Sex of the non-transplanted male and female specimens and their asexually produced buds (upper two rows) and sex of the heterosexual chimaeras and their buds following transplantation. [From Tardent, 1968.]

Male fragments of *H. attenuata,* however, are ineffective after irradiation with 6500 r four days prior to grafting to a female partner [Tardent, 1968]. A change to the opposite direction, i.e., from male to female has been achieved by subjecting whole male polyps to sublethal doses of x-rays [Tardent, 1968] or by exposing them to mustard gas [Figi, 1969].

These few experimental findings strongly suggest that the masculinization of females, as induced by heterosexual parabiosis, is not due to the determining action of diffusing hormone-like substances as had been proposed by Wiese [1953a, b] and Brien [1962, 1963]. Rather, it seems to be due to the migratory invasion [Tardent and Morgenthaler, 1966] of the female by male-determined gonocytes.

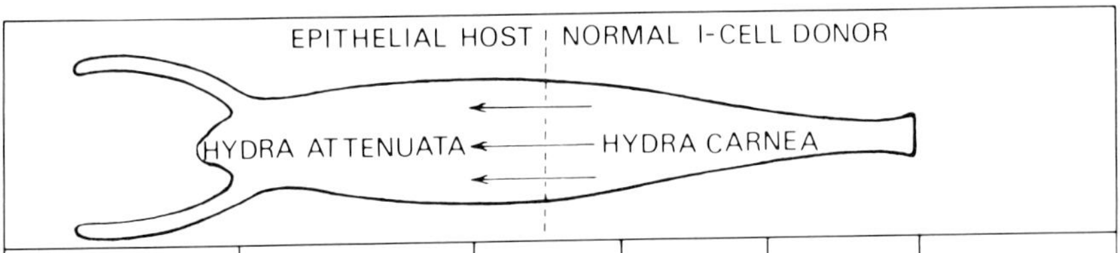

EPITHELIAL HOST | NORMAL I-CELL DONOR

HYDRA ATTENUATA ← HYDRA CARNEA

EPITHELIAL HOST	NORMAL I-CELL DONOR	NUMBER GRAFTED	NUMBER SURVIVED	NUMBER SEXUAL	SEX-RATIO ♂ : ♀
H. ATT. ♀	H. CARN. ♀	16	7	7	0 : 7
H. ATT. ♀	H. CARN. ♂	14	12	10	10 : 0
H. ATT. ♂	H. CARN. ♂	7	6	5	5 : 0
H. ATT. ♂	H. CARN. ♀	17	12	11	1 : 10

Fig. 11. Sex ratios of heterospecific parabiotic chimaeras. Proximal halves of normal female and male *Hydra carnea* (I-cell donors) were grafted to distal halves of epithelial specimens of formerly male and female epithelial *Hydra attenuata* (I-cell acceptors) of which the I-cells had been eliminated by colchicine-treatment. [Modified from Littlefield, 1984a.]

In fact, somatic components seem not to participate in sex determination at all. In a recent study, Littlefield [1984a] has parabiotically introduced I-cells of female *H. carnea* into male epithelial *H. attenuata* whose I-cells, nerve cells, and nematocytes had previously been eliminated by colchicine treatment [Campbell, 1976; Marcum and Campbell, 1978a]. After having been repopulated by migrating I-cells and nematocytes, these epithelial animals which had consisted exclusively of epithelial cells and some endodermal gland cells only gradually reacquired their normal cellular architecture, normal behaviour, and ability to reproduce sexually. Whenever I-cells from female donors were introduced into male epithelial host specimens, the hosts became females, i.e., switched to the sex of the I-cell donor (Fig. 11). Consequently, the soma, at least as far as the epithelial cells are concerned, does not exert a determinant effect upon the germ cells.

In *Hydra*, at least, the interstitial cells, which are precursors of the germ cells, are themselves responsible for sex determination and the observed male dominance. One hypothesis previously proposed [Tardent, 1968] was that the density of the I-cell population might be decisive such that a high number of these cells would lead to maleness and a low number to femaleness. This has been found not to be true, however, as cell counts have failed to detect such differences [Littlefield, personal communication]. We therefore must consider the possibility that each individual's I-cell population contains subpopulations of both predetermined male and female elements.

In simultaneous hermaphrodites (*H. viridis, H. circumcincta*), both elements differentiate simultaneously into spermatozoa and eggs, although in different body regions. In the labile gonochoristic species (*H. attenuata, H. carnea, H. fusca*), only one of the sexes usually is expressed at a time. The decision as to which is expressed may depend upon the ratio of male- and female-determined cells present within the total I-cell population when the polyp becomes gametogenetically active. When the differentiation of the male cells proceeds, it might be also that they exert a direct or indirect effect on the fewer potential oogonia such that their maturation process is inhibited.

Other observations also could be explained by this hypothesis: Masculinization of female specimens by heterosexual parabiosis could be the result of a significant increase of male-determined cells migrating into the female partner, thus changing the balance in favour of spermatogenesis instead of oogenesis. It could be concluded that X-ray or mustard-gas treatments, on the other hand, change the balance in the opposite way, if the additional assumption were made that predetermined male cells are more sensitive to these treatments and therefore suffer higher losses than their female counterparts.

Experimental evidence for the possible existence of I-cell subpopulations has been provided by Littlefield [1984b], who observed that monoclonal antibodies prepared from *Hydra* sperm react only with a fraction of I-cells of sexually inactive specimens. This suggests that these I-cells share certain cell-surface properties with differentiated spermatozoa. If subpopulations of male and female predetermined cells do exist and the functional state of the animal is determined by the relative numbers of such cells, the sex of the individual polyp and its asexual offspring should remain stable as long as the two subpopulations retain their given ratio. The observed "spontaneous" inversions of mature polyps would occur when for some reason the relative frequencies of male and female cells change. Likewise, if a polyp of one sex randomly passes a relatively large proportion of the other subpopulation of I-cells into a developing bud, a change of sex would occur between generations.

This last possibility is supported by our recent investigation where mutant I-cells of *H. magnipapillata* [nem-1 strain of Sugyiama, 1977], which produce abnormal isorhizas (nematocytes), were parabiotically introduced into normal animals (strain 105). Then, for a year, such mosaic specimens as well as their asexual offspring, were monitored as to the relative frequencies of normal and abnormal isorhizas, i.e., nonmutant and mutant I-cells. It was found [unpublished data] that the ratio varied considerably within the growing clones. This indicated that the subsequent buds had received varying proportions of mutant and non-mutant I-cells as theoretically postulated in a previous paper [Tardent, 1984].

This hypothesis, according to which the I-cell population of *Hydra* may harbour subpopulations of predetermined male and female gonocytes, does not, however, exclude the possibility that these cells, despite their respective state of sexual predetermination, are still capable of also becoming engaged in somatic differentiation (nematocytes, nerve, and sensory cells). Within the hierarchy of possible pathways of differentiation, the male or female state of predetermination need be expressed only when the still pluripotent I-cells become engaged in gametogenesis. An earlier decision as to gametic or somatic commitment already would have been made. This interpretation would, of course, be valid only if among the I-cells there were not two distinct and unconvertible subpopulations of I-cells, one being exclusively gametogenic, the other representing a stock of presumptive somatic elements, and if transdetermination were permitted.

THE ORIGIN OF THE GERM CELLS

Historical Background

As an introduction to this topic, it is appropriate to remember that the germ-line concept was formulated by August Weismann [1892] some 100 years ago in his book *Das Keimplasma: Eine Theorie der Vererbung*. He dealt with the question of whether or not structural and functional specialization of somatic cells is accompanied by a loss of genetic potentialities. If so, this would require a separate lineage of cells, the presumptive germ cells, as components of the organism which were protected from such irreversible restrictions in order to be able to fulfill their reproductive function. Weismann's microscopic observations were made on some 35 species of hydromedusae and hydroid polyps [Weismann, 1880a, b, 1883, 1884].

Amongst other things, Weismann [1884] discovered that the "indifferent" cells from which spermatozoa and eggs derive differ morphologically from all other cell types and, of utmost significance, are present long before the polyp or medusa enters a period of gametogenetic activity. In both morphs Weismann distinguished between the anatomical sites where the dormant "indifferent" germ cells ("Urkeimzellen," translated primordial germ cells) are localized and those where gametogenesis ("Reifungsstätte," translated sites of maturation) actually took place. On one hand he was puzzled by the fact that these sites differed from one species to another. On the other hand, he was reassured when he discovered that, at least within the same species, these sites were always the same. Weismann made considerable efforts in trying to explain these variations in phylogenetical terms. As already discussed, he considered the metagenetic species of *Hydrozoa* with a free-swimming medusoid morph (Fig.

3a) as representing the most primitive situation. Here, "Keimstätte" and "Reifungsstätte" happen to be at the same place, e.g., in the ectoderm of the hydromedusa's manubrium (*Anthomedusae*, Fig. 3b) or associated with the radial canals (*Leptomedusae*, Fig. 3c, d).

In agreement with our own view, Weismann considered the gradual reduction of the medusa generation (Figs. 3b–f, 4) as being a secondary evolutionary tendency. Correlated with this trend, the stock of primordial germ cells, and with them the "Keimstätte," were found to be displaced to the more central parts of the polyp or polyp colony, from where the activated germ cells would migrate to the sessile medusoids, gonophores,or sporosarcs ("Reifungsstätte"). This relationship between medusa regression on one hand and the displacement of the site of germ-cell storage on the other is certainly correct. However, its phylogenetic and functional interpretation as proposed by Weismann [1884] are somehow doubtful. Because they are of secondary importance to the main problem, they need not be discussed here [cfr Berrill and Liu, 1948].

The extensively debated question as to whether the primordial germ cells belong to the endoderm or the ectoderm also does not deserve the importance it was given by Weismann because in higher organisms germ cells usually are not attributed to one or another of the three germinal layers. Furthermore, *Cnidaria* mostly lack true gonads. Therefore, the highly mobile oogonia and spermatogonia are forced to move to those places where their trophic requirements can be satisfied.

What stands out in Weismann's remarkable work is the fact that he made a clear distinction between the soma and the germ cells and that he recognized the importance of this scission for the transmission of the unrestricted hereditary information from one generation to another.

In doing so, he proposed an appealing alternative to the somehow confused views of other theoreticians, e.g., von Nägeli [1884], who referred to genetic continuity as an "idioplasm," which, in his opinion, was scattered throughout the body.

Was Weismann correct? Is there indeed such a thing as a germ line in the *Cnidaria*? We must confess that the actual state of knowledge concerning the answer to these questions is about as confusing as it was 100 years ago. For some time, it seemed as if Weismann had founded his germ-line theory on just that group of invertebrates which offered the least possibility for recognizing a clean distinction between the soma of the individual and the postulated lineage of somatically uncommitted germ cells [Berrill and Liu, 1948]. However, as it will be shown below, certain recent findings might again speak in favour of the existence of a germ line in this group.

At this point it seems advisable to examine the extent to which the somatic cells of *Cnidaria* have suffered restrictions of their hereditary potential as compared to a totipotent germ cell.

The Potentialities of Somatic Cells

Detailed studies about the various kinds of cnidarian cells and their kinetics have been performed mainly on members of the genera *Hydra* and *Podocoryne* (*Hydrozoa*). As compared to higher Invertebrates, these animals consist of a fairly small number of different cell types. Some of these somatic cells, for example, the nematocytes [Weill, 1930; Mariscal, 1974; Tardent and Holstein, 1982; Holstein and Tardent, 1984] the cross-striated muscles of the medusa umbrella (Fig. 12) [Chapman et al., 1962; Schmid, 1978; Schmid,

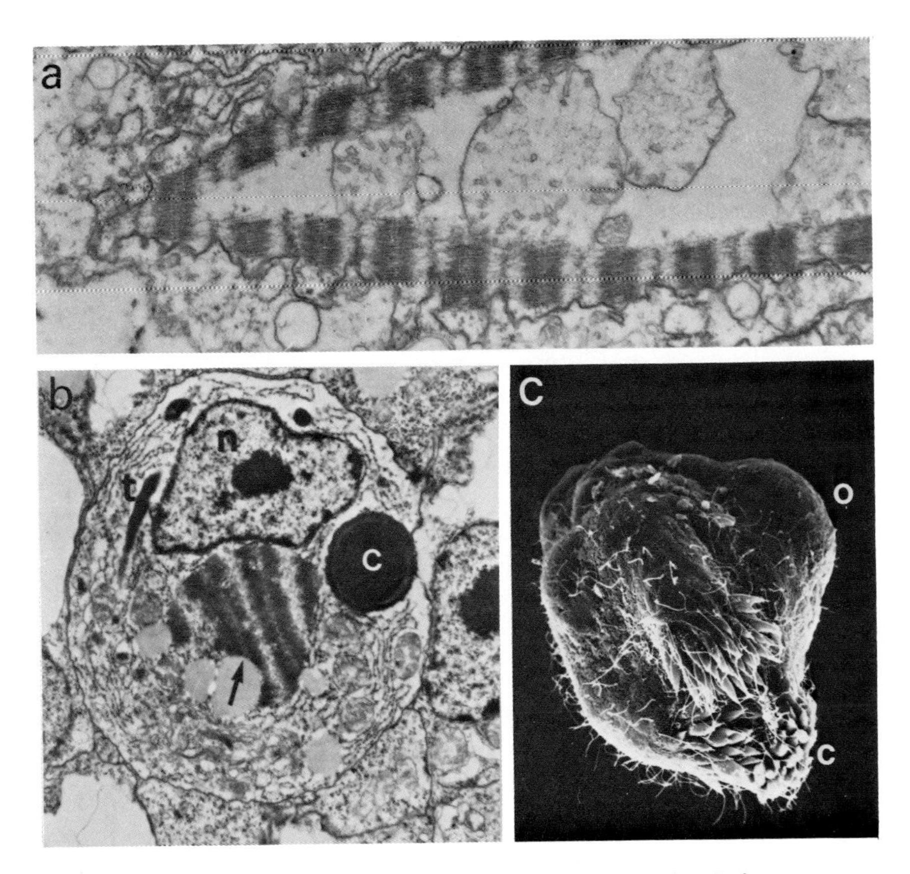

Fig. 12. Transdifferentiation of cross-striated muscles of the hydromedusa *Podocoryne carnea* (Anthomedusae). a) Tangential section through normal muscles of the subumbrellar ectoderm. magnification, ×6000. b) Former muscle cell synthesizing a cnidocyst (C) and its external tubule (t). The arrow points to a portion of persisting myofibrills. magnification, × 13,300. c) Four weeks old manubrium which regenerated from isolated muscles to which I-cell-free endoderm of the subumbrellar plate had been added. The regenerate developed nematocytes (NC) and produced an oocyte (O). C = cnidocyst, syn. nematocyst; n = nucleus; t = section of the external tubule of the nematocyst) (photographs by V. Schmid) SEM magnification, ×555.

et al., 1976], and the nerve- and sensory cells [Tardent and Weber, 1976; Epp and Tardent, 1978; Westfall, 1973] have reached surprising functional specialization and complexity, whereas other cell types are extremely poly-functional. For example, the endodermal digestive cells of *Hydra* are both epithelial-muscular and digestive in nature.

In *Hydra* the ectodermal epithelio-muscular cells and the endodermal digestive cells represent independent, autoreproductive cell lines. Others, such as nematocytes and neurons (Fig. 13), have to be replaced by differentiation from a population of auto-reproductive and multi-potent stem-cells. These are the interstitial cells (Fig. 13) discovered by Kleinenberg [1872]. The

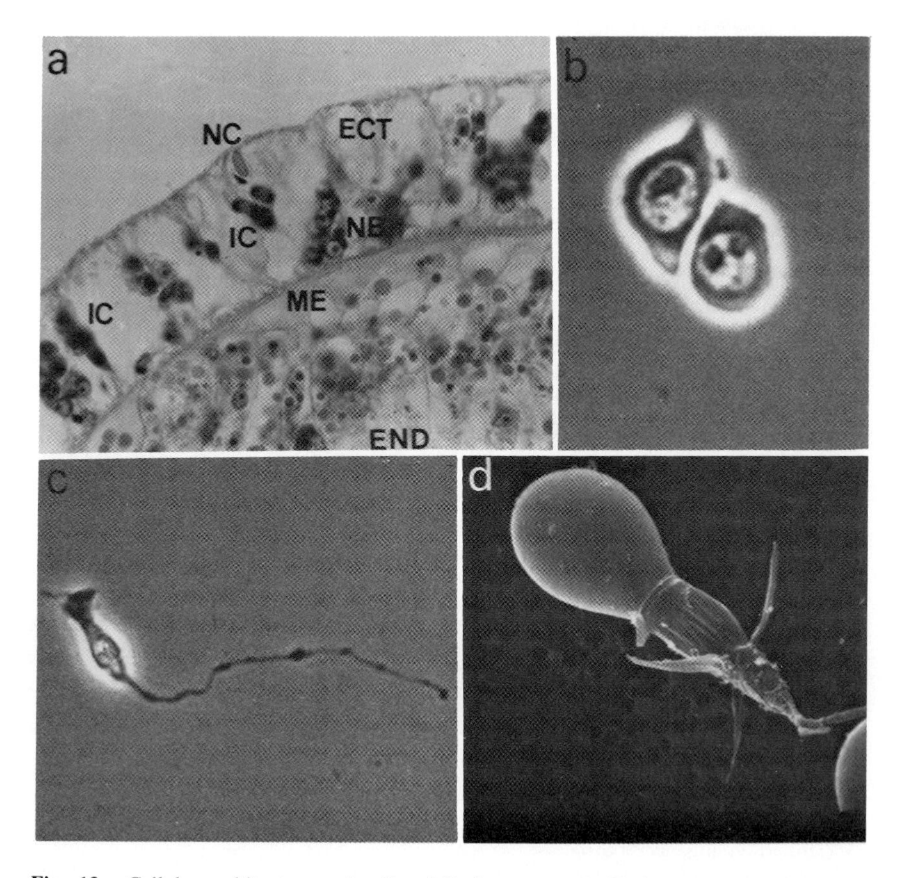

Fig. 13. Cellular architecture and cells of *Hydra attenuata* (Hydrozoa). a) Cross-section through the ecto- and endoderm of the polyps body wall. magnification, ×300. b) Two typical interstitial cells (I-cells) as they appear in a mazeration preparation (phase contrast).magnification, ×1500. c) A bipolar nerve cell (phase contrast). magnification, ×1000. d) Discharged stenotele (nematocyst) [cfr Tardent and Holstein, 1982].

endodermal gland cells apparently fit both these categories, because, although they are capable of multiplying mitotically, worn out clones may be now and then replaced from interstitial cells [Kessler, 1975; Smid and Tardent, 1984].

The question of whether in the course of their differentiation these somatic cells have retained their unrestricted hereditary potential has been answered convincingly only in the case of the cross-striated muscle cells of the umbrella of *Podocoryne carnea* medusa (Fig. 12a). Monotypic isolates of these cells, which are shown to be free of other contaminating cell types, become destabilized after their isolation and engage in a process of dedifferentiation and transdifferentiation [Schmid, 1978; Schmid et al., 1976, 1982; Schmid and Baenninger, 1980]. Even before their myofibrills are broken down, the destabilized cells not only resume mitotic division but also synthesize structures which are typical of other cell types. Their cytoplasm can, for example, contain cross-striated myofibrills and early stages of nematocyte capsules (Fig. 12b) side by side. As a result of such processes, direct transdifferentiation of the formerly monotypic muscular isolate into a rudimentary medusoid manubrium, equipped with the cell types typical for this organ, takes place. So far, none of these small regenerates have been found to produce gametes. Gametogenesis does take place, however, when the muscular isolates are combined with small amounts of live or dead cells taken from the umbrellar endoderm. This endodermal component, itself devoid of interstitial cells, seems to exert an inductive function on the de- and transdifferentiating tissue such that it occasionally develops gametes (Fig. 12c). This ability of striated muscle cells to dedifferentiate and transdifferentiate indicates that the original hereditary potential of at least this specialized cell type has suffered no loss throughout ontogenetic differentiation. Whether other cell types of *Podocoryne* medusa also are able to perform in this way has not yet been demonstrated.

Comparable metaplasiaic events, although not as convincing, were reported in *Hydra*. According to Normandin [1960], the isolated endoderm of *H. oligactis* is capable of regenerating the missing ectoderm and reconstituting a whole polyp "without the demonstrable aid of the interstitial cells." Identical experiments performed on *H. viridis* [Haynes and Burnett, 1963; Burnett et al., 1966] led to the additional observation that the gametes produced by the reconstituted polyps stemmed from dedifferentiated gland cells of the endoderm. These findings led the authors to conclude that in *Hydra* there is no specific germ-cell line segregated from the somatic constituents during embryogenesis. Our attempts to reproduce these observations using these or modified methods for separating the ecto- and endoderm [Smid and Tardent, 1982] have failed. In addition, we confirmed the presence of a small number of interstitial cells in the endoderm [Smid and Tardent, 1984], which were already described by Brien [1966] as "basal cells" and which may well have been the source of the gametes found in the endodermal regenerates [Burnett et al., 1966].

All experiments which undertake to disclose the maximal potentialities of somatic cells must forcibly create exceptional situations to cause these cells to react with such a remarkable degree of plasticity and express their pluri- if not totipotency. If "somatic" cells could, under such conditions, dedifferentiate to such an extent as to become gametes, we still would not know if they would do so under natural conditions. Furthermore, besides somatically committed cells, *Cnidaria* possess a population of autoreproductive, apparently "non-committed" stem-cells (interstitial cells, Fig. 13), some of which differentiate into the germ cells of the polyps or the medusae. Under natural conditions, there would be no need for the cnidarians to use the above-mentioned extravagant pathways that can be induced in the laboratory. However, whether or not these interstitial cells represent the germ line is still controversial and is now discussed.

Interstitial Cells (I-Cells) and Their Origin

The highly mobile [Tardent and Morgenthaler, 1966], spindle- or droplet-shaped interstitial cells (Fig. 13) are believed to be present in all taxonomic groups of the *Cnidaria* but are best known in the *Hydrozoa*. In the fresh water *Hydra*, they populate together with nematoblasts, nematocytes, and nerve cells the intraepithelial spaces of the body's ectoderm. (Fig. 13). From this location they can move into the endoderm by migrating across the acellular mesoglea, which separates the two tissue layers [Smid and Tardent, 1984].

In *Hydra attenuata* the I-cells make up 25–30% of the 120,000 to 200,000 cells of an individual polyp [Bode et al., 1973]. By means of treatment with colchicine, I-cells can be selectively eliminated from the animal [Campbell, 1976; Marcum and Campbell, 1978a, b]. Following such treatment, the polyp not only loses its I-cells but also the nematocytes, the nerve and sensory cells, and in some strains most or all of its gland cells. The epithelial animals thus produced consist only of ectodermal epithelio-muscular cells and endodermal digestive cells. If artificially fed, they are viable, develop buds, and regenerate [Marcum and Campbell, 1978a, b] but never produce spermatozoa or eggs. A normal animal can be produced again if I-cells from a normal donor are parabiotically reintroduced in the epithelial polyp (Fig. 11) [Littlefield, 1984a]. In addition, when properly repopulated with I-cells, an epithelial animal gradually reacquires not only its normal cellular architecture, including nerve cells, nematocytes, and gland cells, but also its usual behaviour [Marcum, personal communication] and the ability to generate gametes.

Together with other findings these facts permit the following conclusions:
1) In *Hydra* neither the ectodermal nor the endodermal cells seem to be

able to replace the missing I-cells, nerve cells, nematocytes, and gland cells by means of transdifferentiation. Neither can one tissue layer regenerate the other when viably separated from one other [Smid and Tardent, 1984]. Therefore, the epithelial layers represent two independent, truly somatic cell lines.

2) Therefore, the I-cells of *Hydra* form another equally independent lineage from which derive all other somatic cell types (neurons, nematocytes, and gland cells) as well as the gametes whenever the animal becomes gametogenetically active.

3) Most significantly for the present discussion, the I-cells can differentiate along either somatic or gametic pathways.

The most crucial problem inherent in this last conclusion is whether cells of the I-cell population are equally totipotent or whether morphologically similar subclones exist which are predetermined in either a somatic or gametic way. This possibility was already discussed in connection with sex determination. If such committed subpopulations really exist, one (or two) of them could be the presumptive gametic (sperm and oocyte) lineage and therefore represent the germ line.

From a morphological and ultrastructural point of view [Hess, 1961; Slautterback, 1961; Lentz, 1966], all I-cells of *Hydra* look alike. Therefore, the question of subpopulations must be resolved by other means. Some indirect evidences speak in favour of only one kind of multipotent I-cell: Except for the tentacles, hypostome, and the basal disc (foot plate), all regions of the polyp's body contain potential germ cells, as any small fragment of the gastric column will, after having reconstituted a polyp, be able to produce gametes. Potential germ cells, therefore, are not confined to the "sexual region" (Fig. 5) of the polyp.

In male and female *Hydra fusca*, gametogenesis brings about a lethal crisis ("crise gamétique") [Brien, 1966] during which the polyp loses all its nematocytes. Once used in feeding, they can no longer be replaced because all I-cells are involved in gametogenesis. If there were separate subpopulations for gametic and somatic purposes, one should expect that the latter would survive and continue to fulfill obligations of somatic replacement regardless of the gametogenic engagement of the others. The recent observations which seem to speak in favour of the existence of predetermined I-cell subpopulations already have been mentioned in connection with the problem of sex determination in hermaphrodites. If in these cases there are indeed separate male and female lineages of I-cells, there could as well be an according distinction between somatically and gametically committed clones of I-cells.

Whatever the final answer to this unresolved problem will be, it is still of interest to know when and how, during embryogenesis, the I-cells are segregated from the cells which are the precursors of the ecto- and endodermal epithelia. The information available on this question is only from studies on a few hydrozoan species, such as *Eleutheria dichotoma* (Fig. 14), *Cladonema radiatum*, *Campanularia johnstoni* [Weiler-Stolt, 1960], *Eudendrium racemosum* [Mergner, 1957b], and *Pennaria* [Martin and Thomas, 1980, 1983]. These studies all demonstrate that cells which have gross morphological and staining properties corresponding to typical cnidarian I-cells appear at an early stage of embryonic development, i.e., soon after the separation of the endoblast from the ectoblast (Fig. 14c). These cells are first found in the endoblast (Fig. 14d), from where they later move into the ectoblast (Fig. 14e). Here, in at least these few species, they remain in the planula larva (Fig. 14g) and throughout the postmetamorphotic stages. Whether or not a small fraction of this initial population remains in the endoblast, thereby representing an independent population, is not clearly shown. It is clear, however, that cnidogenesis (Fig. 14d) also occurs early and involves the primordial I-cells still in the endoblast. Together with the nondifferentiated cells, they migrate to the ectoblast (Fig. 14f). Developed planula-larvae possess a network of nerve and sensory cells [Korn, 1966; Martin and Thomas, 1980] and a complement of glandular elements [Bodo and Bouillon, 1968]. As in adult *Hydra*, these somatic cell types derive from the I-cells. This has been demonstrated by Martin and Thomas [1983] who succeeded in producing viable epithelial planula larvae, i.e., by selectively destroying the interstitial stem cells, were able to produce larvae lacking I-cells, neurons, and nematocytes. These findings suggest that even in early ontogenetic stages I-cells are involved in replacement tasks at a somatic level.

Nothing is known so far about how the first I-cells are generated. They may be formed through differential cell divisions of the fairly large embryonic cells of the endoblast (Fig. 14b) or by a process involving transformation, which would require a significant condensation of their cytoplasmic components. It is clear that the embryo does not feature a particular restricted area in which the I-cells are generated. This would make it difficult for a particular embryonic germ plasm to become incorporated into the developing population of interstitial cells and again would argue against the classical notion of germ-line origin.

SUMMARY AND CONCLUSIONS

Sexual reproduction has been found to occur in all known species of *Cnidaria*. It either alternates or is enacted simultaneously with asexual reproduction. The combination of the two modes of reproduction ensures genetic renewal and variability as well as rapid propagation of successfully adapted genotypes [Tardent, 1984]. With the exceptions of all *Anthozoa* and restricted numbers of *Hydrozoa* and *Scyphozoa*, asexual and sexual reproduction are each practiced by different morphs, the sessile polyps and the free-swimming medusae, respectively.

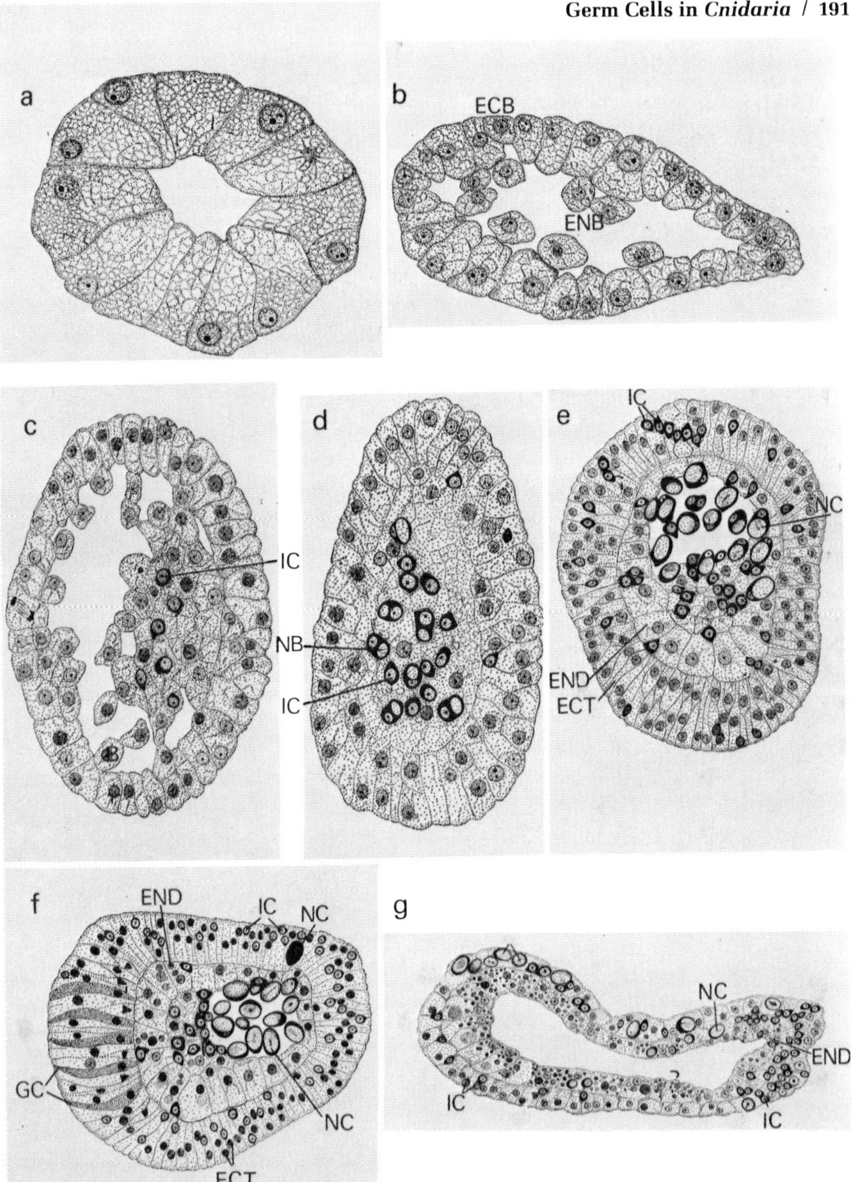

Fig. 14. The appearance of interstitial cells (I-cells) and nematocytes during the early development of *Eleutheria dichotoma* (Hydrozoa, Anthomedusae). Histological sections. a) blastula stage; b) beginning of endoblast formation by multipolar immigration; c–d) first appearances of I-cells and nematoblasts within the endoblast; e–f)migration of I-cells and nematocytes from the endoblast into the ectoblast; g) planula larva before its metamorphosis leading to the founder polyp. ECB = ectoblast; ECT = ectoderm; ENB = endoblast; END = endoderm; GC = gland cells; IC = interstitial cells; NB = nematoblast; NC = nematocyte. magnification a–f, ×862; g, ×609. [Modified from Weiler-Stolt, 1960.]

In all three classes, *Hydrozoa*, *Cubozoa* and *Scyphozoa*, medusa production by polypoid individuals follows fundamentally different pathways [Tardent, 1978]. The products of the different morphogenetic processes (budding in *Hydrozoa*, polyp metamorphosis in *Cubozoa*, strobilation in *Scyphozoa*) also differ considerably from each other (Figs. 1b, 7a, b). Therefore, they may well have independent phylogenetic origins, which would have been the results of different kinds of selective pressures.

The present paper makes suggestions (Fig. 6) as to how and why in *Hydrozoa* the secondary morph, the medusa, could have evolved from the more primitive polypoid level of organization. In agreement with Weismann [1883] and Kühn [1914], the interpretation given recognizes protection and shelter for gametes and embryos as having been the primary function of the hydromedusa. It then would assume that all aberrations from this cycle, as observed in various Hydrozoan species (Fig. 4), are of secondary nature. All the modifications, however, are useful clues in the attempt to reconstruct the early history of this dimorphic cycle. Whenever the medusa persists along with the polypoid morph, it is responsible for the production of the gametes. Thus, the evolution of the pelagic member of the cycle was related primarily to sexual reproduction. The medusa was a kind of "swimming gonad" for the sessile benthic polyp, although trophically independent of it.

Another peculiarity of the *Cnidaria* and their cycles is the generally accepted idea that most polyps are somatically immortal [Brien, 1953; Tardent, 1978], propagating their cells to unlimited numbers of asexually generated homologous offspring. The aging medusae, however, have limited life spans and are therefore somatically mortal. Medusa development occurring at the cost of differentiated and undifferentiated polyp cells [Tardent, 1978] therefore means a transition from somatic immortality to mortality.

This somatic immortality which the *Cnidaria* share only with the *Protozoa* and the *Porifera* requires exceptional qualities at the cellular level. Details about cellular architecture, cell proliferation, and replacement are best known from the fresh water *Hydra*. This polyp is based upon three apparently non-convertible cell lines. Two of these are the two epithelial layers ecto- and endoderm which, in their functional states, proliferate indefinitely. The third line is the multi- if not totipotent stem cells, the interstitial cells (I-cells). This lineage is responsible for the replacement of somatic elements that are subjected to wear and tear and do not represent autonomous cell lines. These are nerve cells, nematocytes, possibly gland cells, although we regrettably do not know if this particular situation found in *Hydra* is true for all cnidarian polyps.

Certain, however, is the fact that all *Cnidaria* contain highly mobile I-cell-like cells known to be the source of the gametes. They are ubiquitous, because individuals which have regenerated from fragments of polyps contain them and are capable of sustaining gametogenesis. Where there is a dimorphic cycle these cells are transferred from the polyp to the medusa.

These properties of such gametogenic stem cells meet the requirements expected of primordial germ cells. Accordingly, Weismann [1880a, b, 1883, 1884] identified them as the representatives of a truly independent germ line. The observations in *Hydra*, however, where these I-cells also produce somatic elements, mean that at least in this case no clear distinction exists between a germ and somatic line. A similar situation exists also in the *Mesozoa* [Lapan and Morowitz, 1972] and the *Porifera* [Kilian, 1980] where the axoblasts and amoebocytes, respectively, are gametogenicly as well as somatogenicly competent. This ambiguity could be the consequence of the still primitive level of metazoan organisation in these groups, where indications for segregation of soma and germ line are evident but where this separation is not fully completed.

The population of stem cells, although consisting of morphologically identical elements, may in reality not be homogenous. There could indeed be subpopulations which differ from each other with respect to their status of predetermination as suggested by recent findings in *Hydra*. Should such distinctions really exist, a considerable degree of flexibility would be necessary in the states of determination of these cells. For example, it would be necessary that in *Hydra fusca*, which after an external stimulus [Brien and Reniers-Decoen, 1949] commits all its I-cells to gametogenesis, I-cells which were determined to become neuroblasts or nematoblasts change their state of predetermination in order to become either oogonia or spermatogonia. This requirement is not unreasonable, as cnidarian cells can undergo much more dramatic transformations, e.g., the transdifferentiation of cross-striated muscle cells [Schmid 1974, 1978; Schmid et al., 1976; Schmid and Bänninger, 1980] which in some cases have even been found to become functional gametes (Fig. 12c). This example is comparable to the transformation of somatic cells as it occurs in *Placozoa* [Grell and Benwitz, 1974] or is suspected also to take place in *Porifera* [Kilian, 1980].

Such observations illustrate the remarkable degree of plasticity which characterizes at least part of the somatic cells of these archaic metazoans. Even so, it is unlikely that, under normal conditions, the cnidarian germ cells ever derive from functional somatic elements. Potentialities which have been revealed under conditions of experimental stress need not necessarily be the case under normal circumstances.

The question of whether or not in *Cnidaria* there is a germ line cannot be answered with an unequivocal "yes" or "no." However, the examination of the facts available about the interstitial cells of this group seems to indicate that these cells, which at an early stage of development segregate from the endoblast (Fig. 14) and eventually produce the gametes, constitute a first step in the direction of a clear-cut segregation of a somatic and germ line. This step has not, however, taken the evolution of this group so far that the potential germ cells may not have additional somatogenic activities.

REFERENCES

Bacci G (1950): Alcuni problemi dell'ermafroditismo negli Invertebrati. Boll Zool Suppl 17:193–212.

Ballard WW (1942): The mechanism for synchronous spawning in Hydractinia and Pennaria. Biol Bull 82:329–339.

Benoît P (1925): L'ovogénèse et les premiers stades de développement chez la myriothèle et chez la tubulaire. Arch zool exp gén 64:85–326.

Berrill NJ, Liu CK (1948): Germplasm, Weismann and Hydrozoa. Q Rev Biol 23:124–132.

Bode H, Berking S, David CN, Gierer A, Schaller C, Trenkner E (1973): Quantitative Analysis of Cell Types during Growth and Morphogenesis in Hydra. Roux's Arch Dev Biol 171:269–285.

Bodo F, Bouillon J (1968): Etude histologique du développement embryonnaire de quelques Hydroméduses de Roscoff: Phialidium hemisphaericum (L.), Obelia sp. Péron et Lesueur, Sarsia eximia (Allman), Podocoryne carnea (Sars), Gonionemus vertens Agassiz. Cah Biol mar 9:69–104.

Boelsterli U (1975): Notes on oogenesis in Tubularia crocea Agassiz (Athecata, Hydrozoa). Pubbl Staz Zool Napoli Suppl 39:53–66.

Brien P (1953): La pérennité somatique. Biol Rev 28:308–349.

Brien P (1962): Contribution à l'étude de la biologie sexuelle. Induction gamétique et sexuelle chez les hydres d'eau douce par des greffes en parabiose. Bull Acad Roy Belg Cl Sci 48:825–847.

Brien P (1963): Contribution à l'étude de la biologie sexuelle chez les hydres d'eau douce. Induction gamétique et sexuelle par la méthode des greffes en parabiose. Bull Biol Fr Belg 97:213–283.

Brien P (1966): "Biologie de la reproduction animale." Paris: Masson.

Brien P, Reniers-Decoen M (1949): La croissance, la blastogénèse, l'ovogénèse chez Hydra fusca (Pallas). Bull Biol Fr Belg 83:295–386.

Brien P, Reniers-Decoen M (1950): Etude d'Hydra viridis (Linnaeus) (La Blastogénèse, la Spermatogénèse, l'Ovogénèse). Ann Soc Roy Zool Belg 81:33–110.

Bührer M (1981): Intraspezifische Verträglichkeit respektive Unverträglichkeit der Gewebe und Zellen von Podocoryne carnea M. Sars (Cnidaria, Hydrozoa) Ph.D. thesis, Zool Inst Univ Zürich, pp 1–47.

Burnett AL, Davis LE, Ruffing FE (1966): A Histological and Ultrastructural study of germinal Differentiation of Interstitial Cells arising from Gland Cells in Hydra viridis. J Morphol 120:1–8.

Campbell RD (1976): Elimination of Hydra interstitial and nerve cells by means of colchicine. J Cell Sci 21:1–13.

Chapman DM, Pantin CFA, Robson EA (1962): Muscle in Coelenterates. Rev Can Biol 21:267–278.

Chia FS, Crawford BJ (1973): Some Observations on Gametogenesis. Larval Development and Substratum Selection in the Sea Pen Ptylosarcus guerneyi. Marine Biology 23:73–82.

Czihak G (1975): "The Sea Urchin Embryo." Heidelberg: Springer-Verlag, pp 1–700.

Epp L, Tardent P (1978): The Distribution of Nerve Cells in Hydra attenuata Pall. Roux's Arch Dev Biol 185:185–193.

Ewer RF (1948): A review of Hydridae and two new species of Hydra from Natal. Proc Zool Soc London 118:226–244.

Figi H (1969): Die Wirkung von Colcemid und Dichloren auf das sexuelle und vegetative Verhalten von Hydra attenuata Pall. Thesis Zool Inst Univ Zürich, pp 1–49.

Franzèn A (1956): On Spermiogenesis, Morphology of the Spermatozoon, and Biology of Fertilization among Invertebrates. Zool Bidr Uppsala 31:355–482.

Glätzer KH (1971): Die Ei- und Embryonalentwicklung von Corydendrium parasiticum mit besonderer Berücksichtigung der Oocyten-Feinstruktur während der Vitellogenese. Helgoländer wiss Meeresunters 22:213–280.

Grell KG, Benwitz G (1971): Die Ultrastruktur von Trichoplax adhaerens FE Schulze. Cytobiologie 4:216–240.

Grell KG, Benwitz G (1974): Elektronenmikroskopische Beobachtungen über das Wachstum der Eizelle und die Bildung der "Befruchtungsmembran" von Trichoplax adhaerens FE Schulze. Z Morph Tiere 79:295–310.

Grell KG, Benwitz G (1981): Ergänzende Untersuchungen zur Ultrastruktur von Trichoplax adhaerens FE Schulze (Placozoa). Zoomorphology 98:47–67.

Hanisch J (1970): Die Blastostyl- und Spermienentwicklung von Eudendrium racemosum Cavolini. Zool Jahrb Anat 87:1–62.

Haynes JF, Burnett AL (1963): Dedifferentiation and redifferentiation of cells in Hydra viridis. Science 142:1481–1483.

Hess A (1961): The Fine Structure of Cells in Hydra. In Lenhoff HL, Loomis WF (eds): "The Biology of Hydra." Coral Gables, Florida: University of Miami Press, pp 1–49.

Holstein Th, Tardent P (1984): An Ultrahigh-Speed Analysis of Exocytosis: Nematocyst Discharge. Science 223:830–833.

Honegger T, Achermann J, Stidwill R, Littlefield L, Bänninger R, Tardent P (1980): Controlled Spawning in Phialidium hemisphaericum (Leptomedusae). In Tardent P, Tardent R (eds): "Developmental and Cellular Biology of Coelenterates." Amsterdam: Elsevier/North Holland Biomed Press, pp 83–88.

Hyman LH (1931): Taxonomic studies on the hydras of North America. III. Rediscovery of Hydra carnea L. Agassiz (1850) with a description of its characters. Trans Am Microsc Soc 50:20–29.

Kessler MI (1975): The origin, maturation and fate of the secretory cells in the gastrodermis of the brown Hydra. PhD Thesis Univ Maine at Orono, pp 1–169.

Kilian EF (1980): Porifera. In Kästner A (ed): "Lehrbuch der speziellen Zoologie Bd 1. Wirbellose Tiere." Stuttgart: Gustav Fischer Verlag, pp 251–288.

Kleinenberg N (1872): Hydra. Eine anatomisch-entwicklungsgeschichtliche Untersuchung. Leipzig: pp 1–90.

Korn H (1966): Zur ontogenetischen Differenzierung der Coelenteraten-Gewebe (Polyp-Stadium) unter besonderer Berücksichtigung des Nervensystems. Z Morph Oekol Tiere 57:1–118.

Kühn A (1914): Entwicklungsgeschichte und Verwandtschaftsbeziehungen der Hydrozoen. 1. Teil Die Hydroiden. Ergebn Fortschr Zool 4:1–284.

Lapan EA, Morowitz HJ (1972): The Mesozoa. Sci Am 227:94–101.

Lentz ThL (1966): "The Cell Biology of Hydra." Amsterdam: North Holland Publ Co, pp 1–199.

Littlefield L (1984a): The Interstitial Cells Control the Sexual Phenotype of Heterosexual Chimaeras in Hydra. Devel Biol 102:426–432.

Littlefield L (1984b): Evidence for a germline in Hydra oligactis males. In Engels W, Clark WH, Fischer A, Olive PJW, Went DF (eds): "Advances in Invertebrate Reproduction." Vol 3. Amsterdam: Elsevier Science Publishers, p 608.

Mangan J (1909): The entry of Zooxanthellae into the ovum of Millepora and some particulars concerning the medusae. Quart J Micr Sci 53:697–709.

Marcum BA, Campbell RD (1978a): Development of Hydra lacking nerve and interstitial cells. J Cell Sci 29:17–33.

Marcum BA, Campbell RD (1978b): Developmental roles of epithelial and interstitial cell lineages in hydra: Analysis of chimaeras. J Cell Sci 32:233–247.

Mariscal RN (1974): Nematocysts. In Muscatine L, Lenhoff HM (eds): "Coelenterate Biology." New York: Academic Press Inc, pp 129–178.

Martin V, Thomas MB (1980): Nerve Elements in the Planula of the Hydrozoan Pennaria tiarella. J Morphol 166:27–36.

Martin V, Thomas MB (1983): An SEM analysis of early development in Pennaria tiarella. Amer Zool 23:1014.

Mergner H (1957a): Cnidaria. In Reverberi G (ed): "Experimental embryology of marine and fresh-water invertebrates." Amsterdam: North Holland Publ Co, pp 1–84.

Mergner H (1957b): Die Ei- und Embryonalentwicklung von *Eudendrium racemosum* Cavolini. Zool Jahrb Anat Ontog 76:63–164.

Moore GPH, Dixon KE (1972): A Light and Electronmicroscopical Study of Spermatogenesis in Hydra cauliculata. J Morphol 137:483–501.

Munck ML (1983): Spermatogenese bei Hydra carnea. Thesis Zool Inst Univ Munich, pp 1–65.

Nagao Z (1965): Studies on the development of Tubularia radiata and Tubularia venusta (Hydrozoa). Publ Akkeshi Mar Biol Stat 15:6–35.

Nägeli C von (1884): "Mechanisch-physiologische Theorie der Abstammungslehre." München/Leipzig.

Normandin DK (1960): Regeneration of Hydra from the Endoderm. Science 132:678.

Nyholm KG (1943): Zur Entwicklung und Entwicklungsbiologie der Ceriantharien und Aktinien. Zool Bidr Uppsala 22:85–248.

Schmid V (1974): Structural alterations in cultivated striated muscle cells from Anthomedusae (Hydrozoa). A metaplasiaic event. Exp Cell Res 86:193–198.

Schmid V, Schmid B, Schneider B, Stidwill R, Baker G (1976): Factors Affecting Manubrium-Regeneration in Hydromedusae (Coelenterata). Roux's Arch Dev Biol 179:41–56.

Schmid V (1978): Striated Muscle: Influence of an Acellular Layer on the Maintenance of Muscle Differentiation in Anthomedusa. Dev Biol 64:48–59.

Schmid V, Baenninger R (1980): Manubrium Regeneration "in Vitro." In Tardent P, Tardent R (eds): "Developmental and Cellular Biology of Coelenterates." Amsterdam: Elsevier/North Holland Biomedical Press, pp 353–360.

Schmid V, Wydler M, Alder H (1982): Transdifferentiation and Regeneration in Vitro. Dev Biol 92:476–488.

Schmidt H, Zissler D (1979): Die Spermien der Anthozoen und ihre phylogenetische Bedeutung. Zoologica 129:1–96.

Slautterback DB (1961): Nematocyst Development. In Lenhoff HL, Loomis WF (eds): "The Biology of Hydra." Coral Gables, Florida: University of Miami Press, pp 77–129.

Smallwood WM (1899): A contribution to the morphology of Pennaria tiarella McCrady. Am Natur 33:861–870.

Smid J, Tardent P (1982): The Influence of Ecto- and Endoderm in Determining the Axial Polarity of Hydra attenuata Pall. (Cnidaria, Hydrozoa). Roux's Arch Dev Biol 191:64–67.

Smid J, Tardent P (1984): Migration of I-cells from Ectoderm to Endoderm in Hydra attenuata Pall. (Cnidaria, Hydrozoa) and their subsequent differentiation. Dev Biol 6:469–477.

Sugyiama T, Fujisawa T (1977): Genetic Analysis of Developmental Mechanism in Hydra. I. Sexual reproduction of Hydra magnipapillata and isolation of mutants. Dev Growth Differ 19:187–200.

Szollosi D (1964): The structure and function of the Centrioles and their satellites in the jellyfish Phialidium gregarium. J Cell Biol 21:465–479.

Tardent P (1966): Experimente zur Frage der Geschlechtsbestimmung bei Hydra attenuata Pall. Rev Suisse Zool 73:481–492.

Tardent P (1968): Experiments about Sex Determination in Hydra attenuata Pall. Dev Biol 17:483–511.
Tardent P (1978): Coelenterata, Cnidaria. In Seidel F (ed): "Morphogenese der Tiere, Lieferung 1:A-I." Stuttgart: Gustav Fischer Verlag, pp 69–415.
Tardent P (1984): The Significance of the Metagenetic Life Cycles of Cnidaria for Genetic Diversification and Adaptability. In Engels W, Clark WH, Fischer A, Olive PJW, Went DF (eds): "Advances in Invertebrate Reproduction Vol 3." Amsterdam: Elsevier Science Publishers, pp 269–278.
Tardent P, Morgenthaler U (1966): Autoradiographische Untersuchungen zum Problem der Zellwanderungen bei Hydra attenuata Pall. Rev Suisse Zool 73:468–480.
Tardent P, Leutert R, Frei E (1968): Untersuchungen zur Taxonomie von Hydra circumcincta Schulze 1914 und Hydra ovata Boecker 1920. Rev Suisse Zool 75:983–998.
Tardent P, Weber Ch (1976); A Qualitative and Quantitative Inventory of Nervous Cells in Hydra attenuate Pall. In Mackie GO (ed): "Coelenterate Ecology and Behavior." New York: Plenum Press, pp 501–512.
Tardent P, Holstein T (1982): Morphology and morphodynamics of the stenotele nematocyst of Hydra attenuata Pall. (Hydrozoa, Cnidaria). Cell Tissue Res 224:269–290.
Weiler-Stolt B (1960): Ueber die Bedeutung der interstitiellen Zellen für die Entwicklung und Fortpflanzung mariner Hydroiden. Roux's Arch Dev Biol 152:398–455.
Weill R (1930): Essai d'une classification des nématocystes des Cnidaires. Bull Biol Fr Belg 64:141–156.
Weismann A (1880a): Zur Frage nach dem Ursprung der Geschlechtszellen bei den Hydroiden I. Zool Anz 3:226–233.
Weismann A (1880b): Ueber den Ursprung der Geschlechtszellen bei den Hydroiden II. Zool Anz 3:367–370.
Weismann A (1883): "Die Entstehung der Sexualzellen bei Hydromedusen." Jena: Gustav Fischer Verlag, pp 1–295.
Weismann A (1884): Die Entstehung der Sexualzellen bei den Hydromedusen. Biol Centralbl 4:12–32.
Weismann A (1892): "Das Keimplasma. Eine Theorie der Vererbung." Jena: Gustav Fischer Verlag, pp 1–628.
Weissman A, Lentz TL, Barrnett RJ (1969): Fine Structural Observations in Nuclear Maturation during Spermiogenesis in Hydra littoralis. J Morphol 128:229–240.
Werner B (1958): Die Verbreitung und das jahreszeitliche Auftreten der Anthomeduse Rathkea octopunctata M. Sars, sowie die Temperaturabhängigkeit ihrer Entwicklung und Fortpflanzung. Helgoländer wiss Meeresunters 6:137–170.
Werner B (1973): New Investigations of Systematics and Evolution of the Class Scyphozoa and the Phylum Cnidaria. In Tokioka T, Nishimura S (eds): "Recent Trends in Research in Coelenterate Biology." Seto Mar Biol Lab 20:35–61.
Werner B (1984): Cubozoa. In Kästner A (ed): "Lehrbuch der speziellen Zoologie Bd. 1,2. Teil." Stuttgart: Gustav Fischer Verlag, pp 106–133.
Westfall JA (1973): Ultrastructural evidence for neuromuscular systems in Coelenterates. Amer Zool 13:237–246.
Widersten B (1965): Genital organs and fertilization in some Scyphozoa. Zool Bidr Uppsala 37:45–58.
Wiese L (1953a): Geschlechtsverhältnisse und Geschlechtsbestimmung bei Süsswasserhydroiden. Zool Jahrb Abt Zool 64:55–83.
Wiese L (1953b): Ueber die Bestimmung und Realisation des Geschlechts bei Süsswasserpolypen. Naturwissenschaften 6:189–192.
Yoshida M (1959): Spawning in Coelenterates. Experientia 15:11–12.
Zihler J (1972): Zur Gametogenese und Befruchtungsbiologie von Hydra. Roux's Arch Dev Biol 169:239–267.
Zürrer D (1983): Untersuchungen zur Oogenese von Hydra carnea Agassiz (Cnidaria, Hydrozoa). Thesis Zool Inst Univ Zürich, pp 1–62.

The Origin and Evolution of Sex, pages 199–211

Bkm Sequences and Their Conservation

Kenneth W. Jones

Department of Genetics, University of Edinburgh, Scotland

INTRODUCTION

In eukaryotes, chromosomal mechanisms play an important role in both the control of developmental gene expression and evolution. The involvement of chromosomal factors in these processes is most evident in the case of chromosomes that are specialized for sex determination. However, consistent with our ignorance about how developmental genes are controlled and how evolution occurs, why and how sex chromosomes evolved is also unknown. In this article, new evidence and ideas are discussed which may have some bearing on the evolution of sex chromosomes and their role in the evolutionary process.

Examples of specialized sex chromosomes are common in both male (XY/XX) and female (ZW/ZZ) heterogamety throughout higher eukaryotes, including plants, insects, and vertebrates. This indicates that the preconditions for evolving sex-related chromosomal specialization are widespread amongst eukaryotes. The specialization of a chromosome for sex determination involves the loss of most of its other genetic functions. In some species, for example, the bandicoot (*Isoodon obesculus*), the Y chromosome is absent from somatic tissues [Hayman and Martin, 1965], and in others, such as the mole cricket *Gryllotalpa fossor* [Rao and Arora, 1978], it is absent altogether. Curiously, this phenomenon of sacrificing genetic functions in the cause of sex determination does not appear to have been evolutionarily disadvantageous. Quite the contrary, species in which specialised sex chromosomes have evolved, in general, tend to be relatively evolutionarily advanced. For example, in the vertebrates, species of the lower orders, such as the fishes and amphibia, for the most part lack specialized sex chromosomes, whereas in land-dwelling species of reptiles and mammals, sex chromosomes are well differentiated [reviewed in Ohno, 1967]. This tendency is evident even within a single line of descent. Thus, in the snakes, the relatively primitive species of constrictors have no visibly specialized sex chromosomes, whereas the more evolutionarily advanced poisonous species all have a well-defined ZW/ZZ system of chromosomal sex determination [reviewed in Bull, 1980]. Similarly, amongst birds, the anatomically relatively primitive ratites, which include the ostriches, moas, and emus, lack the specialized W sex chromosomes that are found in more recently evolved species [Tagaki et al., 1972; Tagaki and Sasaki, 1974]. It appears that the evolution of chromosomal mechanisms of sex determination has gone hand in hand with accelerated evolution of somatic specialization. This suggests either that more rapid evolution frequently resulted in sex chromosome specialization or that the emergence of sex chromosomes themselves somehow accelerated the evolutionary process. One aspect of a change to a more rapid evolutionary rate would be an increase in variability upon which natural selection can operate. Intuitively, it seems unlikely that a chromosomal (i.e., a more stable) sex-determining system, in itself, would have contributed to a faster pace of evolution. Therefore, if sex chromosomes have influenced evolutonary rates, the mechanism is more likely to be incidental to their sex-determining functions. In this context, it recently has been found in many species that specialized sex chromosomes contain a high concentration of particular types of repeated sequences which have been referred to as Bkm [Singh et al., 1976, 1979, 1980]. This article will summarise the work which led to the characterisation of Banded Krait minor satellite sequences and discuss their possible significance for some of the questions outlined.

ISOLATION OF W CHROMOSOME REPEATED DNA (BKM)

The possibility of isolating sex chromosome DNA quantitatively, in order to investigate its structure and functions, was originally indicated from the fact that the W chromosome in snakes is entirely heterochromatic and stains differentially with giemsa in the C-banding technique [Singh and Ray-Chaudhuri, 1975]. Heterochromatin with these characteristics was known in mice and humans [Pardue and Gall, 1970; Jones, 1970; Jones et al., 1972] to indicate locations of simple sequence, or satellite, DNA, which can be recovered by differential centrifugation. Analytical ultracentrifugation of the DNAs obtained from male and female Banded krait snakes therefore was used to define a minor female-specific satellite DNA which was referred to as Bkm. This was shown by in situ hybridization to be relatively specific for the W chromosome. However, Bkm amounted to less than 1% of the genomal DNA, whereas the W chromosome accounts for between 5% and 10%. Moreover, the pattern of labeling obtained by in situ hybridization of a Bkm probe

TABLE I. BKM Sequences in Primitive and Advanced Snake Species

Species	Family	Counts/min hybridized			Ratio of Male/Female counts	Genomal % of W chromosome
		Male	Female	Control		
Eryx johni johni	Boidae	717	744	119	1.03	—
Python reticulatus	Boidae	—	722	—	—	—
Xenopeltis unicolor	Boidae	—	678	—	—	—
Ptyas mucosus	Colubridae	1852	4165	116	2.25	8
Bungarus caeruleus	Elapidae	1861	6681	166	3.59	11
Bungarus fasciatus	Elapidae	1108	1742	59	1.57	4

Equal amounts of male and female genomal DNA bound to nitrocellulose filters was hybridized with a ^{32}p-labeled uncloned Bkm satellite DNA probe and compared with a control of the same amount of bacterial DNA. Note the relatively low hybridization values obtained with DNA from members of the primitive family Boidae compared with the hybridization obtained with DNA from the species representing the more highly evolved families. Also note the higher hybridization values of the female DNA of these latter families, reflecting the contribution of Bkm DNA from the W chromosome.

The data shown were abstracted from Singh et al. [1976, 1980]. The hybridization values for *P. reticulatus* and *X. unicolor* taken from a different set of data have been normalized against the values obtained for *P. mucosus* in both sets of data.

to female snake chromosomes extended over the entire W chromosome. This showed that the Bkm sequences are interspersed with other DNA on this chromosome. Subsequently, the DNAs of males and females of several families of snakes, including those of family Boidae and representatives of the more advanced families Colubridae, Elapidae, and Viperidae, were analysed for the presence of Bkm-related sequences by quantitative filter hybridization [Singh et al., 1976, 1980]. This approach showed that Bkm sequences are not confined to the W chromosome. Some of the data from these experiments are shown in Table I.

In summary, these studies showed, first, that Bkm-related DNA is conserved throughout all snake families; second, that it is also present in males, despite the lack of a demonstrable DNA satellite, or of any prominent in situ localization on chromosomes; third, that even the males of species with chromosomal sex determination have approximately two to three times more of these particular sequences compared to either sex in species lacking chromosomal sex determination; fourth, that the W chromosome in all snakes contains as much, or sometimes more, Bkm DNA than is found in toto in the remainder of the genome. The absolute amount of hybridization in females of different species shown in Table I reflects the relative length of the W chromosome. The females of some snakes, for example, *Notechis scutatus*, have been found to have as much as 14 times the amount of Bkm-hybridizable DNA found in constrictors. From these data, it can be concluded that there is a striking correlation between the presence of specialized sex chromosomes carrying Bkm and changes in the abundance of these same sequences in the DNA of the genome as a whole.

THE CONSERVED SEQUENCES OF THE BKM SATELLITE

Unlike most satellite DNAs, which tend to be relatively unconserved, Bkm satellite sequences are conserved in a wide range of eukaryotes [Jones and Singh, 1981]. Without exception, all eukaryotic DNAs tested showed hybridization with a probe of uncloned Bkm satellite DNA. So far, 63 species have been surveyed, representing virtually the entire spectrum of eukaryotes, including dinoflagellates, yeast, sea urchins, coelenterates, insects, plants, and the major groups of vertebrates. The sequence homologies responsible for this hybridization have been characterised in the case of snakes, Drosophila, and mouse by sequencing Bkm-related clones recovered from genomic DNA libraries. One consensus sequence found in these comparisons comprises uninterrupted repeats of the tetranucleotide GATA [Singh et al., 1984]. Another clone, referred to as p1581, has been isolated by Bkm probing of a mouse cDNA library, but it is not yet fully characterised. Clone p1581 does not hybridize with GATA, but it contains sequences that are conserved over the same extensive range of eukaryotic DNAs. The GATA and p1581 sequences together account for most of the hybridization patterns previously seen on Southern blots using an uncloned Bkm satellite probe. Their quantitative distribution varies in different species' DNAs and in the sex chromosomes of different species. Snake W chromosomes contain substantial amounts of both sequence types, whereas the W chromosomes of certain birds show a higher concentration of p1581-related DNA relative to GATA repeats (unpublished). The sex-specific pattern found in mouse DNA by Bkm probing [Singh et al., 1981] mainly reflects the high concentrations of GATA repeats on the Y chromosome. These repeats are located in the region of the murine Y chromosome, which is responsible for sex determination [Singh and Jones, 1982]. The proximal end of the Drosophila melanogaster X chromosome also contains a high local concentration of GATA repeats [Singh et al., 1984]. Clone p1581, on the other hand, hybridizes indistinguishably to male and female mouse DNA by Southern blotting. It appears, therefore, that Bkm satellite DNA comprises at least two consensus sequences that are common in the genomes of eukaryotes and often concentrated on sex chromosomes. Sequencing studies on mouse, snake, and bird DNAs reveal that one possible p1581 consensus comprises long stretches alternating purine-pyrimidine dinucleotides [Walker, unpublished]. However, probes consisting purely of this consensus sequence do not account entirely for the hybridization patterns obtained with a probe of p1581, indicating that there is another consensus sequence yet to be found. Evidence suggesting a function for some Bkm-related sequences comes from studies of transcription.

TRANSCRIPTION OF THE BKM SEQUENCES IN MICE

Clone p1581 was isolated from a cDNA library and can be presumed on these grounds to originate from a transcribed region of the mouse genome. Consistent with this, RNA extracted from different tissues and organs of the mouse hybridize with p1581, on Northern blots, in a pattern which varies somewhat according to the tissues of origin (unpublished). GATA repeats also are transcribed, as has been shown by probing the same northern blots of RNA with a single-strand GATA probe [Singh et al., 1984]. The combined hybridization patterns of p1581 and GATA probes can be reproduced by probing the same blots with Bkm satellite DNA, suggesting that these two cloned probes represent the major transcribed conserved Bkm sequences in mice. Interestingly, the sequences of Bkm-related subclones from genomic libraries of mouse and Drosophila contain GATA repeats in open reading frames [Singh et al., 1984]. However, evidence of translation of these sequences has not yet been sought. Paradoxically, therefore, sequences which are abundant on developmentally silent sex chromosomes also form part of the transcribed regions of developmentally expressed genes.

POTENTIAL CONTROLLING FUNCTIONS OF BKM SEQUENCES

Transcription of GATA DNA has been found in several unrelated species. In all cases so far examined, probes comprising the repeats of the complementary sequence CTAT hybridize to mRNA at a much reduced level compared to GATA probes [Singh et al., 1984]. This indicates that the GATA repeats are most frequently orientated in the sense strand of genes. Transcription of the complementary CTAT strand would yield RNA in which there are translational stop codons in all reading frames, in both directions. Perhaps significantly, therefore, GATA repeats within individual genomic clones have all been found to exhibit the same strand polarity. This finding is consistent with selective pressure to eliminate non-sense strand repeats of this sequence. Evidently, if CTAT sequences became inserted into transcribed regions of chromosomes, such regions would very possibly be rendered translationally defunct. GATA/CTAT, therefore, is potentially a highly effective mutagenic sequence whose normal function might be in relation to controlling the polarity of gene expression. Anti-sense RNA can inhibit gene expression in eukaryotes [Izant and Weintraub, 1984]. However, the extent to which this type of control is utilized in eukaryotes still has to be established. These potential negative controlling effects of the GATA sequence may explain why GATA sequences seem to have accumulated in high concentration on developmentally inactivated specialized sex chromosomes. However, whether this reflects or explains their inactivity is a matter for speculation.

One of the likely p1581 consensus sequences appears to consist of alternating purine-pyrimidine dinucleotide repeats. If confirmed, this suggests a possible function for these particular Bkm sequences in B- to Z-conformation shifts of DNA. Z-DNA is a left-handed alternative conformation of the right-handed double helix [Wang et al., 1979, 1981] which occurs under physiological conditions [Haniford and Pulleyblank, 1983]. The Bkm sequences represented by p1581 therefore could potentially be implicated in a very wide range of controlling functions. For example, transcriptional enhancers contain potential Z-form DNA [Nordheim and Rich, 1983], and eukaryotic transcription can be inhibited by potential Z DNA sequences in vitro [Clarkson et al., 1981; Hipskind and Clarkson, 1983; Santoro et al., 1984]. Also, it has been shown that sequences with the capacity to form Z-DNA are transcribed in vivo [Santoro and Costanzo, 1983]. Z-DNA impairs the formation of normal chromatin structure [Nickol et al., 1982], involves binding of novel proteins [Nordheim et al., 1982], and leads to a lower affinity for regulatory DNA-binding proteins [Fried et al., 1983]. Relevant to the fact that W chromosomes contain high concentrations of Bkm sequences related to p1581, it recently has been shown that the X chromosome of Drosophila also contains significantly more sequences which hybridize to a probe of alternating purines and pyrimidines, poly(dC.dA)-poly(dG.dT), than do the autosomes [Pardue et al., 1984]. Such domains of alternating purine-pyrimidine dinucleotides therefore may be particularly significant in the context of sex chromosomes in general. Perhaps relevant in this regard is the fact that X-linked dosage compensation in females of *Drosophila* involves a modulation of the levels of transcription of X-linked genes [reviewed by Baker and Belote, 1983]. The fact that the W sex chromosomes which, on morphological grounds, appear to be at a relatively early stage in their evolution contain high concentrations of sequences with potential functions in the control of gene expression, is particularly interesting in view of the general idea that these particular chromosomes are genetically degenerate.

MODELS FOR SEX CHROMOSOME EVOLUTION

The finding of concentrations of Bkm on the Y and W chromosomes is relevant to the unanswered question of how these chromosomes evolved. The classical model of the evolution of specialized sex chromosomes is based on Muller's original idea [Muller, 1914] that the genetic degeneration of the Y chromosome was initiated by a reduction in crossing over with the X chromosome. The genetically active but recombinationally deficient Y chromosome then was envisaged gradually to become genetically inactive due to the accumulation of loss of function mutations under the influence of selection. However, Fisher [1935] criticised this idea on the grounds that a difference in the rate of accumulation of mutations between the Y and X chromosomes would not be expected. Subsequent calculations show that there could be a somewhat higher rate of chance fixation of deleterious recessives on the Y chromosome [Nei, 1970]. However, the debate about whether the evolutionary divergence of Y and X, or Z and W chromosomes, can be satisfactorily accounted for on a Mullerian model continues [see Charlesworth, 1978, for further discussion]. Sex chromosomes in snakes are particularly relevant to this debate in that, unlike in Drosophila or in mammals whose sex chromosomes have stimulated previous models of sex chromosome evolution, examples of most of the stages in the evolution of these chromosomes exist in living species. Moreover, whether specialised sex chromosomes are present, and the extent of their elaboration, correlates positively with other indices of evolutionary advancement. Snake families have been classified broadly on anatomical grounds into primitive, intermediate, and advanced evolutionary types [reviewed in Bull, 1980]. The primitive condition is represented by the snakes of the family Boidae, which also lack differentiated sex chromosomes. The most evolutionarily advanced snakes are represented by the Viperidae, which exhibit highly heteromorphic ZW bivalents, indicating an advanced stage in the evolution of the W chromosome. Snakes of the family Colubridae are intermediate between these extremes, in that cranial anatomy associated with adaptations for prey capture has not evolved to the same extent as that of the Viperidae [Jansen and Foehring, 1983]. Consistent with this pattern, in which there is a correlation between anatomical specialisms and sex chromosome evolution, in the Colubridae, the ZW bivalent is very often homomorphic, or nearly so [Singh, 1972; Bull, 1980]. Given the propensity of sex chromosomes to diverge rapidly in evolutionary time, it is suggested that these chromosomes are in a relatively early stage in their evolution in Colubrid snakes. According to the classical hypothesis, the processes of genetic degeneration of W chromosomes might be expected to have proceeded less far in Colubrid snakes compared with Viperid snakes chromosomes. However, it is apparent that in Colubrids, as in all other snakes, Bkm sequences are interspersed throughout the W chromosome, showing that there has already been considerable evolution at the DNA sequence level. Moreover, in both the Colubridae and the Viperidae, the entire W chromosome, like the inactive mammalian X chromosome, forms a heteropycnotic sex chromatin body in somatic tissues [Singh et al., 1976], signifying inactivation. These aspects of the W chromosome are difficult to explain on the conventional selectionist model and are more consistent with the idea that somatic inactivation of the entire W chromosome became established at a relatively early stage in W chromosome evolution, before major genetic or morphological divergence had time to occur. This concept inverts the Mullerian model in suggesting that, rather than failure to cross over being responsible for *genetic* inactivation, *chromosomal* inactivation caused the interference with crossing over and the rapid subsequent divergent evolution of sex bivalents. This obviously implies an evolutionarily relatively abrupt mechanism, and it has been suggested [Jones, 1983a,b] that this may have been similar to that which triggers developmental X chromosome inactivation in mammals. If this idea is essentially correct, a relatively singular mutational cause of evolutionary inactivation of the sex chromosome is implied. Because this inactivation is obviously consistent with the sex-determining functions of these chromosomes, it has been suggested to be controlled by the sex genes themselves [Jones, 1983b, 1984]. Consistent with this idea, inactivation and activation behaviour of sex chromosomes is conspicuously associated with germ cells and with early stages of embryogenesis, and similar behaviour of the sex chromosomes is evident in the females of insects, snakes, and mammals. In snakes in situ hybridization, using a Bkm probe, confirmed that the W chromosome remains somatically condensed but decondenses as a whole in the postmeiotic oocyte and becomes associated with a prominent nucleolus [Singh et al., 1979]. Transcription involving morphological changes in the W chromosome during oogenesis also has been described in the Lepidopteran *Ephestia* [Traut and Scholz, 1978]. Thus, such behaviour appears to be a common aspect of female heterogamety. However, the inactivated mammalian X chromosome also reactivates during oogenesis [reviewed by Gartler and Riggs, 1983]. This common behaviour pattern is consistent with the sex/fertility-determining role of the X and W chromosomes in oogenesis. Consistently, in male heterogamety the Y chromosome becomes activated during spermatogenesis [Hess and Meyer, 1968; Hennig et al., 1974, 1983], whereas the X chromosome apparently inactivates at this stage [Monesi, 1971]. Failure of the X chromosome to inactivate at this time is associated with male infertility [Lifschytz and Lindsley, 1972], suggesting that the condensation cycle is obligately connected with sex gene function. A model which attempts to account for how the condensation behaviour of the chromosome became linked to sex gene expression has been recently put forward [Jones, 1983a,b]. This envisages that mutations which established genetic control of sex in some cases also caused these genes to take control of genes involved in chromosomal condensation. This proposed mechanism has been called chromosomal "hijacking." Because chromosomal behaviour thereafter was controlled by the cycle of expression of the sex gene which it carries, somatic chromosomal inactivation signifies the somatic inertness of the sex genes, and chromosomal reactivation in the germ line indicates activation of these genes in connection with sex determination. The essential failure of crossing over in the heterogametic sex in reptiles and mammals then is seen to have arisen because the sex bivalent carries functionally divergent sex alleles which impose different cycles of condensation on the Z and W and X and Y chromosomes, respectively. The Y and W chromosomes therefore are doomed to genetic degeneration in all but their sex-determining functions, which survive because they control the entire process. This model obviously challenges the view that the essential significance of X-chromosome inactivation is to establish dosage compensation between males and females. Because only one X chromosome remains active in a given cell, the model also has new implications for the mode of function of alleles of sex genes carried by the X chromosome.

The accumulation of Bkm sequences on the W chromosome as an early consequence of its inactivation presumably reflects the abrupt cessation of selective pressure on gene function, which permitted the rapid evolution of resident sequences. Why these particular sequences evolved on sex chromosomes in snakes, birds, and mammals is unclear, but it may reflect their relative instability and/or appropriateness of function with respect to the altered control of the sex determining chromosome.

EVOLUTIONARY IMPORTANCE OF SEX CHROMOSOMES

Whatever the precise details of the evolution of the W chromosome may have been, it is a fact that its high concentration of Bkm sequences is echoed in a similar abundance of these sequences in the genome as a whole. This constitutes the first molecular evidence linking general genomal changes with the evolution of sex chromosomes. Although we have no direct evidence, it seems probable, on the grounds that the W chromosome constitutes a concentrated potential source of Bkm, that the general increase in Bkm DNA in the genome was derived from the W chromosome by sequence transposition. If this was so, it suggests that specialized sex chromosomes constitute a hitherto unsuspected potential source of evolutionary change. This is based on the fact that a major proportion of spontaneous mutations have been shown to be correlated with insertional events associated with sequence transposition. This is clear from recent work on Drosophila genes such as bithorax [Bender et al., 1983], white [reviewed by Rubin, 1983], notch [Kidd et al, 1983; Artavanis-Tsakonas et al., 1983] and scute [Carramolino et al., 1982]. In principle, depending on its size and insertion site, the insertion of any DNA sequence directly within the coding sequence or promotor region could interfere with gene function, but, as discussed, GATA sequences might be especially mutagenic in this context. Although it is less clear how an insertion lying outside the gene might affect its expression, in principle, it is foreseeable that more subtle effects could result. For example, instances have been shown in both yeast [reviewed by Roeder and Fink, 1983] and Drosophila [Modolell, et al., 1983] in which the insertion of a transposable sequence caused the expression of a gene to be influenced by an unlinked gene; a phenomenon with possible implications for the "hijack" model for sex-chromosome inactivation. Also, sequences which are potentially capable of DNA B-form to Z-form conformational shifts have been implicated in the control of crossing over, transcriptional enhancement, and the promotion of instability amongst domains of repeated genes [reviewed by Rich et al., 1984]. Alternating purine-pyrimidine dimers are characteristically unstable. For example, the protein coding domains of MHC genes which exhibit the greatest degree of polymorphism are those containing alternating purine pyrimidine-rich regions [Tykocinski and Max, 1984]. Gene conversion involving repeats of alternating purines and pyrimidines also has been implicated in the high incidence of polymorphism in the H-2 class 1 major histocompatibility locus of the mouse [Loh and Baltimore, 1984]. The interspersion of sequences such as those which are present in p1581 consensus sequences, therefore, might have had pleiotropic consequences affecting both gene expression and recombination. If sequence transposition is potentiated in metabolically active chromosomes [Rogers, 1984], it is probable that sequence evolution involving sex chromosomes will have its most significant effects in the germ line of the heterogametic sex, where they are decondensed and reactivated. This would have the maximum effect in generating diversity. In consequence, species which evolved sex chromosomes subsequently may have been exposed to a considerably increased incidence of mutation due to unbridled sequence evolution on these chromosomes in the heterogametic sex [Jones, 1984]. Foreseeably, such a novel genotypic mechanism could have contributed to a marked acceleration of evolution and may be the explanation of why, in general, species with specialized sex chromosomes are amongst the more advanced life forms. The emergence of sex chromosomes thus can be seen potentially as one significant explanation of the great radiations of new species at certain evolutionary periods, particularly amongst vertebrates. A corollary of this general model might be that evolutionary rates eventually would tend to slow down as the specialized sex chromosome became reduced in size and/or evolved relatively functionless families of unconserved repeated DNAs. In this sense, if there is any evolutionary future, it may belong to species in which, like the snakes, sex chromosomes are in a relatively early stage of their evolution.

EUKARYOTES AND THE EVOLUTION OF SEX

If the evolution of sex chromosomes has been responsible for increased evolutionary rates by the mechanisms discussed, it follows that the underlying processes of sequence interspersion have played a significant role throughout evolution and that the Bkm satellite defines some of the sequences which are implicated. Bkm probes hybridize apparently universally to the DNA of a very wide spectrum of eukaryotes but have not been detected in prokaryotes under the same conditions of hybridization. We may therefore speculate that the consensus sequences of the Bkm satellite form part of a class of DNA which may have played a significant role in the evolution of the eukaryotes as a group. Sequences potentially capable of altering chromatin functional structure at some distance (p1581 consensus) and/or exhibiting strand polarity affecting translation (GATA), together with transposable DNA sequences in general, presumably evolved in primitive oganisms which contained redundant DNA sequences which buffered the genome against inactivation of essential genes. To the extent that these were in chromosomal domains rather than in plasmids, they may have contributed to selection favouring more complex chromosomes and larger genome size characteristic of eukaryotes. Some of these initially nonessential chromosomal domains containing potential Z-DNA may have facilitated altered gene expression under changing environmental conditions. In this context, it is perhaps worth pointing out that sex ratio in many organisms is determined by environmental factors [reviewed in Bull, 1980]. Simple multicellular organisms would have better accommodated mutations by means of metabolic complementation between cells. The mutational modification of cell membranes which caused cells to adhere also would be a precondition for sexual processes. The evolution of eukaryotes, with their propensity for sexual differentiation, therefore can be envisaged as one outcome of adaptations to the challenge posed by DNA sequences of the type comprising the Bkm satellite. The widespread conservation of Bkm sequences may reflect the finite variety of DNA sequences that possess properties relevant to these processes.

REFERENCES

Artavanis-Tsakonas S, Muskavitch MAT, Yedvobnick B (1983): Molecular cloning of Notch, a locus affecting neurogenesis in Drosophila melanogaster. Proc Natl Acad Sci USA 80:1977–1981.

Baker B, Belote JM (1983): Sex determination and dosage compensation in *Drosophila melangaster*. Ann Rev Genet 17:345–393.

Bender W, Akam M, Karch F, Beachy PA, Pfeifer M, Spierer P, Lewis EB, Hogness DS (1983): Molecular genetics of the bithorax complex in Drosophila melanogaster. Science 221:23–29.

Bull JJ (1980): Sex determination in reptiles. Q Rev Biol 55:3–20.

Carramolino L, Ruiz-Gomez M, Guerrero M, Campuzano S, Modelell J (1982): DNA map of mutations at the scute locus of Drosophila melanogaster. EMBO J 1:1185–1191.

Charlesworth B (1978): Model for evolution of Y chromosomes and dosage compensation. Proc Natl Acad Sci USA 75:5618–5622.

Clarkson SG, Koski RA, Corlet J, Hipskind RA (1981): Influence of 5′ flanking sequences on tRNA transcription in vitro. In Brown DD, Fox CF (eds): "Developmental Biology Using Purified Genes." New York: Academic Press, pp 463–472.

Fisher RA (1935): The sheltering of lethals. Am Nat 69:446–499.

Fried MG, Wu H-M, Crothers DM (1983): CAP-binding to B and Z forms of DNA. Nucleic Acids Res 11:2479–2494.

Gartler SM, Riggs AD (1983): Mammalian X chromosome inactivation. Ann Rev Genet 17:155–190.

Haniford DB, Pulleyblank DE (1983): Facile transition of poly[d(TG).d(CA)] into left-handed helix in physiological conditions. Nature 302:632–634.

Hayman DL, Martin PG (1965): Genetics 52:1201–1206.

Hennig W, Huijser P, Vogt P, Jäckle H, Edstrom J-E (1983): Molecular cloning of the microdissected lampbrush loop DNA sequences of Drosophila hydei. EMBO J 2:1741–1746.

Hennig W, Meyer GF, Hennig I, Leoncini O (1974): Structure and function of the Y chromosome of Drosophila hydei. Cold Spring Harbor Symp Quant Biol 38:673–683.

Hess O, Meyer GF (1968): Genetic activities of the Y chromosome in Drosophila during spermatogenesis. Adv Genet 14:171–218.

Hipskind RA, Clarkson SG (1983): 5′ flanking sequences that inhibit in vitro transcription of a Xenopus laevis tRNA gene. Cell 34:881–890.

Izant JG, Weintraub H (1984): Inhibition of thymidine kinase gene expression by anti-sense RNA: a molecular approach to genetic analysis. Cell 36:1007.

Jansen DW, Foehring RC (1983): The mechanism of venom secretion from Duvernoy's Gland of the snake Thamnophis sirtalis. J Morphol 175:271–277.

Jones KW (1970): Chromosomal and nuclear location of mouse satellite DNA in individual cells. Nature 225:912–915.

Jones KW (1983a): Evolutionary conservation of sex-specific DNA sequences. Differentiation (Suppl) 23:56–59.

Jones KW (1983b): Evolution of sex chromosomes. In Johnson MH (ed): "Development in Mammals," Vol 5. New York: Elsevier, pp 297–319.

Jones KW (1984): The evolution of sex chromosomes and their consequences for the evolutionary process. In Bennet MD, Gropp A, Wolf U (eds): London: Allen and Unwin, London pp 241–255.

Jones KW, Prosser J, Corneo G, Ginelli E, Bobrow M (1972): Satellite DNA, constitutive heterochromatin and human evolution. In "Modern Aspects of Cytogenetics: Constitutive Heterochromatin in Man." Stuttgart-New York: F. K. Schattauer Verlag, pp 45–61.

Jones KW, Singh L (1981): Conserved sex-associated repeated DNA in vertebrates. In Dover G, Flavell R (eds): "Genome Evolution." Systemics Association, Special Volume 20. London: Academic Press, pp 135–154.

Kidd S, Lockett TJ, Young MW (1983): The Notch locus of Drosophila melanogaster. Cell 38:135–146.

Lifschytz E, Lindsley DL (1972): The role of X-chromosome inactivation during spermatogenesis. Proc Natl Acad Sci USA 69:182–186.

Loh DY, Baltimore D (1984): Sexual preference of apparent gene conversion events in MHC genes of mice. Nature 309, 639–640. Modelell J, Bender W, and Meselson M (1983): Drosophila melanogaster mutations supressible by the supressor of Hairy-wing are insertions of a 7.3 kilobase mobile element. Proc Natl Acad Sci USA 80:1678–1682.

Modolell J, Bender W, Meselson M (1983): Drosophila melanogaster mutations suppressible by the suppressor of hairy wing are insertions of a 7.3 kilobase mobile element. Proc Nat Acad Sci USA 80:1678–1682.

Monesi V (1971): Chromosome activities during meiosis and spermiogenesis. J Reprod Fertil [Suppl] 13:11–17.

Muller HJ (1914): A gene for the fourth chromosome of Drosophila. J Exp Zool 17:325–336.

Nei M (1970): Accumulation of non-functional genes on sheltered chromosomes. Am Nat 104:311–322.

Nickol J, Behe M, Felsenfeld G (1982): Effect of B-Z transition in poly (dG-m^{5}dC).poly (dG-M^{5}dC) on nucleosome formation. Proc Natl Acad Sci USA 79:1771–1775.

Nordheim A, Tesser P, Azorin F, Kwon YH, Miller A, Rich A (1982a): Isolation of Drosophila proteins that bind selectively to left-handed Z-DNA. Proc Natl Acad Sci USA, 79:7729–7733.

Nordheim A, Rich A (1983): The sequence $(dC\text{-}dA)_n.(dG\text{-}dT)_n$ forms left handed Z-DNA in negatively supercoiled plasmids. Proc Natl Acad Sci USA 80:1821–1825.

Ohno S (1967): "Sex chromosomes and sex-linked genes." Berlin: Springer-Verlag.

Pardue ML, Gall JG (1970): Chromosomal localization of mouse satellite DNA. Science 168:1356–1358.

Pardue ML, Nordheim A, Moller A, Weiner LM, Stollar BD, Rich A (1984): Z-DNA and chromosome structure. In Bennet MD, Gropp A, Wolf U (eds): "Chromosomes Today, Vol 5." London: Allen & Unwin, London pp 34–45.

Rao SRV, Arora P (1978): Insect sex chromosomes: Part 1. Differential response to 5 bromodeoxyuridine of two X chromosomes in females of the mole cricket Gryllotalpa fossor (Scudder). Indian J Exp Biol 16:870–872.

Rich A, Nordheim A, Wang AH-J (1984): The chemistry and biology of left-handed Z-DNA. Annu Rev Biochem 53:791–846.

Rogers J (1984): The origin and evolution of retroposons. Int Rev Cytol [Suppl] 17. (in press)

Roeder GS, Fink GR (1983): Transposable elements in yeast. In Shapiro JA (ed): "Mobile Genetic Elements." New York: Academic Press, pp 299–328.

Rubin GM (1983): Dispersed repetitive DNAs in Drosophila. In Shapiro JA (ed): "Mobile Genetic Elements." New York: Academic Press, pp 329–361.

Santoro C, Costanzo F (1983): Stretches of alternating poly(T-dG) with the capacity to form Z-DNA are present in human liver transcripts. FEBS Lett 155:69–72.

Santoro C, Costanzo F, Ciliberto G (1984): Inhibition of eukaryotic tRNA transcription by potential Z-DNA sequences. EMBO J 3:1553–1559.

Singh L (1972): Evolution of karyotypes in snakes. Chromosoma 38:185–236.

Singh L, Purdom IF, Jones KW (1976): Satellite DNA and the evolution of sex chromosomes. Chromosoma 59:43–62.

Singh L, Purdom IF, Jones KW (1979): Behaviour of sex chromosome associated satellite DNAs in somatic and germ cells in snakes. Chromosoma 71:167–181.

Singh L, Purdom IF, Jones KW (1980): Sex chromosome associated satellite DNA: evolution and conservation. Chromosoma (Berl.) 79:137-157.

Singh L, Purdom IF, Jones KW (1981): Conserved sex chromosome associated nucleotide sequences in eukaryotes. Cold Spring Harbor Symp Quant Biol 44:805-813.

Singh L, Ray-Chaudhuri SP (1975): Location of C-band in the W sex chromosome of common Indian krait *Bungarus caeruleus* (Schneider). Nucleus 18:163–166.

Singh L, Jones KW (1982): Sex reversal in the mouse (Mus musculus) is caused by a recurrent non-reciprocal crossover involving the X and an aberrant Y chromosome. Cell 28:205–216.

Singh L, Phillips C, Jones KW (1984): The conserved nucleotide sequences of Bkm which define Sxr in the mouse are transcribed. Cell 36:111–120.

Traut W, Scholz D (1978): Structure replication and transcriptional activity of the sex-specific heterochromatin in a moth. Exp Cell Res 113:85–94.

Tagaki N, Itoh M, Sasaki M (1972): Chromosome studies in four species of Ratitae (Aves). Chromosoma 36:281–291.

Tagaki N, Sasaki M (1974): A phylogenetic study of bird karyotypes. Chromosoma 46:91–120.

Tykocinski ML, Max EE (1984): G-C dinucleotide clusters in MHC genes and in 5′ demethylated genes. Nucleic Acids Res 12:4385–4396.

Wang AH-J, Quigley GJ, Kolpak FJ, Crawford JL, van Boom JH, Rich A (1979): Molecular structure of a left-handed double helical DNA fragment at atomic resolution. Nature 282:680–686.

Wang AH-J, Quigley GJ, Kolpak FJ, van der Marel G, van Boom JH, Rich A (1981): Left-handed double helical DNA: variations in the backbone conformation. Science 211:171–176

The Origin and Evolution of Sex, pages 213–220

Allorecognition and Microbial Infection: Roles in the Evolution of Sex and Immunity

Virginia L. Scofield

Department of Microbiology and Immunology, School of Medicine, University of California, Los Angeles, Los Angeles, California 90024

INTRODUCTION

The major histocompatibility complex (MHC), the master system of vertebrate adaptive immunity, was discovered first in the mouse as the H-2 system, which is the locus of control for graft rejection. Products of MHC loci now are known, however, to be the primary participants in an elaborate surveillance system that functions for host defense against microbial pathogens [Snell, 1968; Klein, 1983]. MHC genes also appear to operate in kin recognition by olfactory cues [Yamakazi et al., 1983], and related or linked loci also appear to be involved in differentiation and fertilization events [Brickell et al., 1983; Lyon, 1984]. This gene cluster seems, therefore, to be the "identity card" of the vertebrate organism [Dausset, 1981].

Studies with the colonial tunicate *Botryllus* have uncovered a surprisingly vertebrate-like system for tissue-based kin recognition [Scofield et al., 1982b]. Colonies of *Botryllus* live crowded together and compete for limited space. Cell-cell recognitions determine whether contacted colonies fuse with or reject one another; likewise, they govern recognition between gametes and, perhaps, sibling co-settlement near the mother colony [Oka, 1970; Grosberg, personal communication]. These discriminations are controlled by a single Mendelian locus at which a large number of alleles segregate in natural populations. This polymorphism is maintained by gametic self-incompatibility systems linked or identical to the fusibility genes [Oka, 1970; Scofield et al., 1982b]. In their control of cell-mediated transplantation rejections, kin recognition functions and linkage to genes affecting fertilization, *Botryllus* fusibility genes bear a strong family resemblance to loci of the MHC [Scofield et al., 1982a]. Because nucleotide sequences hybridizing to cloned MHC-immunoglobulin (MHC-Ig) superfamily genes are present and expressed in *Botryllus* blood cells [Danska, Weissman, and McDevitt, unpublished], it seems likely that gametic and somatic self-recognition genes in protochordates represent ancient functions of primitive MHC genes. In this light, it is most interesting to note that MHC genes in mammals also elicit graft rejections, and that this function is tightly linked to their primary adaptive function in immune defense against microbial infection. This correspondence between two apparently disparate recognition systems (which employ identical molecules) may hold the central clue to the history of individuality and sexual reproduction in metazoan organisms. The multilevel control of kin recognition by both the *Botryllus* and vertebrate histocompatibility systems adds to the available DNA hybridization evidence arguing for their homology.

NATURAL TRANSPLANTATIONS, MICROBES, AND IMMUNITY

Genetics of Transplantation

Botryllus colonies are common organisms in bays and yacht harbors worldwide. Each colony is a clone of individuals derived by asexual budding from a founder, or *oozooid*. Different colonies grow crowded together under conditions of fierce space competition. When the growing edges of a colony approach one another from opposite directions (the result of growth around spherical surfaces), contact is followed by rapid fusion of their blood vessels and establishment of circulation between them. When the contacted surfaces are from different, unrelated colonies, the same reciprocal transplantation leads to rejection and reseparation. This behavior is controlled by a gene system that can be studied in the laboratory [Scofield et al., 1982b].

When two *Botryllus* colonies are crossed sexually, four types of progeny result (Fig. 1). If the parents are AB and CD, for example, the offspring will be AC, AD, BC and BD as determined by fusion experiments between them (Fig. 1) that show a) that wild colonies are heterozygous at one locus for fusion, and b) that two colonies sharing one allele can fuse [Oka and Watanabe, 1957]. There are 40–100 alleles for fusibility in most *Botryllus* populations [Scofield et al., 1982b]. This degree of polymorphism is seen elsewhere only in the major transplantation loci of the MHC, the related immunoglobulin genes, and the self-incompatibility systems of plants [Oka, 1970; Burnet, 1971].

Gamete Incompatibility and Tissue Recognition

Tunicates are hermaphroditic, and many species are self-sterile [Lillie, 1919]. In *Botryllus,* self-sterility appears to be governed by fusibility genes or by loci closely linked to them [Oka, 1970; Scofield et al., 1982b]. Crosses

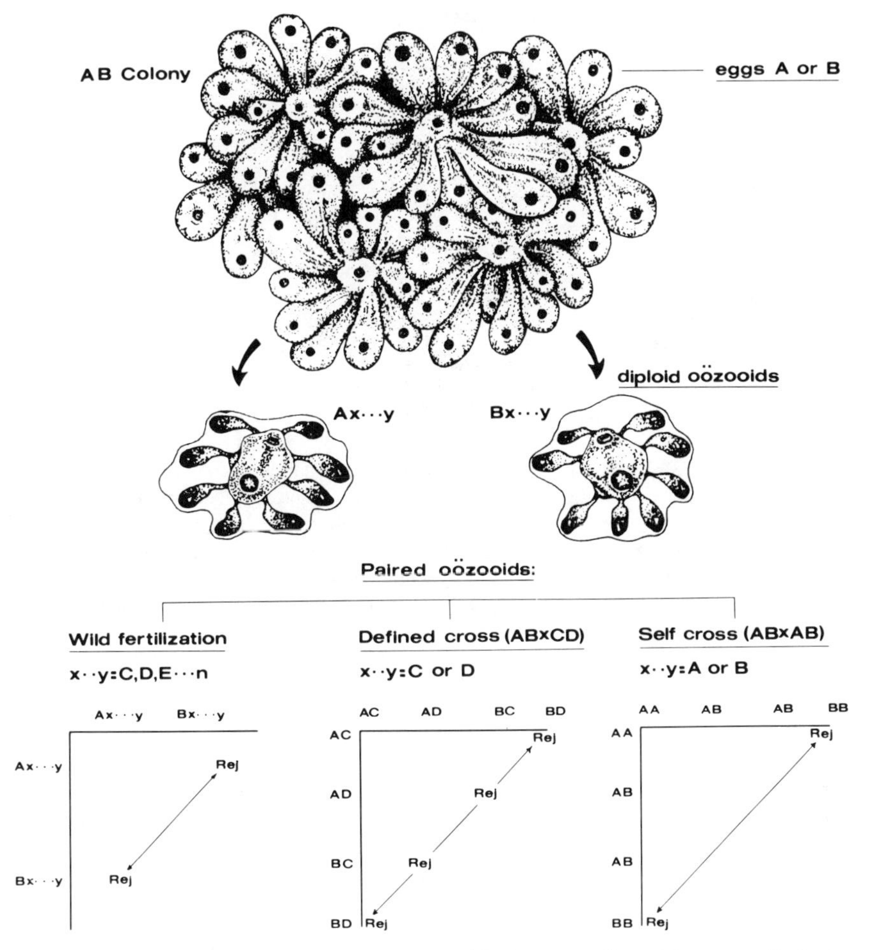

Fig. 1. Genetics of fusibility in *Botryllus* by oozooid microassay. Oozooid progeny of colonies fertilized in the natural environment (left panel) or of defined or self-crosses done in the laboratory (center and right panels) are paired and scored two days later for fusion and rejection. Expected percentages of fused and rejected pairs are predicted according to a one-locus, multiple-allele model in which two oozooids sharing one allele can fuse. Observed frequencies were close to those expected for each type of cross. See text for details.

between sibling colonies sharing one fusion allele show a segregation distortion in their progeny that reflects an inability of gametes carrying shared alleles to fertilize (Fig. 2). Removal of the egg envelope with proteases abolishes this specificity [Oka, 1970]; as in other ascidians, therefore, the self-incompatibility recognition elements are in this diploid, maternally de-

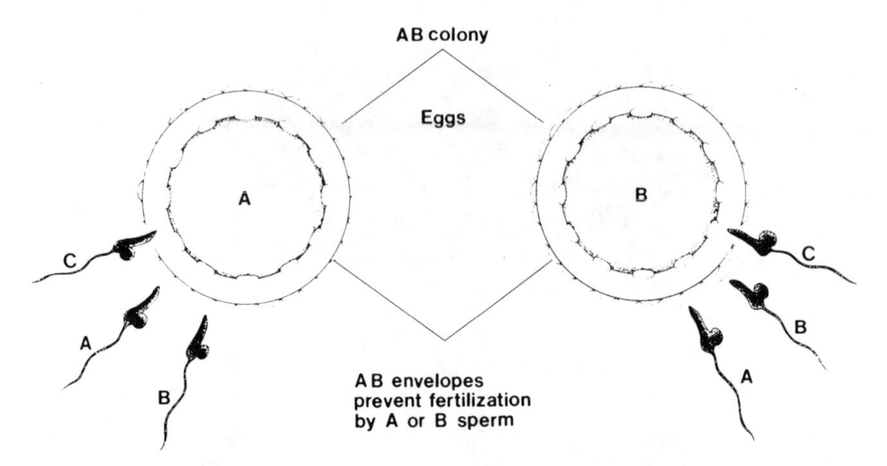

Fig. 2. Genetic control of fertilization in *Botryllus* parallels that for colony fusion. Eggs from an AB colony cannot be fertilized by A or B sperm but can be fertilized by sperm carrying other fusibility alleles (C, above). Removal of the diploid, maternally derived egg envelopes abolishes this pattern and allows fertilization by sperm of any type. It appears that a haploid fusibility pattern on the sperm is recognized by a diploid one on the egg envelopes. Failure of reaction between gametes sharing alleles, and activation (acrosome reaction leading to fertilization) between allogeneic ones may reflect similar recognition events to those occurring between mixed blood cells. Syngeneic and semiallogeneic blood cells are indifferent to one another, whereas allogeneic ones bind and kill each other. For details, see text.

rived structure [Rosati and DeSantis, 1978]. Differential binding and activation of non-self sperm may be by soluble histocompatibility elements in the egg chorion like those present in the tunic [Scofield, in press; Koyama and Watanabe, 1982]. The apparently reciprocal recognitions between allogeneic gametes may involve the same mechanisms as the binding and cell activation leading to lytic reactions between allogeneic blood cells in vitro [Tsuchida and Scofield, in preparation] and to rejections *in vivo*.

Initial reactions between fully allogeneic or semiallogeneic *Botryllus* colonies are all-or-none; transplantation leads either to rejection or fusion within hours of first contact between healthy colonies. Within 1–2 weeks after semiallogeneic fusions, however (e.g., AB to BC, Fig. 3), the blood cells of one or the other partner seem to disappear from the common circulation [Taneda and Watanabe, 1983]. If the fused partners are oozooids, one of them may disappear altogether (Fig. 3) in a reaction that, for unknown reasons, spares certain epithelial tissues [Scofield, unpublished results]. Similiar reactions have been reported to occur between fused semiallogeneic colonies of *Botryllus scalaris* in Japan [Saito and Watanabe, 1982]. These semiallogeneic reactions are similar in their tissue specificity to the resorption of old zooids that occurs at the end of each blastogenic (budding) cycle in all normal

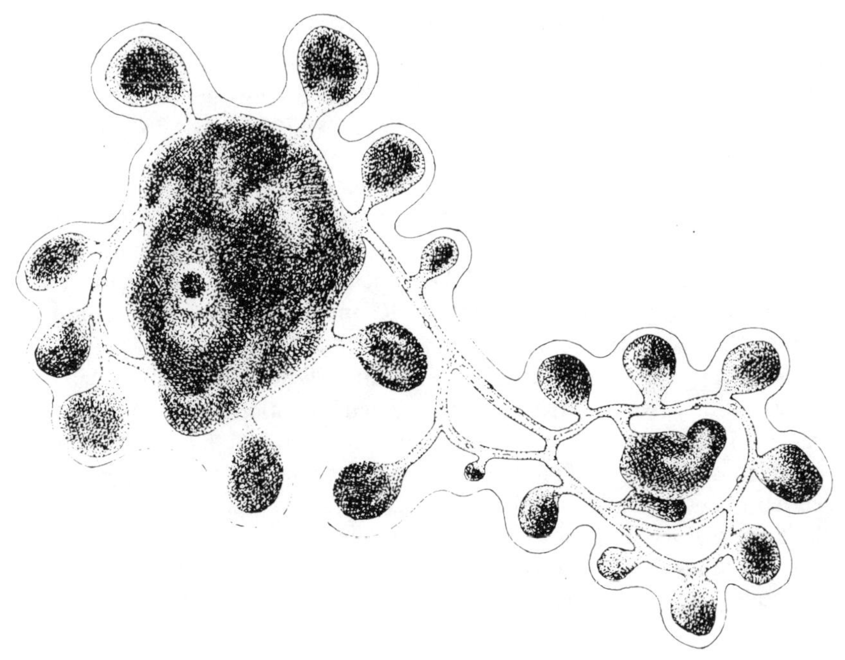

Fig. 3. Illustration of resorption of one partner in a pair of fused semiallogeneic oozooids. One oozooid (to the right), together with its bud, shrinks and eventually disappears. The ampullae and blood vessels usually remain intact. See text for details.

Botryllus colonies. This suggests that this histocompatibility system, like the mammalian one, is also involved in developmental events. In the *Botryllus* case, the programmed senescence of old zooids appears to involve blood cell-mediated destruction of self cells [Harp et al., 1985].

Because of sibling cosettlement, *Botryllus* colonies tend to grow near colonies with which they can fuse. Despite the large number of alleles usually present in North American populations, therefore, about 20% of contacted colony pairs in the Monterey, CA population are fused [Scofield et al., 1982b]. The capacity to fuse (at least initially) with siblings may persist because of the significant protective effects of such fusions; even with eventual resorption of one partner, each member of such a pair could be said to benefit because of shared genes between them. In support of this idea, we find that the "winners" of such interactions are larger and healthier than are single oozooids of the same age [Tsuchida and Scofield, unpublished].

The MHC: Parallels with Protochordate Allorecognition

Vertebrate members of the phylum *Chordata* universally possess an adaptive immune system operated by the related MHC and immunoglobulin (Ig) gene families [Williams and Gagnon, 1982]. The Ig system appears to have evolved from ancestral MHC-like genes whose products, like MHC glycoproteins in contemporary species, were specialized for binding to microbes. This binding, which may depend upon lectin-ligand interactions [Stewart et al., 1982], occurs on the surfaces of *antigen-presenting* cells, which "process" microbial components and display them, in association with MHC elements, to recognizing T and B lymphocytes. In this context, MHC glycoproteins appear to be seen by the T and B cell antigen receptors as "altered-self"; every antigen recognition, therefore, also involves an allorecognition-like component [Kaye and Janeway, 1984]. Bacteria, viruses, and multicellular parasites frequently mimic (or adopt) MHC antigens to escape host immune attention [Simpson et al., 1983; Mann et al., 1983; Hirata et al., 1973], likewise, every vertebrate can reject allografts, showing a universal capacity for recognition of small MHC differences and alterations. The dual involvement of MHC genes in allorecognition and antimicrobial defense reflects the simultaneous presence of allogeneic challenge and microbial infection as selective forces operating on them throughout their evolution. It seems possible, then, that the individuality-maintaining functions of MHC genes evolved precisely because variation *per se* at these loci (possession of rare alleles by individuals) confounds these coevolving antagonists [Bodmer, 1972].

CONCLUSIONS

Variability, MHC Genes, and the Evolution of Sex

For any individual of any species, it can be argued that the most dangerous (and most rapidly changing) component of any habitat is its complement of species-specific microbial pathogens. These usually have a short generation time and can adapt quickly to the unique biochemistries of eukaryotic hosts. If the effectiveness of associative recognition of any given microbe varies with different allotypic MHC gene products, extreme polymorphisms for MHC loci may benefit individuals, because microbes would tend to be adapted to the class I and II molecules of the more common haplotypes. It has been argued that possession of rare alleles at MHC loci carries an adaptive advantage for this reason [Bodmer, 1972; Travers et al., 1984].

In the colonial tunicates, there is linkage or identity between gene regions that determine gamete identity and govern individual tissue integrity. This functional linkage between fertilization and histocompatibility is reminiscent of the linkage between the mouse H-2 genes and the neighboring t-chromatin, whose contained loci affect sperm structure and fertilization [Lyon, 1984]. In mammals, the same cell surface glycoproteins that mark individual identity are specialized for binding to pathogens. It seems possible from these relationships that infectious challenge may have provided the first selective forces that led to sexual reproduction and the amazing variability (and individuality) it allows. This model has been put forward, in a slightly different form, as the "Red Queen" hypothesis [Van Valen, 1973].

While colony specificity systems are present in most colonial tunicate species [Koyama and Watanabe, 1982], they vary widely in the numbers of alleles involved and the manner in which they are expressed. Likewise, gametic self-incompatibility systems are not universally present within the same genus (or species) [Lillie, 1919]. This kind of variation suggests that genes controlling these discriminations are preadapted for rapid change, as are yeast mating-type genes [Klar et al., 1981]. As in the yeast systems, MHC gene polymorphisms appear to result in part from *gene conversion* events [Weiss et al., 1983]. By generating new, functional variants of proteins centrally involved in host defense, gene conversion could augment sexual reproduction and conventional recombination to speed up evolution at these critically important loci. In this role, gene conversion has reached its highest expression in the immunoglobulin genes [Loh and Baltimore, 1984], where the somatic diversification machinery for antibody production allows vertebrates to respond to invading disease organisms by a kind of somatic evolution and selection [Klein, 1983]. The MHC and immunoglobulin genes together illustrate eloquently the parallel evolutionary relationships between immunity, infection, recombination and reproduction.

The genetic structures of the vertebrate and protochordate histocompatibility systems show that tissue recognition and sexual reproduction are functionally and evolutionarily interlinked. As host-parasite relationships have always been central actors in MHC evolution, so also might they have been instrumental in the evolution of individuality and its maintenance by sexual reproduction.

REFERENCES

Bodmer WF (1972): Evolutionary significance of the HC-A system. Nature 237:139–145.

Brickell PM, Latchman DS, Murphy D, Willison PW, Rigby PWJ (1983): Activation of a Qa/Tla class I major histocompatibility antigen gene is a general feature of oncogenesis in the mouse. Nature 366:756–760.

Burnet FM (1971): "Self-recognition" in colonial marine forms and flowering plants in relation to the evolution of immunity. Nature 232:230–236.

Dausset J (1981): The major histocompatibility complex in man: past, present and future concepts. Science 213:1469–1474.

Harp JA, Weissman IL, Tsuchida CB, Scofield VL (1985): Programmed senescence of old individuals in colonial protochordates involves autoreactive blood cells. Submitted for publication.

Hirata AA, McIntyre FC, Terasaki PI (1973): Crossreaction between human transplantation antigens and bacterial lipopolysaccharide. Transplantation 15:441–449.

Kaye T, Janeway CA (1984): The Fab fragment of a directly activating monoclonal antibody

that precipitates a disulfide-linked heterodimer from a helper T cell clone blocks activation by either allogeneic Ia or antigen and self-Ia. J Exp Med 159:1397–1412.
Klar AJS, Strathern JN, Hicks JB (1981): Regulation of transcription in expressed and non-expressed mating type cassettes of yeast. Nature 289:239–241.
Klein J (1983): "Immunology: The Science of Self-Nonself Discrimination." New York: John Wiley.
Koyama H, Watanabe H (1982): Colony specificity in the ascidian *Perophora sagamiensis*. Biol Bull 162:171–183.
Lillie FR (1919): "Problems of Fertilization." Chicago: University of Chicago Press.
Loh DY, Baltimore DB (1984): Sexual preference of apparent gene conversion events in MHC genes of mice. Nature 309:639–640.
Lyon MF (1984): Transmission ratio distortion in mouse t-haplotypes is due to multiple distorter genes acting on a responder locus. Cell 37:621–628.
Mann DL, Popovic M, Sarin P, Murray C, Reitz MS, Strong DM, Haynes BF, Gallo RC, Blattner RA (1983): Cell lines producing human T cell lymphoma virus show altered HLA expression. Nature 304:58–60.
Oka H, Watanabe H (1957): Colony-specificity in compound ascidians as tested by fusion experiments. Proc Jap Acad 33:657–659.
Oka H (1970): Colony specificity in compound ascidians: the genetic control of fusibility. In Yukawa H (ed): "Profiles of Japanese Scientists." Tokyo: Kodansha.
Rosati F, DeSantis R (1978): Studies of fertilization in ascidians I: self-sterility and specific recognition between gametes of *Ciona intestinalis*. Exp Cell Res 112:111–119.
Saito Y, Watanabe H (1982): Colony specificity in the compound ascidian *Botryllus scalaris*. Proc Japan Acad 58 (Series B): 105–108.
Scofield VL (1985): Control of fusibility in colonial tunicates. In Cell Receptors and Cell Communication in Invertebrates. Greenberg A, Cinader B, eds. Elsevier, in press.
Scofield VL, Schlumpberger JM, Weissman IL (1982a): Colony specificity in the colonial tunicate *Botryllus* and the origins of vertebrate immunity. Am Zool 22:411–423.
Scofield VL, Schlumpberger JM, West LA, Weissman IL (1982b): Protochordate allorecognition is controlled by a MHC-like gene complex. Nature 295:499–503.
Simpson AJG, Singer D, McCutchan TF, Sacks DL, Sher A (1983): Evidence that schistosome MHC antigens are not synthesized by the parasite but are acquired from the host as intact glycoproteins. J Immunol 131:962–976.
Snell GD (1968): The H-2 locus of the mouse: comparative genetics and polymorphism. Folia Biol (Praha) 14:335–358.
Stewart J, Glass EJ, Weir DM (1982): Macrophage binding of *Staphylococcus albus* is blocked by anti-I region antibody. Nature 298:852–854.
Taneda Y, Watanabe H (1983): Effects of X-irradiation on colony specificity in the compound ascidian *Botryllus primigenus* Oka. Dev Comp Immunol 6:665–673.
Travers PT, Blundell C, Sternberg MJE, Bodmer WF (1984): Structural and evolutionary analysis of HLA-D region products. Nature 310:235–240.
Van Valen L (1973): A new evolutionary law. Evolutionary Theory 1:1–30.
Weiss EH, Mellor A, Golden L, Fahrner K, Simpson E, Hurst J, Flavell R (1983): The structure of a mutant H-2 gene suggests that the generation of polymorphism in H-2 genes may occur by gene conversion-like events. Nature 306:792–795.
Williams AF, Gagnon J (1982): Neuronal cell Thy-1 glycoprotein: homology with immunoglobulin. Science 216:696–703.
Yamakazi K, Beauchamp GK, Wysocki CJ, Bard J, Thomas L, Boyse EA (1983): Recognition of H-2 types in relation to the blocking of pregnancy in mice. Science 221:186–188.

The Origin and Evolution of Sex, pages 221–256

The Origin and Early Evolution of Germ Cells as Illustrated by the Volvocales

Graham Bell

Biology Department, McGill University, 1205 Ave Docteur Penfield, Montreal, Quebec, Canada H3A 1B1

Few groups of organisms hold such a fascination for evolutionary biologists as the Volvocales. It is almost as if these algae were designed to exemplify the process of evolution. They do not merely exhibit a rich variety of sexual and asexual reproduction, homothallism and heterothallism, monoecy and dioecy; the same could be said of other groups. They also display, within a narrow taxonomic compass, extremes of somatic and sexual organization which normally could be found only in entirely unrelated groups: their somatic organization may be unicellular at one extreme, or multicellular with a macroscopic, functionally differentiated plant body at the other; their sexual reproduction may be isogametic, with no distinction of male and female, but it is in some forms oogametic, with a massive immotile ovum and a tiny motile sperm. These algae, therefore, offer an unparalleled opportunity to describe and interpret the evolution of fundamental features of biological organization and have been the subject of phylogenetic speculation from Haeckel's time down to the present day. This essay represents an attempt to improve on my predecessors in two ways. First, I shall replace the qualitative arguments of previous accounts with a quantitative treatment based on regression analysis. Second, I shall attempt to synthesize trends in multicellularity, sexuality, and gamete dimorphism by explaining them all as consequences of a single more fundamental trend: increase in size. In particular, I hope to show how sexual and asexual germ cells come to evolve as a consequence of increasing size.

ALLOMETRIC VARIATION

The sizes of different structures or the rates of different processes often vary together, and their relationship is conveniently expressed as their common dependence on body size. The study of allometry, brought into prominence by J.S. Huxley [1932] and J.T. Bonner [1965], recently has attracted an increasing amount of attention, culminating in the review by Peters [1983]. Two allometric rules which seem firmly established as empirical generalizations are important to the analysis below.

The first is that logarithmic transformation is very broadly applicable: it normalizes frequency distributions, linearizes regressions, and stabilizes the variance around the regression line. The reason for the success of the log transform is not known with certainty, though it may be due to the multiplicative, or autocalytic, nature of many biological processes. The success itself is beyond doubt, and almost all my analyses will be regressions of logarithmically transformed variates.

The second rule is a principle of similitude, identified by Peters, which relates metabolic rates to body size. Let K be the total production or movement of matter or energy in unit time by an organism of mass V; then K and V are generally related by a power function

$$K = \text{constant} \times V^b \qquad (\text{cal sec}^{-1})$$

in which the exponent b has a value of about 0.75. Thus we can write the linear equation

$$\log K = \text{constant} + 0.75 \log V$$

This has the immediate corollary that production per unit mass (i.e., a rate of increase, or ratio of production to biomass) generally decreases with body size, with an exponent of about -0.25:

$$\log K - \log V = \text{constant} - 0.25V \qquad (\text{cal g}^{-1} \text{ sec}^{-1})$$

It is not strictly necessary for my purposes that $b = 0.75$, but only that $b < 1$; nevertheless, I have found that 0.75 and -0.25 are good descriptions of data which span many orders of magnitude of body size V.

I have measured body size in units of cubic microns ($1\ \mu^3 = 10^{-18} m^3$) because, when expressed as a logarithm, this unit varies between 0 and 10 for small organisms. For unicells, cell volume V was measured by taking midpoints of the published ranges of length and breadth and assuming the cell to be spheroidal. For colonies, the volume is calculated as the sum of protoplast volumes and does not include intercellular material or the space inside hollow colonies, so that values for unicells and colonies are comparable. The raw data on which the analyses are based and the statistics of the regressions I shall refer to can be obtained from the author.

My principal source of information for allometric data has been von Huber-Pestalozzi [1961], with some use of Ettl's monograph on *Chlamydomonas* [Ettl, 1976]. For *Volvox,* however, I have used the review by Smith [1944] and the primary literature: Shaw [1916, 1919, 1922a, 1922b, 1922c], Pocock [1933a, 1933b], Rich and Pocock [1933], Iyengar [1933], Darden [1966], Kochert [1968], Starr [1970a, 1970b, 1971a, 1971b], vande Berge and Starr [1971] and Karn et al. [1974]. Reviews of sexual and asexual reproduction can be found in Smith (1944), Wiese [1976], Coleman [1979] and Kochert [1982].

The frequency distribution of body size of unicells and colonies within the order Volvocales is shown in Figure 1. For unicells, the modal value of V is about 3 (i.e., 1000 μ^3), but they span about four orders of magnitude in size, from very small cells of less than 1 (i.e., 10 μ^3) to large forms close to 5 (i.e., 100,000 μ^3). This enormous range characterizes not only the order as a whole (Fig. 1A) but even the single genus *Chlamydomonas* (Fig. 1B). Colonial forms are of course even larger, and have total chloroplast volumes extending from about 3 to nearly 7. The overall range of size within the group is thus about a millionfold, from unicells such as *Nephroselmis* and *Platychloris,* which are not much bigger than a large bacterium, up to the larger species of *Volvox,* easily visible to the naked eye.

WHY GET BIGGER?

The lower limit of size is presumably set by physical constraints on the space required by eukarote organelles such as nucleus and chloroplast. The advantages of being larger than this minimal size are not known, but a recurring and plausible speculation is that larger cells are less likely to be eaten by filter-feeding animals.

The smallest filter feeders are rotifers. *Brachionus rubens* feeds chiefly on particles of between 14 and 270 μ^3; the corresponding range for *B. calyciflorus* is 1750–4200 μ^3 [Pilarska, 1977]. In *B. plicatilis* the relationship between the maximum diameter of ingested particles (y) and the length of the lorica (x) is $y = 0.09x - 0.03$ [Hino and Hirano, 1980], giving a maximum volume of roughly 3000 μ^3 of particles ingested by a 200μ-long rotifer. The corresponding relationship for cladocerans was estimated to be $y = 0.022x + 4.9$ by Burns [1968]. This suggests that very large cladocerans 4mm long could ingest particles of up to about 400,000 μ^3, but the maximum values observed in feeding trials with algae are much smaller. Kryutchkova [1974] found that *Daphnia* took particles of up to about 8000

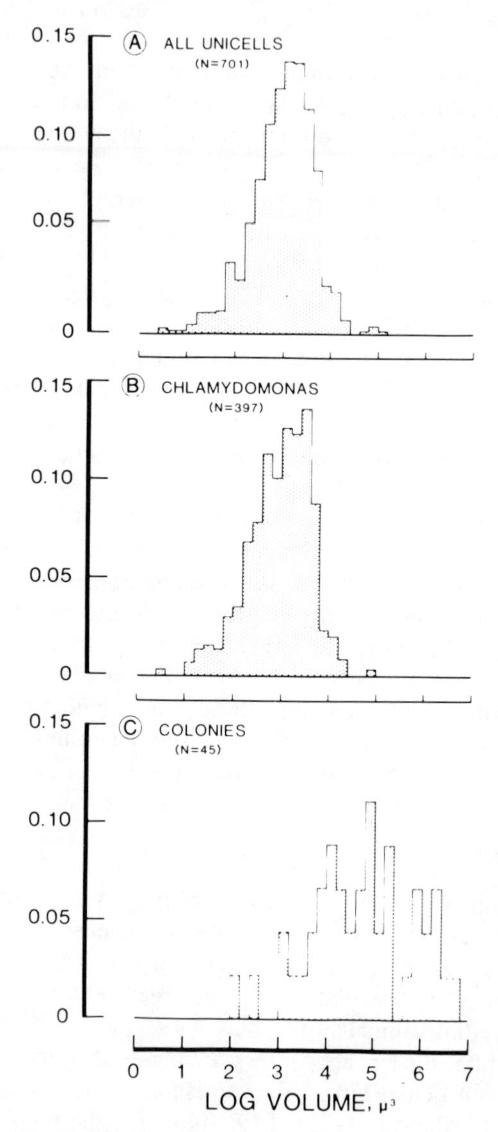

Fig. 1. Frequency distribution of size in Volvocales. A, all unicells; B, *Chlamydomonas* only; C, all colonies, Source: von Huber-Pestalozzi.

μ^3, with smaller cladocerans such as *Bosmina* ingesting particles of up to about 1700 μ^3. Peak filtration rates for *Daphnia* are between 1700 and 4200 μ^3 [Pilarska 1977]. Copepods are the largest common planktonic filter feeders but still prefer smaller particles: *Diaptomus oregonensis* has peak filtration rates for particles of 245–490 μ^3 [McQueen, 1970], while other species take particles of up to 8200 μ^3 [Kryutchkova, 1974].

Large cells are thus less likely to be eaten by filter feeders, and algae with volumes in excess of about 10,000 μ^3 are virtually immune from grazing. This is close to the upper limit of size for unicells, but most colonies will effectively escape predation, especially since their actual volumes are greatly inflated by the presence of large amounts of intercellular material and the presence of a large internal space.

It is possible, therefore, that large size evolves as a means of increasing survival rates in permanent aquatic habitats. Some comparative evidence which supports this conclusion is given below, but I do not pretend that the fundamental advantage of large size in planktonic algae has yet been demonstrated unequivocally. The fundamental disadvantage, however, can be identified with some confidence.

ALLOMETRY OF RATES OF INCREASE

The rate of increase under favourable conditions is a ratio of production to biomass, and—in accordance with the principle of similitude described above—it therefore declines as body size increases with an exponent of about -0.25. The most consistent estimate we have of this rate is the limiting rate of increase r_o, attained as population density approaches zero. The value of r_o obtained in the laboratory depends on the conditions of culture and will vary with temperature, light intensity, nutrient concentrations, bacterial contamination, and so forth. The maximal rate of increase (r_{max}, the value of r_o which cannot be exceeded) is impossible to measure, as we cannot exclude the possibility that there exists some combination of culture variables under which any given value of r_o will be exceeded. Nevertheless, we can define r_{max} operationally as the greatest value of r_o observed when strenuous attempts have been made by experienced observers to optimize the conditions of culture.

Blueweiss et al. [1978] show that r_{max}, defined in this way, declines at the expected rate as body size increases; translated into units of cubic microns, their result is that

$$\log r_{max} = 1.52 - 0.26 \log V$$

over nearly 20 orders of magnitude of V, from prokayotes to large vertebrates. Similar results have been obtained at the lower end of this scale, for small eukaryotes, by Fenchel [1974], Taylor and Shuter [1981], and Banse [1982]. However, the number and taxonomic variety of unicells included in these surveys is severely restricted (only 5 to 25 species, mostly ciliates), and I have collated data for about 150 species of unicells and small multicellular eukaryotes in an attempt to falsify the proposed rule.

The data for this analysis can be obtained from the author. Figure 2 is a scatter plot of the results, which are organized according to major taxonomic and structural groups in Figure 3. We can draw two major conclusions. The first is simply that for the majority of taxa the principle of similitude is confirmed and, indeed, greatly strengthened for organisms in the lower third

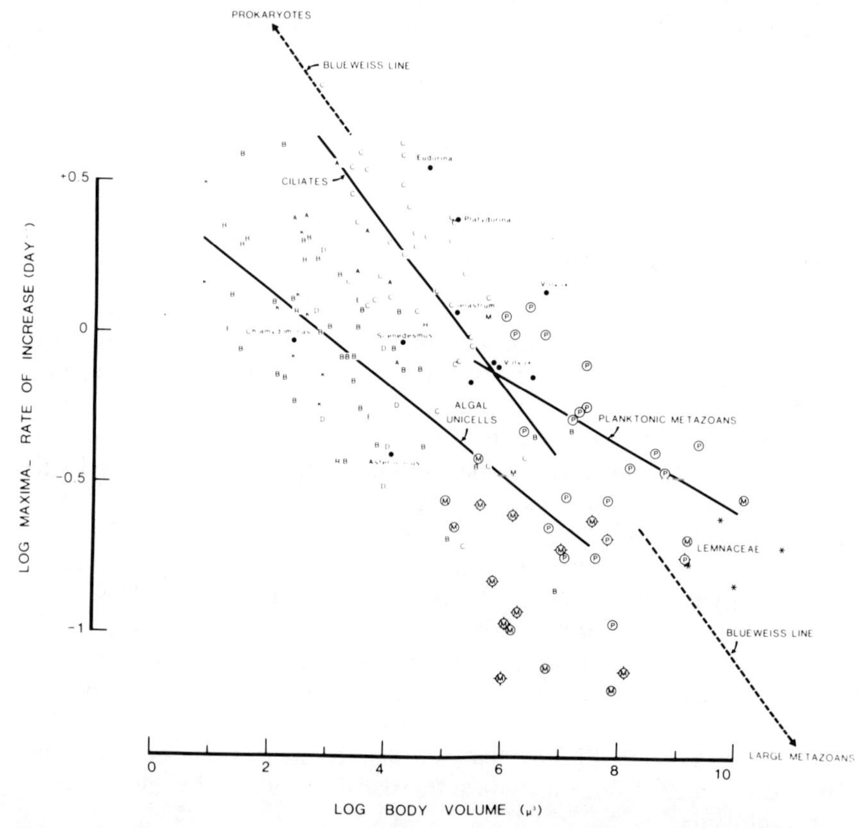

Fig. 2. Allometry of the rate of increase for unicells and small multicellular organisms. Taxa are identified as follows: A, amoebas; B, Bacillariophyceae (diatoms); C, Ciliata; D, Pyrrhophycophyta; E, Euglenophycophyta; H, Heliozoa; K, Chlorophycophyta; M, benthic metazoans; P, planktonic metazoans; R, Cryptophycophyta; Y, Myxophyceae. Colonial algae are identified by solid circles, and named, as is *Chlamydomonas;* metaphytes (Lemnaceae) are identified by asterisks. All metazoan values are distinguished by a circle surrounding the letter. Those organisms for which this circle is pecked are exclusively sexual, and to obtain values comparable with the other, asexual, organisms they should be multiplied by two (i.e., add 0.3 to the log value graphed here). The three solid lines are the least-squares regressions for planktonic metazoans (P), ciliates (C) and algal unicells (B, D, E, and K). The broken line is the regression obtained by Blueweiss et al. (1978), using a smaller number of species over a greater range of body size. Note that the data for ciliates, planktonic metazoans, colonial Volvocales and metaphytes is organized around the Blueweiss line, but that for algal unicells is not.

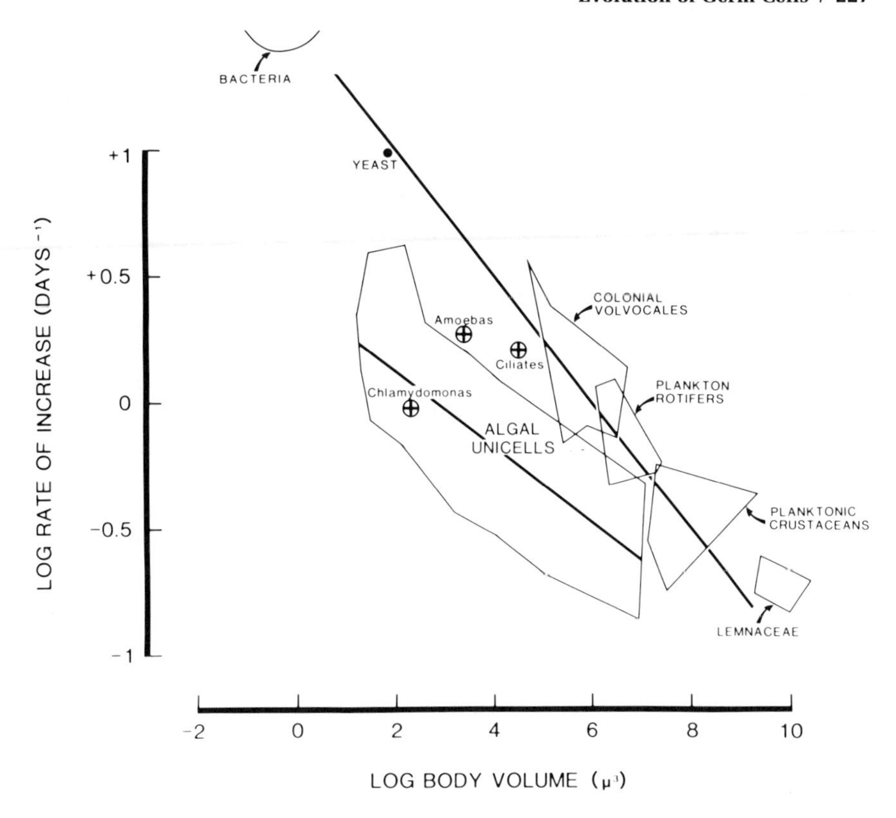

Fig. 3. A simpler version of Figure 2, designed to bring out the similarities and differences between taxa. The data appear to be organized around two regressions, shown here as thick lines. The upper is the Blueweiss line, extending from prokaryotes to large vertebrates, and providing a good prediction of the location of protozoans, colonial Volvocales, planktonic metazoans, and metaphytes. The thin lines are the envelopes of the data points shown in Figure 2; crosses within circles are bivariate means. Yeast and bacteria are roughly located for purposes of illustration. The lower line is the regression for algal unicells, clearly separate from the main Blueweiss regression. The bivariate mean for *Chlamydomonas* shown here (and in Fig. 2) refers to 30 species from the University of Texas collection, growing in soilwater medium at 22°C under constant illumination in my laboratory; I was unable to find adequate data in the literature.

of the range of body size; the points are clearly organized around a slope of -0.25 with an elevation close to that obtained by Blueweiss et al. [1978]. The second conclusion is original and much more surprising: algal unicells are an exception to the general rule, lying below the main regression line. In particular, the fairly large data sets available for diatoms and ciliates make it clear that the diatoms have a much lower capacity for increase at given size. For forms of average size, this difference approaches an order of magnitude. The data sets for other algal unicells are much smaller, but there is no indication whatsoever that they lie above the diatom regression. However, the colonial Volvocales do not share the low rates of increase of their unicellular relatives; they lie on the main regression. Moreover, the smallest metaphytes also lie along the main regression. Multicellular plants and ciliate protozoans thus evade the physiological penalty which is for some reason levied on unicellular autotrophs. How does this situation arise?

THE EFFECT OF DIVIDING LABOUR

In this section I shall suggest that the unexpectedly high rates of increase shown by colonial algae are made possible by the division of labour between somatic and germ cells. A host of previous authors has identified the division of labour as the crucial step in the evolution of multicellular organisms, but so far as I am aware the benefit obtained by dividing labour between cells has not been demonstrated qualitatively, nor expressed quantitatively, nor has its nature been elucidated.

I have attempted to demonstrate the effect of dividing labour between somatic and germinal function in Figure 2. The crucial comparisons are as follows.

a) Heterotrophic metazoans are described by the same regression line as autotrophic metaphytes and colonial Volvocales (i.e., case 6 = case 2). There is, therefore, no effect of the mode of nutrition itself on the rate of increase. All the forms included in this comparison are more or less planktonic, the metaphyte Lemnaceae being floating plants with aquatic roots; benthic metazoans almost invariably have lower rates of increase than planktonic forms (hydras and the nematode *Caenorhabditis* are exceptions), and there are other effects associated with sexual versus asexual and vegetative versus oviparous reproduction. As no benthic algae are involved in the comparisons, however, I have omitted any discussion of these secondary effects.

b) Colonial algae (and small metaphytes) have a greater capacity for increase than unicells of comparable volume (i.e., case 6 > case 5). This suggests the existence of a physiological advantage associated with the division of labour between somatic and germinal cells within a multicellular structure.

c) Ciliates have a greater capacity for increase than amoebas (including the heliozoan *Actinophrys*) of comparable volume (i.e., case 3 > case 1). This suggests the existence of a physiological advantage associated with the division of labour between somatic and germinal nuclei within a single cell.

A statistical evaluation of these inferences is given in the legend to Table 1. The technique used is multiple linear regression, using log r_{max} as the dependent variable, log V as a continuously distributed independent variable, and the mode of nutrition and the presence or absence of cellular and nuclear

TABLE I. Rate of Increase of Organisms as a Function of Mode of Nutrition and Organization

Case	Mode of nutrition	Nuclear div lab	Cellular div lab	Taxa	Position
1	0	0	0	amoebas	slightly below
2	0	0	1	metazoans	on line
3	0	1	0	ciliates	on line
4	0	1	1	—	—
5	1	0	0	algal unicells	below
6	1	0	1	colonial algae, metaphytes	on line
7	1	1	0	—	—
8	1	1	1	—	—

Coded variables are: mode of nutrition, 0 = heterotrophic, 1 = autotrophic; nuclear division of labour, 0 = absent, 1 = present; cellular division of labour, 0 = absent, 1 = present. Final column is rough indication of position of taxa on plot of rate of increase vs somatic volume, relative to the Blueweiss line (see Figs. 2 and 3). Multiple regression using log rate of increase as the dependent variable yields:

Variable	Coefficient	t	p
log somatic volume	−0.1745	10.56	<0.0001
mode of nutrition	−0.1189	1.86	0.065
nuclear div lab	+0.3326	4.35	<0.0001
cellular div lab	+0.3573	4.02	<0.0001

division of labour as coded independent variables. The effects of body size and cellular and nuclear division of labour are very highly significant. There is no effect of the mode of nutrition at P = 0.05, though the computed probability of 0.065 is sufficiently small to suggest that autotrophic organisms might have a somewhat lower capacity for increase if more data were available. I conclude that the division of labour between cells or between nuclei is associated with a greater capacity for increase when the effect of size is removed. The effect of a nuclear division of labour is intriguing, but because it does not occur in the Volvocales, I shall restrict my discussion to the effect of a cellular division of labour. The next section will attempt to identify the physiological basis of this effect.

THE ADVANTAGES OF DIVIDING LABOUR

According to Adam Smith [1776], the increase in productivity gained by dividing labour is due to the increased dexterity resulting from concentration on a single task, to saving the time that would otherwise be spent in passing from one task to another, and from the relative ease of inventing specialized machines. The last possibility may have an important evolutionary analogue, if by creating specialized castes of cells it is more likely that an improvement in the performance of one of these castes can be achieved by unit genetic change. However, I shall consider only short-term physiological advantages and attempt to show how the kinetics of resource acquisition and utilization lead to an increased dexterity in performing a narrow range of functions. My argument hinges on two propositions:

a) First, that the Michaelis-Menton equation is an adequate description of the way in which rates of reaction change with substrate concentration. This equation is used routinely for describing metabolic processes, and in particular those processes such as the uptake of phosphorus from solution, which are directly relevant to the problem.

b) Secondly, that there is negative feedback between the quantity of a resource accumulated within a cell and the rate of its further accumulation; in particular, that the rate of photosynthesis decreases with the quantity of photosynthate acquired and more generally that the rate of nutrient uptake decreases with internal nutrient concentration. For photosynthesis at least, this is still a controversial issue [see reviews by Neales and Incoll, 1968; Pinto, 1980]. If the proposition is correct, we can use the Michaelis-Menton equation to describe the way in which the rate of acquisition of a resource varies with the steepness of the concentration gradient between the external medium and the cytoplasm.

The Michaelis-Menton equation states that the rate of uptake of a resource R from some external source across a concentration gradient will be given by

$$R = C.R_{max}/(C + K_s)$$

where C is the external concentration, R_{max} is the limiting rate of uptake when the external concentration is very great, and K_s, the so-called half-saturation constant, is the value of external concentration at which $R = R_{max}/2$. This description suggests that there are two ways in which the rate of uptake might be increased: by steepening the concentration gradient (effectively, increasing the mean value of C) or by homeostasis (decreasing the variance of C).

The former possibility is intuitively transparent. If an organism cannot alter the external concentration of nutrients or the intensity of light, it must steepen the concentration gradient by reducing the internal concentration. It can do this by locking up the products of synthesis into compounds or structures which interfere relatively little with the reactions which contribute to their formation, as for instance starch is stored in pyrenoids within volvocacean cells. Even better, however, would be to get rid of these end products altogether. A multicellular organism could accomplish this by having some cells (somatic cells) specialized as sources and others (germ cells) specialized as sinks. An isolated, starved unicell will take up nutrients (or fix carbon) at a rate proportional to the concentration gradient, initially high but continually decreasing until a sufficiently high product:substrate ratio virtually abolishes

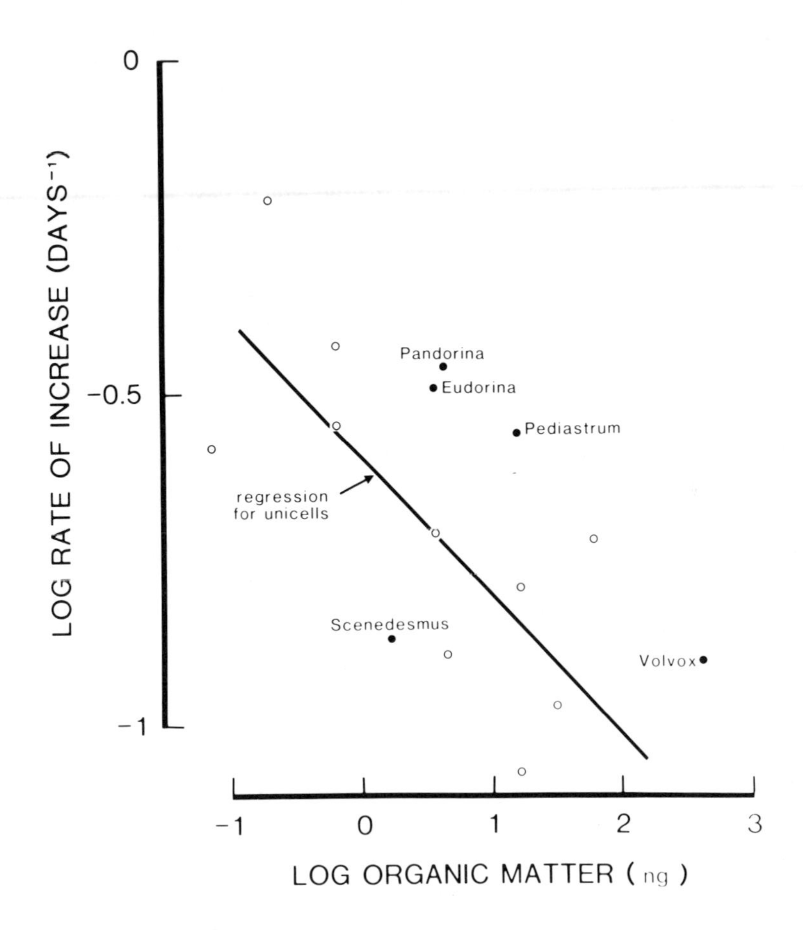

Fig. 4. Growth rates of various green algae in culture, from Moss [1973]. Hollow circles are unicells; solid circles with names are coenobia. *Scenedesmus* does not depart significantly from the regression line, which was calculated for unicells only, but the large coenobia are significantly above the line. Note that the x-variate is quantity of organic material (as reported by Moss), for which I could not find a satisfactory conversion to cell volume. Results of a multiple regression analysis in which a dummy variable "coloniality" was used to separate coenobia from unicells (0 = unicell, 1 = coenobium, *Scenedesmus* excluded) were as follows:

	independent variable:	
	log organic matter	coloniality
regression coefficient	−0.199	+0.253
± standard error	±0.046	±0.098
t(P)	4.33 (<0.001)	2.58 (0.026)

F for this regression was 9.95, P = 0.003.

the reaction. A similar unicell coupled to a catabolically inactive partner can pass the products of synthesis to its partner and thus maintain a high rate of synthesis by maintaining a steep concentration gradient. I shall call this the "Gradient-steepening" hypothesis.

The second possibility is a little more complicated. Suppose that we compare two isolated unicells cultured in media which have the same mean concentration, but one is constant whereas the other varies. In the constant medium, the rate of uptake is given by the standard Michaelis-Menton equation $R = C\,R_{max}/(C + K_s)$. In the variable medium, the concentration may be either $(C - x)$ or $(C + x)$ at any moment in time, with equal probability. The expected rate of uptake is then

$$E(R) = \frac{R_{max}}{2}\left[\frac{(C - x)}{C - x + K_s} + \frac{(C + x)}{C + x + K_s}\right].$$

It is easy to show that $R > E(R)$ for any $x > 0$; thus, the unicell cultured in the more stable environment has the greater rate of uptake on average. More generally, if the mean external concentration is C and its variance s_C^2, the expected rate of uptake is approximately

$$E(R) = \frac{\overline{C}\,R_{max}}{\overline{C} + K_s}\left[1 - s_C^2\left(\frac{R_{max}/\overline{C} - 1}{\overline{C} + K_s}\right)\right],$$

which is a monotonically decreasing function of variance s_C^2. Any capacity for homeostasis will therefore raise expected rates of uptake or synthesis. (I suspect, though I have not proven, that this is true for any uptake function with negative second and third derivatives.) Now, consider any multicellular organism whose cells are not completely isolated from one another. Each somatic cell will act simultaneously as source and sink to its neighbours, as substances will tend to pass from regions of high to regions of low concentration. The cells within a colony will therefore experience less variation in concentration gradients and will exhibit a greater total rate of uptake than a comparable number of isolated unicells. I shall call this the "Homeostasis" hypothesis.

EVIDENCE FOR A PHYSIOLOGICAL ADVANTAGE ASSOCIATED WITH DIVIDING LABOUR

Direct evidence bearing on these hypotheses is difficult to find, and the detailed micro-physiological measurements needed to test them probably have not been made. Later, I shall predict the results of some of these measurements and hope that this will stimulate interest in the physiology of colonial algae.

The first concern was to find independent confirmation of the greater capacity for increase of colonial forms, as the number of colonial species represented in Figure 1 is small and the rates of increase are poorly estimated. Moss [1973] collected 15 species of chlorophyte algae, including unicellular and colonial forms, and measured their rates of growth at different levels of temperature and nitrogen concentration. The highest growth rates he found for each species are not nearly maximal (perhaps because light intensity was relatively low), but they can be used to check the generality of the r_{max} regression. They are shown as a function of volume in Figure 5. Again, large colonies of *Eudorina* and *Volvox* lie significantly above the regression line for unicells, with the small coenobia of *Scenedesmus* being indistinguishable from unicells of the same size. It seems likely, therefore, that the higher growth rate of colonies is a reliable and repeatable observation.

The only detailed physiological measurements I have found which enable colonies to be compared directly with unicells concern uptake of phosphorus, an important process because phosphorus is thought to limit algal production in fresh water [Schindler, 1975]. How do we expect the performance of colonies and unicells to differ? The following argument is summarized in Figure 6. At very low external phosphate concentrations, cells will take up phosphorus from solution at some characteristic rate. As the external concentration is increased, the rate of uptake increases but does so more and more slowly; thus, the first derivative of the rate of uptake with respect to external concentration is maximal when the external concentration is very low. The limiting value (that approached as the external concentration approaches zero) is equal to R_{max}/K_s, and its logarithm is therefore equal to $\log R_{max} - \log K_s$. We have no reason to expect this to differ systematically between unicells and colonies, as at very low concentrations the absolute value of the rate of uptake will be very low and feedback inhibition negligible; so we can check the fundamental similarity of phosphorus uptake kinetics in unicells and colonies by predicting that both will fall along the same regression line when $\log R_{max} - \log K_s$ is plotted on $\log V$. This prediction is supported by observation (Fig. 6A): The limiting rate of uptake is strongly mass-scaled, and the same scaling rule describes both unicells and colonies. The difference that we do expect to observe is that colonies should be able to maintain higher rates of uptake at greater external concentrations, through possessing greater R_{max} or greater K_s or both. Thus, if we plot $\log R_{max} + \log K_s$ on $\log V$, we expect that the colonies will lie above the regression line describing the unicells. This prediction is supported also (Fig. 6B): $R_{max} \times K_s$ is only weakly mass-scaled, and the average value for colonies is significantly, and substantially, in excess of that for unicells. This seems to provide confirmation that the physiological benefits assumed to underlie the greater capacity for increase of colonial forms do exist and can be measured.

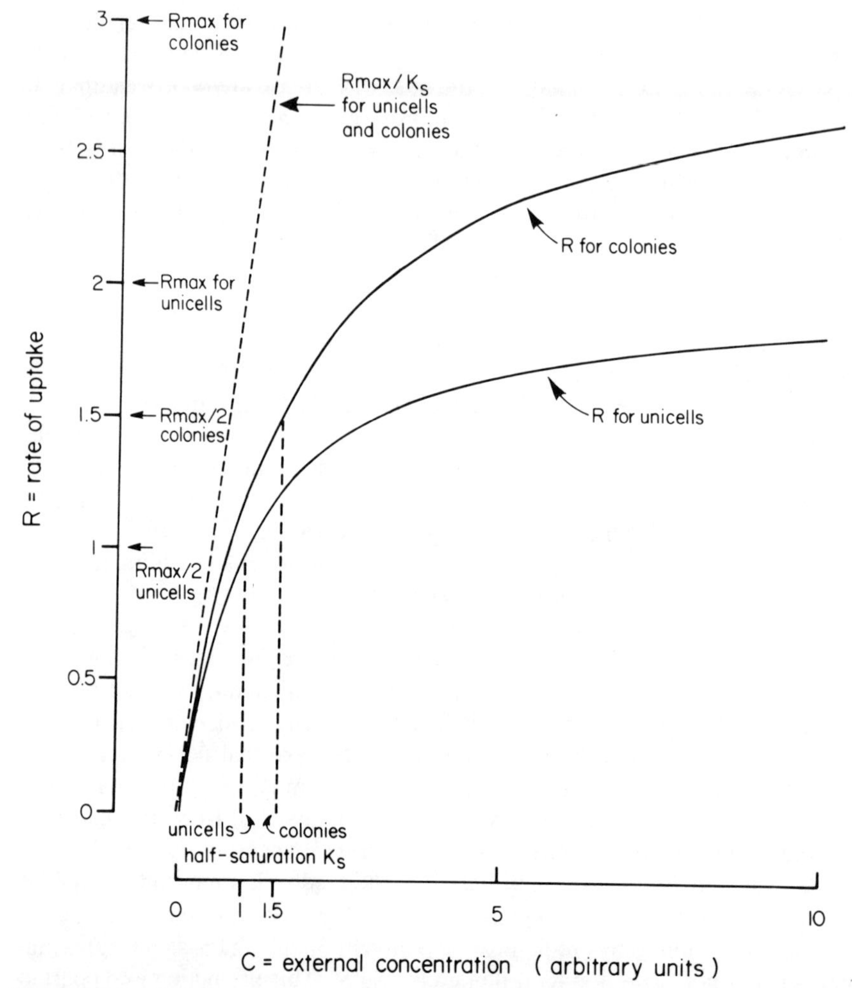

Fig. 5. The Michaelis-Menton equation used to describe the expected uptake kinetics of unicells and colonies. The two curves are example of the function $R = CR_{max}/(C + K_s)$, where R is the rate of uptake, C the external concentration, R_{max} the limiting rate of uptake and K_s the half-saturation constant (see text). Colonies and unicells are expected to perform equally well at very low external concentrations (ie R_{max}/K_s does not differ between colonies and unicells), but colonies should maintain a higher rate of uptake at higher external concentrations (ie $R_{max} \times K_s$ greater for colonies than for unicells).

The higher growth rate of colonies could be procured either by a shorter development time or by a greater production of offspring (or, of course, both) relative to unicells of the same size. Unfortunately, I have not been able to obtain adequate information on development times in the literature, and this possibility cannot be explored further. However, the allometric data provides

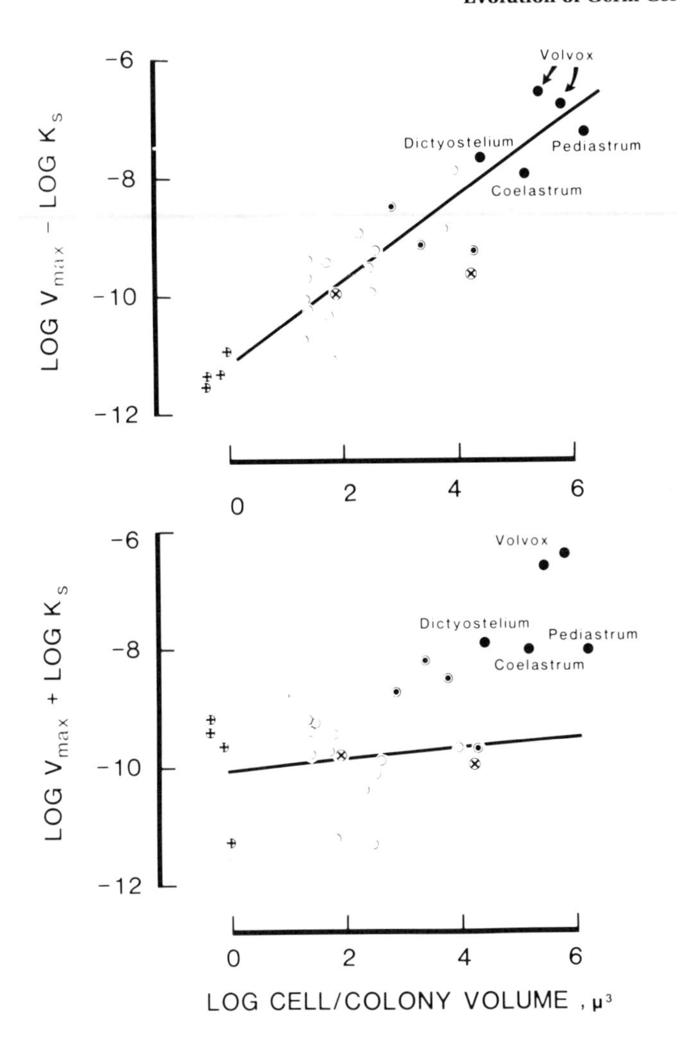

Fig. 6. Test of the predicted physiological difference between colonies and unicells, using phosphorus uptake data. The upper graph is the allometry of log (R_{max}/K_s); the regression equation is:

$$y = 11.16 + 0.70x \ (r^2 = 0.58, P = 0.0025).$$

---

extensive information on offspring size and shows that the total production of offspring by colonies does exceed that by unicells (Fig. 7). Moreover, the offspring of colonial forms are relatively smaller than those of unicells (Fig. 7). Quite apart from any difference in development rate, therefore, colonies produce a greater bulk of smaller offspring than do unicells of comparable

The lower graph is the allometry of log ($R_{max} \times K_s$):

$$y = 10.06 + 0.09x \ (r^2 = 0.01, P = 0.74).$$

The standardized residuals (in units of standard errors) for large coenobia are:

	dependent variable:	
	log (R_{max}/K_s)	log (R_{max}x K_s)
Coelastrum microsporum	−0.78	+1.94
Dictyosphaerium pulchellum	+0.84	+2.14
Pediastrum boryanum	−0.88	+1.82
Volvox aureus	+1.48	+3.66
Volvox globator	+0.51	+3.86
mean ± sd	+0.234 ± 1.033	+2.684 ± 0.991

These colonies therefore lie on the unicell regression for log (R_{max}/K_s) but above it for log ($R_{max} \times K_s$). The two sets of residuals are significantly different from one another ($t_8 = 3.83$, $P < 0.01$).

size, and their greater capacity for increase can be attributed, at least in part, to this.

COLONY ORGANIZATION IN VOLVOCALES

In this section I shall try to identify the crucial stages in the evolution of coloniality in the Volvocales and to account in detail for the success of the colonial habit.

Most of the Volvocales are active, photosynthetic, biflagellate unicells, typified by *Chlamydomonas*. Many species are sometimes or often palmelloid, having a variable number of cells embedded in a gelatinous matrix-like currants in a bun. This appears to be a resistant stage and is of no direct relevance to the evolution of multicellularity. More interesting are algae such as *Raciborskiella, Dangeardinella,* and the Spondylomoraceae, which usually comprise irregular clumps of variable numbers of motile cells. These forms are of great interest in representing the first step towards the evolution of multicellularity, confirming a tendency even among normally unicellular taxa toward a more or less permanent association of cells (e.g., *Chlamydomonas botryopara* and *C. pulvinata*). I do not know what significance can be attached to such loose and variable aggregations other than the mere consequence of being bigger than isolated cells and thus experiencing lower rates of removal by filter feeders.

The smallest regular colonies (coenobia) are sheets or solid globoids of 4–16 cells, most of which are in contact with one another (e.g., *Gonium, Platydorina*). There is no division of labour: Each cell functions first as a somatic cell and later as a sexual or asexual germ cell. Their regularity of

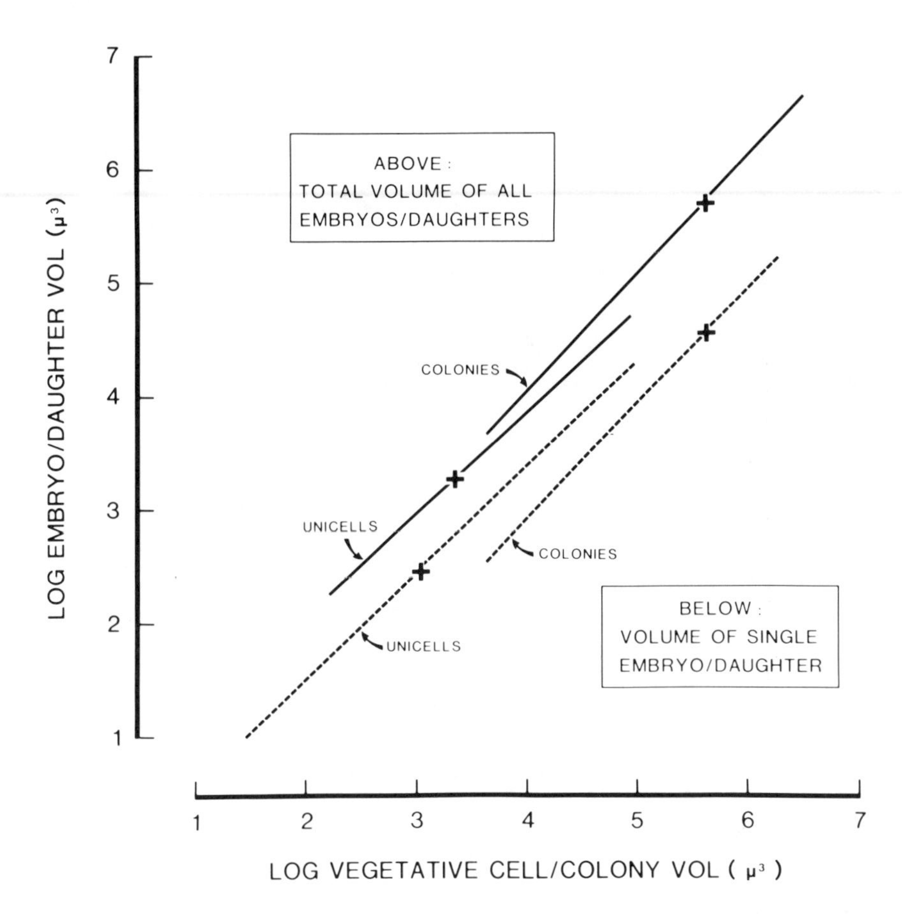

Fig. 7. Allometry of reproductive allocation in unicellular and colonial Volvocales. Broken lines are regressions of embryo protoplast volume (or daughter cell volume), solid lines of total volume of embryos or daughter cells. In this and subsequent figures the lines represent expected values of the y-variate over the observed range of the x-variate, the cross marking the bivariate mean.

form reduces the variance in size characteristic of irregular colonies and perhaps represents the outcome of stabilizing selection, in this case a balance between some force such as reduced predation favouring large size and the countervailing physiological disadvantage of a smaller surface area:volume ratio. It is in principle possible that some degree of homeostasis might be achieved by diffusion between neighbouring cells, but in motile colonies this seems unlikely, and small non-volvocacean coenobia such as *Scenedesmus* do not in fact have a greater rate of increase than unicells of the same size (Figs. 1 and 5).

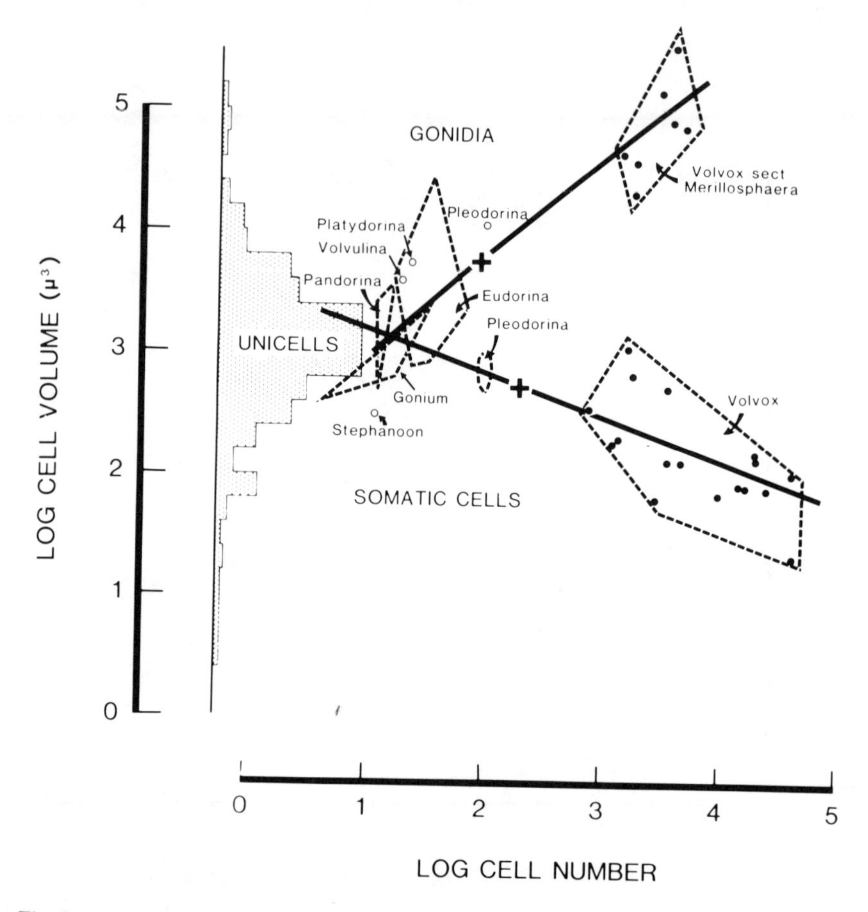

Fig. 8. Increasing specialization of somatic and reproductive cells in larger colonies, as shown by their divergence in size. Histogram at left is frequency distribution of cell volume in unicells. Solid lines are regressions for gonidia (upper) and somatic cells (lower). Broken lines are envelopes of data for taxa indicated. Hollow circles are data for single species of small colonies; solid points are data for *Volvox*.

In somwhat larger forms such as *Eudorina*, with 32–64 cells, an important innovation appears: the internalization of a space, the colony lumen. This increases the apparent size of the colony far beyond the combined volume of its constituent cells or protoplasts. The very general occurrence of hollow colonies therefore reinforces the notion that a great deal of importance attaches to mere increase in bulk, independently of any physiological correlate of increased size. A second consequence, however, is of much greater significance for the further development of the colonial habit. The colony lumen is an internal environment which can be regulated by the organism itself. Once a lumen has been created, the diffusion or active transport of substances into it from the somatic cells can be regulated by feedback, creating a homeostatic organ of enormous potential. Paradoxically, one of the most important structures of the volvocacean colony may be the space in the middle.

The first signs of a division of labour appears in hollow coenobia of 64 or more cells, with the anterior cells being somatic and the posterior somatic at first and reproductive later. This grade of organization is reached in *Eudorina illinoisensis* and *Pleodorina*. The appearance—and thereafter the universal occurrence—of a division of labour beyond a certain threshold minimum number of cells reflects a very basic economic principle. It was argued by Adam Smith, and accepted by all later writers, that the primary determinant of the appearance and degree of the division of labour is the extent of the market: specialization cannot pay in small communities. More precisely, it seems that the number of functionally distinct levels within an organization is a power function of the size of the organization with an exponent of less than unity, so that a plot of log division of labour on log organization size would be a straight line with a slope of less than unity. Quantitative accounts of the division of labour are very scarce, however, and the only statistical description I have found is that of the structure of U.S. unemployment offices by Blau [1974; 1977; e.g., Fig. 17.2 of Blau, 1974]. According to Wilson [1971], caste differentiation is generally a function of colony size in social hymenopterans and termites; Bonner [1965] plots the number of cell types as a function of body size in animals and plants. There is, then, a very general but so far poorly described tendency for the degree of division of labour to increase, at a decreasing rate, as the size of an organization increases. For any particular sort of organization, therefore, there is a threshold size at which it becomes profitable to divide labour: For colonies of volvocacean flagellates, this threshold occurs of about 64 cells, and the initial division is made between somatic and germinal function.

However, this threshold is crossed fully only by the rather larger coenobia (comprising about 10^3 cells) of *Volvox* section *Merillosphaera*. The crucial step taken by these organisms is the creation of a separate caste of reproductive cells and their displacement into the colony lumen. A caste of asexual germ cells, the gonidia, are distinguishable from sterile somatic cells early in colony development, being created by unequal cell division in the 16-cell embryo. Thereafter, the gonidia, which soon sink down below the level of the somatic cells and enter the lumen, increase markedly in size whereas the somatic cells remain more or less the same size. It is certain, therefore, that the production of the somatic cells must somehow be channelled to the developing gonidia. I suggest that this is accomplished by the transport of nutrients and photosynthate acquired by the somatic cells into the lumen, where it can be taken up by the gonidia. This realizes a twofold advantage. The somatic cells can use the lumen as a sink into which to discharge their products and thus maintain as steep as possible a concentration gradient with the external medium, while at the same time the gonidia are bathed in a nutrient medium of more or less constant composition and thus realize the benefits of homeostasis.

The gap between *Pleodorina* and *Volvox* is not unbridgeable; indeed, the isolation of developmental variants of species within both genera suggests that it might be possible eventually to connect the two levels of organization with a more or less continuous series of intermediate forms. Gerisch [1959] found that the somatic cells of *Pleodorina californica* occasionally divide to produce small spheroids which give rise to normal strains. Starr [1970] described a stable mutant of *Volvox carteri* in which the somatic cells are reproductive; on backcrossing to the normal type, strains with sterile and reproductive somatic cells segregated in roughly equal proportions. Vande Berg and Starr [1971] isolated a mutant of *Volvox powersii* in which only the anteriormost cells were sterile, all the rest being reproductive. Both *Volvox* mutants formed unusually small colonies of 256–512 cells, illustrating again the dependence of the division of labour on the size of the colony.

The relationship between the extent of the market and the degree of division of labour in Volvocales is shown in Figure 8. The very smallest colonies show no differentiation between somatic and germ cells, all of their cells being similar and close to the modal size of unicellular forms. As colony size increases, the somatic cells become smaller and the gonidia become larger, showing a tendency towards progressive specialization of the two cell types. Eventually, the difference between the two becomes so extreme that the somatic cells are smaller than all but the smallest unicells, whereas the gonidia are as large as the largest unicells. This picture holds for *Merillosphaera* but breaks down for other sections of *Volvox* which, although they have even larger colonies, actually have smaller gonidia (Fig. 9). What this represents, however, is a change in the mode of embryogenesis and a further advance in organization.

In *Merillosphaera* the gonidium enlarges to maximal size before cleavage; growth then ceases, so that successive cleavages produce twice as many cells, each of which is half as big. In the other sections of *Volvox*, the gonidium begins cleavage when still quite small, but the cells enlarge between cleavages. This may represent a considerable economy, if it is mechanically easier to cleave a small volume of cytoplasm. Some indication of the scale of this economy is given by Figure 10. In young embryos of *Merillosphaera* species, the difference between log somatic volume and log gonidial volume is about 2.3, i.e., the gonidia are some 200 times larger than the somatic cells at this stage. In adult colonies this difference increases to about 2.7, so that growth of the gonidia relative to that of the somatic cells is $2.7 - 2.3 = 0.4$, i.e., a factor of about 3. Data from *Volvox rousseletii* and *V. capensis* suggest that in the section *Euvolvox* the corresponding difference is a factor of about 300. The conversion of somatic production into embryo cytoplasm is thus much more efficient in *Euvolvox* than in *Merillosphaera*.

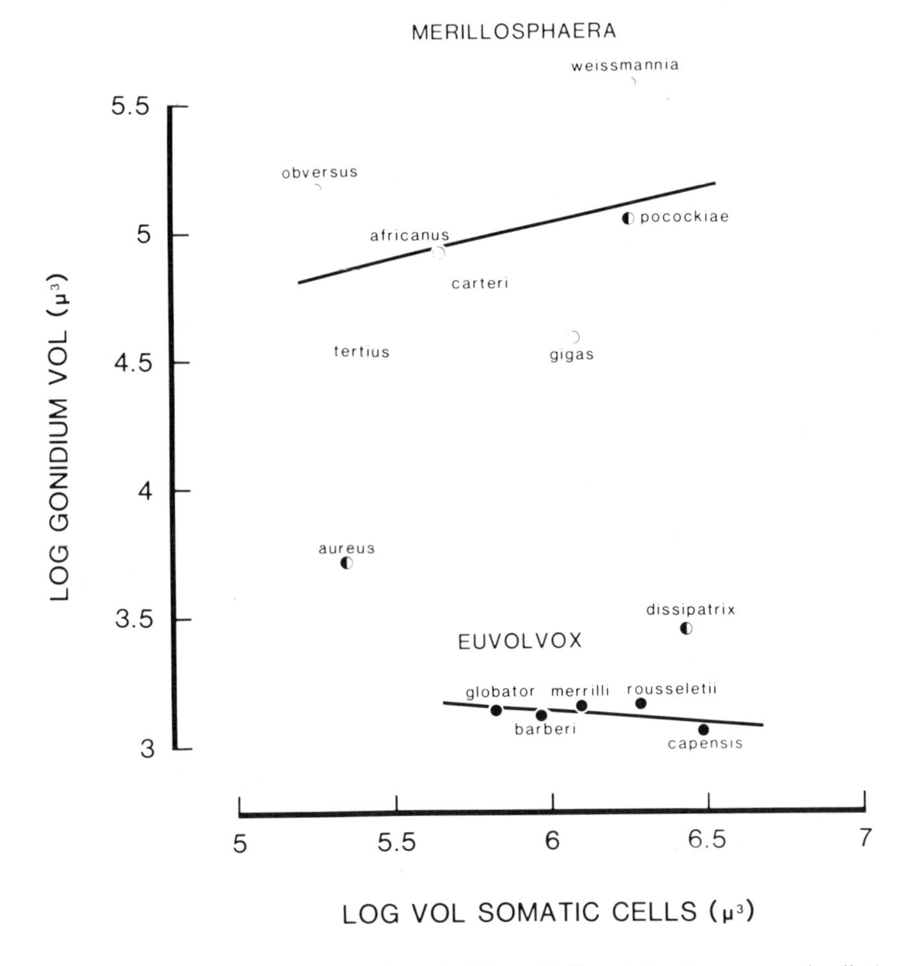

Fig. 9. Dimorphism of gonidium volume in *Volvox*. Hollow circles (upper regression line) are large gonidia for species in section *Merillosphaera*, lacking cytoplasmic connections. Solid circles are small gonidia of species in section *Euvolvox*, with broad cytoplasmic connections. Half-shaded circles are species in two other sections, which have gonidia of intermediate size and fine cytoplasmic connections.

The increased efficiency is made possible by virtue of another economic principle recognised by Adam Smith: in markets of given size, the degree of division of labour will depend upon the ease of communication. Clearly, more efficient communication with distant individuals in effect increases the extent of the market. In all species of *Volvox* (and in *Eudorina*), the cells of the young embryo are connected to one another by rather broad cytoplasmic strands, formed as the result of incomplete cell division. In *Merillosphaera*, these connections are lost during inversion of the daughter colony; in other

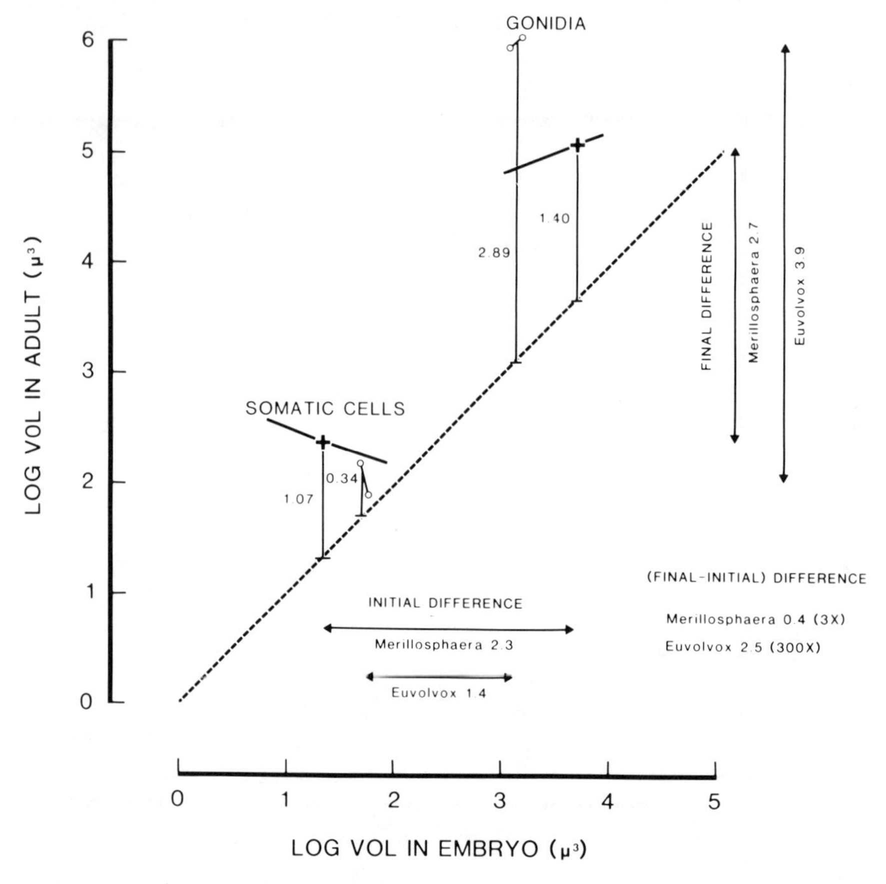

Fig. 10. Relative growth of gonidia and somatic cells in *Volvox*. Broken line is the line of equality (volume of cell type in embryo is equal to volume in the adult). Heavy solid lines are regressions for *Merillosphaera* species, with crosses marking the bivariate means. Hollow circles, connected by a line, indicate *V. rousseletii* and *V. capensis*, in the section *Euvolvox*. Because of their different mode of embryogenesis (see text), the points for "gonidial" volume are total protoplast volume of an embryo. By dropping perpendiculars from mean values to the line of equality, we can estimate the growth in size of both cell types in either section. This shows that the growth of the reproductive cell or tissue relative to that of the somatic cell is much greater in *Euvolvox* than in *Merillosphaera*.

sections of *Volvox,* however, they persist in the adult. In *Volvox aureus* they are initially broad enough to allow the passage of organelles such as mitochondria, but later become stretched by the growth of the intercellular matrix, narrow to 0.1–0.3 μ, and contain only endoplasmic reticulum and ribosomes [Bisalputra and Stein, 1966]. Somatic cells are attached to each of their neighbours by between one and three of these strands. In the section *Euvolvox,*

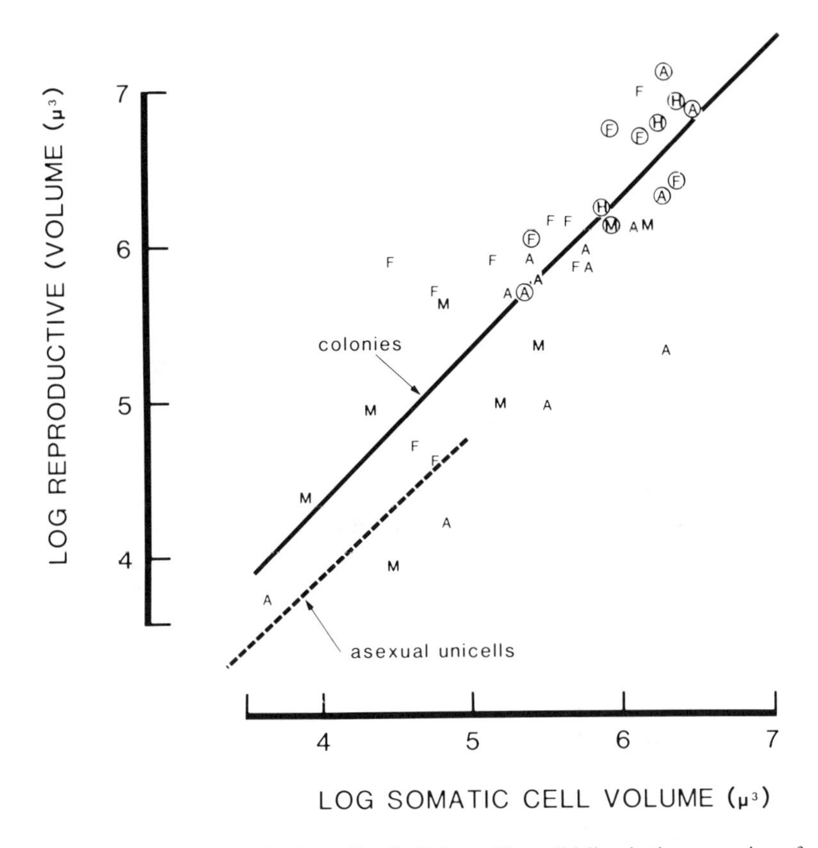

Fig. 11. Allometry of reproductive effort in *Volvox*. The solid line is the regression of all data for colonies: plotted points are shown as: A, total gonidial volume; F, total zygote volume; M, total sperm volume; H, sum of total zygote volume and total sperm volume for monoecious colonies. Points surrounded by a circle refer to species with cytoplasmic connections between cells. The regression of total daughter cell volume for unicells (broken line) is shown for comparison.

the cytoplasmic connections remain broad throughout the life of the colony, so that the somatic cells when seen from above are shaped like little starfish; in *V. rousseletii* the chloroplast extends into these strands. In *V. globator* the stands increase in thickness as the colony matures, and reach their greatest thickness when the gonidia begin to divide [Pocock, 1933a]. In advanced species of *Volvox,* therefore, the somatic cells are linked together as a tissue whose function is to provision the gonidia. The immediate effect of this linkage will be to even out any local inequalities in resource status and thus create a much greater capacity for homeostasis.

At the same time, the somatic cells are linked directly to the gonidia; indeed, the connections between somatic cells and gonidia are generally coarser and more numerous than those between neighbouring somatic cells. This cytoplasmic continuity between soma and germ is maintained throughout embryogenesis, until inversion. In *Euvolvox,* Pocock [1933b] has described the situation in some detail. Each gonidium in the embryo is surrounded by a ring of 9–12 somatic cells, to each of which it is connected by stout cytoplasmic strands. After birth the gonidia continue to enlarge, lose their flagella, and sink inwards, drawing out the connecting strands. As the gonidium sinks down, it draws with it the circle of somatic cells to which it is connected, forming a dimple on the surface of the colony. When the gonidium cleaves, all the cells of the developing embryo retain a direct cytoplasmic connection with the somatic cells of the mother colony up to the 16-cell stage, after which only the cells bordering the phialopore are directly connected to the maternal soma. The connection between mother and daughter is finally broken only when the daughter colony inverts, shortly before its release and the subsequent dissolution of the mother colony. These cytoplasmic connections linking mother to daughter must increase enormously the power of the growing embryo to act as a sink for resources acquired by the maternal somatic tissue, thus maintaining as steep as possible a concentration gradient between the colony and the external medium.

The scheme outlined above depends on the ability of the cytoplasmic strands of *Euvolvox* to translocate resources between neighbouring somatic cells and between somatic cells and gonidia. Although this has never been investigated, an extremely suggestive observation is reported by Pocock [1933b]. She describes cells in *V. rousseletii* which are similar to normal somatic cells but larger and richer in food reserves. These cells appear to lack cytoplasmic connections, and Pocock makes the very cogent suggestion that, having been cut off from their neighbours, they enlarge because of an accumulation of resources which can no longer be efficiently translocated. Pocock also remarks that although the somatic cells of *V. rousseletii* enlarge somewhat after birth, they cease to grow when the gonidia begin to divide and then actually decrease in size, especially in the posterior region of the colony, where the gonidia are situated.

I have been unable to show conclusively that the advance in colonial organization and made of embryogenesis displayed by *Euvolvox* is associated with a greater capacity for increase, as there is no adequate comparative data on development periods. Some preliminary but suggestive information on total reproductive output is given in Figure 11. The total volume of reproductive tissue—asexual embryos, zygotes, or sperms—is greater in colonies than in unicells of comparable size, as already established for purely asexual reproduction, and the data for *Euvolvox* lie somewhat above the general

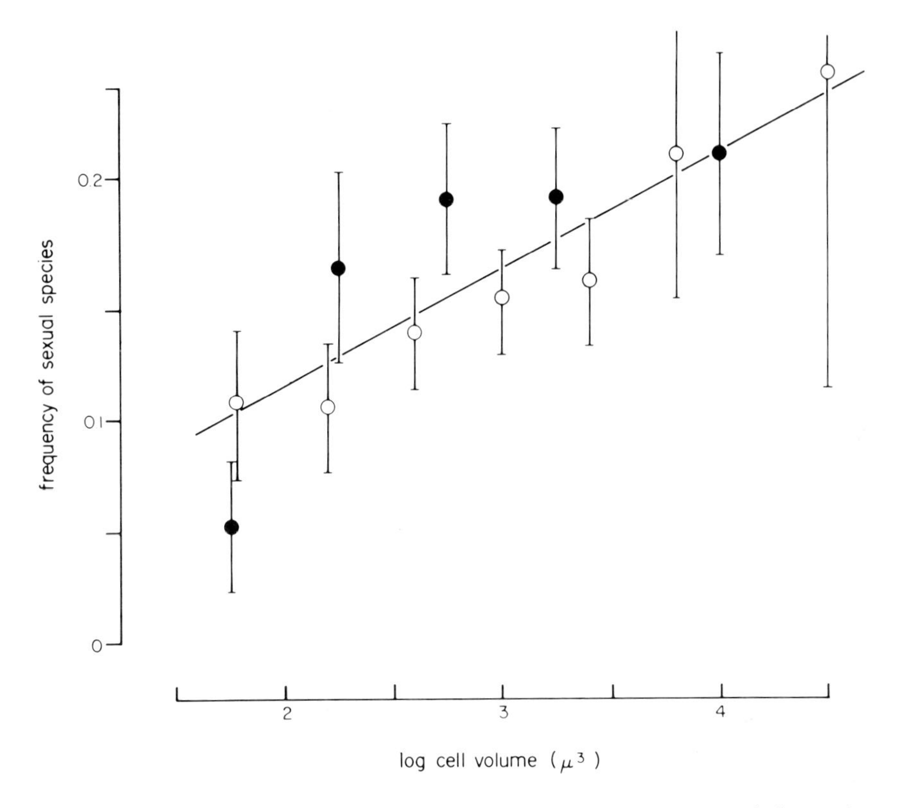

Fig. 12. Frequency of sex in relation to size in unicellular Volvocales. Open circles are data for all unicells (n = 845 species), from von Huber-Pestalozzi; regression of frequency of sexual species (at increments of 0.4 units of log cell volume) is

$$y = 0.0141 + 0.0484x \ (r^2 = 0.92, P < 0.001).$$

Closed circles are data for *Chlamydomonas* alone (n = 452 species), from Ettl, plotted at intervals of 0.5 units of log cell volume; regression is

$$y = -0.0105 + 0.0639x \ (r^2 = 0.69, P < 0.02)$$

Regression of combined data (shown on graph) is

$$y = 0.0103 + 0.0523x \ (r^2 = 0.75, P < 0.001)$$

but data sets may overlap, so independence of points is doubtful.

regression line for colonies. However, these data are at the limit of reliability, and a more careful laboratory investigation of this pattern would be welcome.

Finally, the ability of the gonidia to act as resource sinks in all sections of *Volvox* is probably further enhanced by their mode of development. In *V. carteri,* for example, they are formed by an unequal fission of each cell in the anteriormost two tiers of cells in the 16-cell embryo [Kochert, 1968; Starr, 1970], and further cell proliferation maintains them more or less equidistant from one another. In *V. weissmania,* the arrangement of the gonidia is "almost mathematically regular" according to Shaw [1922c]. The spacing of the gonidia as far as possible from one another will decrease competition between them and thus maximize the rate of uptake of the assemblage as a whole.

These speculations about the function of colony architecture could be tested by a variety of physiological measurements. Two experiments with *Merillosphaera* species are fairly straightforward: 1) if the fluid held in the colony lumen is extracted, it will be found to contain much greater concentrations of carbohydrate and nutrients than the external medium; 2) a pulse of labelled phosphate (or any other macronutrient) added to the culture medium will be detected first in somatic protoplasts, then in the lumen, and finally in the gonidia. Furthermore, pulsed label will pass more rapidly to the gonidia in *Euvolvox* species than in *Merillosphaera* species. If it is possible to administer a pulse of label to a small region of the colony surface, then in *Euvolvox* this label should subsequently become translocated to neighboring somatic cells, eventually draining into the gonidia even if administered as far from the gonidia as possible. In *Merillosphaera* there will be little translocation to neighboring somatic cells, the label instead moving into the lumen. Microsurgery, if feasible, would abolish this distinction; severance of the cytoplasmic connections in *Euvolvox* would prevent translocation of label to neighbouring regions and would cause the isolated group of cells to increase in size.

The creation of an internal space, the increase in the number of somatic cells, the improved communication between somatic cells and between somatic cells and gonidia, and perhaps also the spacing of the gonidia with respect to one another, are the crucial features which make possible the large capacity for increase of the larger coenobial Volvocales. Their joint effect is shown dramatically by *Euvolvox,* where although the cells of the embryo must double in volume between each cell division, they nevertheless cleave every hour or less. This is almost an order of magnitude faster than an isolated unicell of comparable size and structure. It is made possible by the emergence of the germ cell as a separate functional category and the consequent sterilization of the soma.

SEX AND SIZE

Whereas asexual reproduction by the mitotic formation of daughter cells or autocolonies is the prevalent mode of reproduction in the Volvocales, sexuality is widespread. It has been reviewed recently by Wiese [1976], Coleman [1979], and Kochert [1982], and the induction of sexual differentiation in *Volvox* by specific and highly potent pheromones is a favourite research topic [e.g., Kochert, 1968; Starr, 1969; 1970a; Starr and Jaenicke, 1974; Meredith and Starr, 1975; Pall, 1974]. I shall describe only the allometry of sex and in the next section its bearing on the evolution of morphologically distinct male and female cells.

In multicellular animals, obligate parthenogenesis is rare and tends to be confined to small organisms such as bdelloid rotifers and chaetonotoid gastrotrichs, which are comparable in size with the largest unicells. The correlates of asexual reproduction have been described extensively by Bell [1982], who mentions the tendency for asexuality to be associated with small size and interprets this by speculating that small organisms inhabit ephemeral habits in which competition for resources is minimal because high population density for prolonged periods is experienced rarely and which therefore puts a premium on a cheap method of rapid multiplication. This interpretation can be investigated quantitatively among the Volvocales. It is well known that sex is nearly universal among colonial forms, especially among the larger colonies, although it has not yet been described in many of the unicells. There is therefore a broad tendency for sex to be associated with large size, running parallel with the same tendency among animals. A much riskier prediction is that sex will continue to be associated with large size when the analysis is restricted to unicells. The argument runs as follows:

1) High rates of increase are favoured in ephemeral habitats; but

2) Small size is associated with high rates of (asexual) multiplication; therefore,

3) Unicells inhabiting ephemeral habitats will tend to be smaller than those in more permanent habitats; moreover,

4) Sex is favoured only in permanent habitats, where diversification functions to reduce competition between sibs; therefore,

5) Larger unicells are more likely to possess sexual reproduction. Propositions 2, 3, and 5 can be tested empirically by direct measurement. Proposition 2 is already known to be true, as r_{max} decreases with cell size in ciliates, amoebas, and various taxa of algal unicells (Fig. 1). Proposition 3 cannot be tested with the data to hand, and I have been unable to find reliable habitat descriptions for adequate numbers of volvocacean unicells. It is easily

possible, however, to distinguish species which have only been collected by rewetting dry soil samples from those which have been found also in the water column of ponds and lakes. On average, the former group should include a greater proportion of species living in highly ephemeral habitats which often dry out completely and should therefore be smaller in size. This one-tailed hypothesis can be tested by comparing the logarithms of cell volume in the two categories, yielding

a) log V = 3.115 ± 0.618 (N = 168) for aquatic samples

b) log V = 2.898 ± 0.546 (N = 34) for soil samples for 202 unicells catalogued by von Huber-Pestalozzi; these data have t = 1.90, for which the one-tailed P = 0.03. Moreover, there is a tendency for the unicells from soil samples to produce more daughter cells (log midpoint daughter cell number = 0.576 ± 0.194 versus 0.542 ± 0.149 for the aquatic samples) despite their smaller size, which suggests that they may have greater r_{max} independently of their smaller size; however, this effect is not formally significant (t = 1.48). Proposition 3 is therefore supported by these data, though a more detailed set of habitat descriptions is required before it can be accepted confidently.

Two tests of the crucial proposition 5 were made. Both involved counting the total number of unicells T(x) in some interval x of cell volume and identifying the number among these S(x) for which sexual reproduction has been described. The ratio S(x)/T(x) is then an estimate of the conditional probability of possessing sexual reproduction for unicells of size x, and it is predicted that this probability will be an increasing function of x. The two data sets used were: 1) species from all Volvocacean families, from von Huber-Pestalozzi; and 2) species from the genus *Chlamydomonas* alone, from Ettl's monograph. The results are shown in Figure 12; in both cases the frequency of sexual forms increases with cell volume.

The pattern discerned by comparative analysis among multicellular animals, and on a broad scale between unicells and colonies in the Volvocales, is therefore confirmed quantitatively for unicells alone, even within the confines of a single genus.

SEXUAL GERM CELLS

The appearance and specialization of a caste of asexual germ cells is accompanied by the parallel development of sexual germ cells. In unicells and small colonies, all cells are first somatic and then, given appropriate conditions, differentiate into gametes. In larger colonies certain cells become differentiated as sexual initials, and in *Euvolvox* they retain cytoplasmic continuity with the somatic layer just as the gonidia do. The sexual germ cells therefore realise the same advantages of a division of labour as do the gonidia (cf Fig. 11).

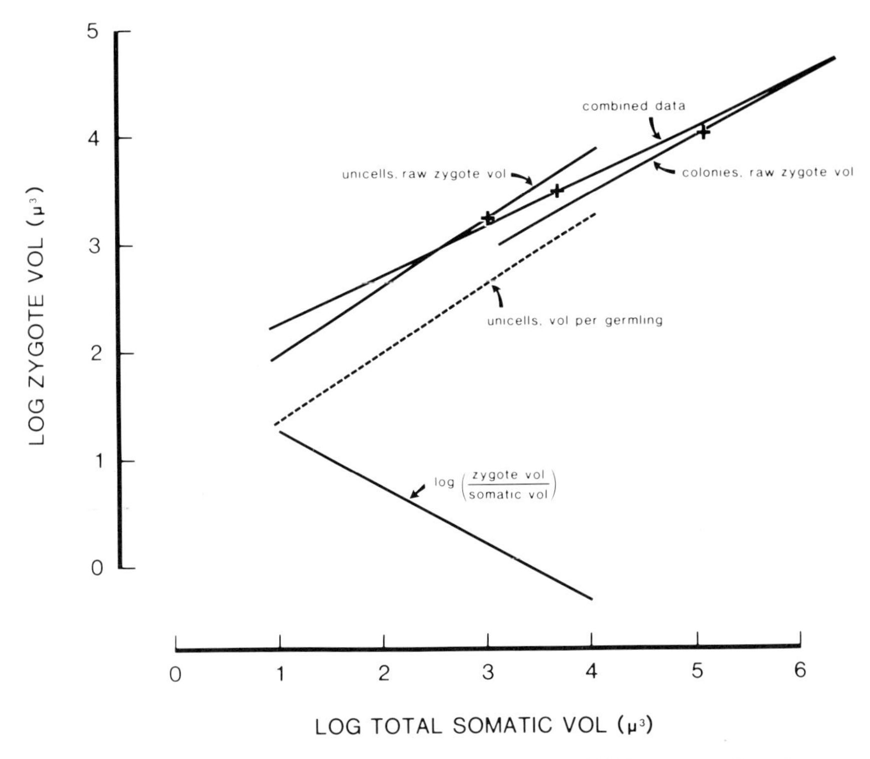

Fig. 13. Allometry of zygote volume in Volvocales. Three solid lines at top of graph are raw zygote volume for unicells, for colonies, and for both combined. The broken line is the expected volume per zygote germling for unicells, assuming that all four meiotic products emerge, and taking no account of the zygote wall. The lower solid line illustrates how the ratio of zygote volume to somatic volume decreases both for unicells and for colonies.

Most sexual unicells and some colonies are isogametic, producing gametes which have different gender but identical or very similar size and structure. In other forms, however, the gametes are dimorphic. Gamete dimorphism normally is accompanied by a clear separation into male and female categories, but in a few cases gamete size is uncorrelated or only loosely correlated with gender, and fusion may occur between equal or unequal partners. The evolution of gamete dimorphism and thus the origin of male and female categories was first explained satisfactorily by Parker et al. [1972] and later elaborated by Bell [1978, 1982], Charlesworth [1978], and Maynard Smith [1978]. Very briefly, the argument goes as follows. Other things being equal, it is better to produce small gametes as more can be synthesized from a given mass of material. Suppose, however, that the viability of zygotes increases more than proportionately with zygote size. Then small gametes will continue to be favoured because they can be produced in great numbers, whereas large gametes will also be favoured because they give rise to large and thus highly viable zygotes; gametes of intermediate size will be least fit. The effect of size on zygote viability may thus create disruptive selection for gamete size, and gamete dimorphism will evolve as a result.

A quantitative test of this hypothesis hinges on the allometry of zygote volume. Large size will be crucial to the vegetative success of the zygote germling when it must subsidize considerable growth or differentiation with stored reserves before becoming self-supporting. We expect, therefore, that unicells, with relatively little postzygotic growth, will tend to be isogametic, whereas multicellular algae will often be heterogametic, producing distinct microgametes and macrogametes. This trend is apparent in many algal series and especially clear in the Volvocales, where the female gametes of large colonial forms are massive immobile cells; the male gametes are reduced to small, highly motile sperm [Knowlton, 1974; Bell, 1978]. It may be the case also that forms living in ephemeral habitats which require large zygotes in order to achieve rapid growth during brief periods of flooding tend to be heterogametic [Madsen and Waller, 1983].

Because the Volvocales display a continuum of degrees of gamete dimorphism from complete isogamy to fully developed oogamy, we can use the allometric data for more precise and riskier tests of the hypothesis. The allometry of zygote volume is shown in Figure 13. Both in unicells and in colonies, zygote volume increases with somatic volume, the exponent being about 0.60. (In passing, I should mention that a remarkably similar regression is obtained for the volume of asexual cysts: A single allometric rule therefore applies to all resistant propagules, whether produced sexually or asexually.) Unicells produce rather larger zygotes than colonies of comparable size, but all four meiotic products usually emerge from unicell zygotes as germlings; colonies generally produce only a single germling per zygote, so that germling volume is, if anything, somewhat less for unicells than for colonies, especially as colonies arising from zygotes are generally smaller than asexually produced colonies. The important point, however, is that zygote volume increases less steeply than linearly with somatic volume. This is not a pecularity of Volvocales; in Figure 14 I have abstracted some data from Peters [1983] to show that an exponent of about 0.75 provides an excellent description of the relationship between zygote size and body size across some 15 orders of magnitude in body size, from *Chalamydomonas* to ostriches. As the exponent is less than unity, the ratio of zygote volume to somatic volume falls as somatic volume increases (Fig. 13): larger forms produce relatively small zygotes, whether they are unicells or colonies. We expect, therefore, that the crude association between gamete dimorphism and increased bulk will continue to hold when unicells and colonies are analysed separately. The truth of this prediction is demonstrated in Figure 15. This shows that isogametic unicells

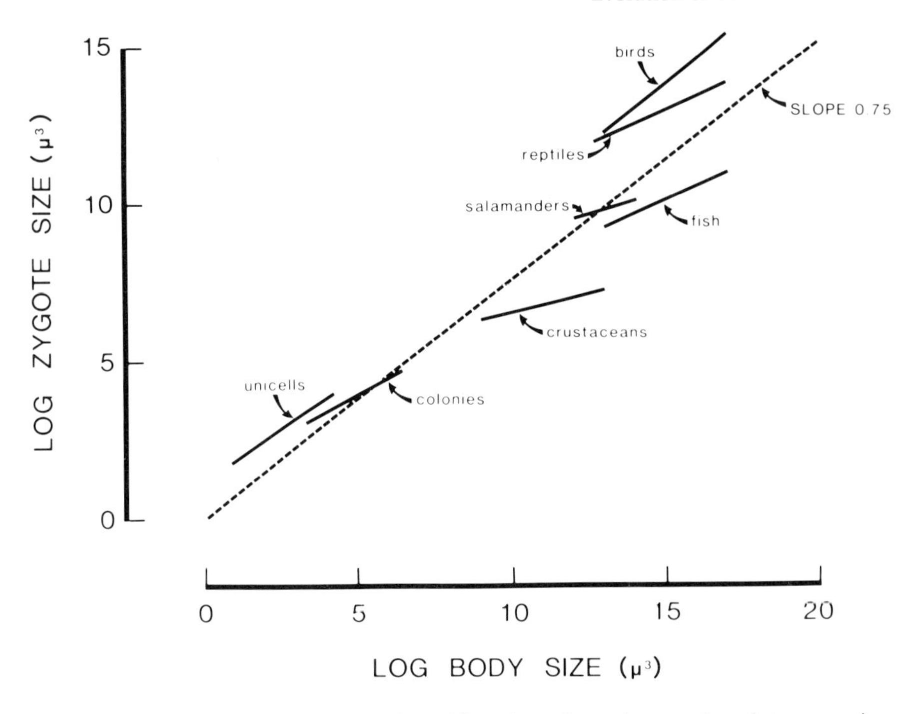

Fig. 14. Allometry of zygote volume in a wide variety of organisms, to show that a regression with a slope of about 0.75 provides a description of data from algal unicells to vertebrates. Regression equations for animals from Peters [1983]; for algae from Figure 13.

are smaller than heterogametic unicells, and isogametic colonies are smaller than heterogametic colonies. Moreover, the degree of gamete dimorphism, measured as a continuous variable by taking the ratio of sizes of female and male gametes, is an increasing function of size when heterogametic unicells and colonies are analysed separately. This gives conclusive support to the general association of gamete dimorphism with large size and suggests that Parker's functional interpretation of gamete dimorphism is substantially correct.

MICROGAMETE PACKAGING

The differentiation of gametes in Volvocales has one final twist. The microgametes of heterogametic unicells are released from the androgonium as separate cells; in colonial forms, however, they are organised into platelets or spheroids and, indeed, develop in much the same way as daughter colonies. When released from the parental envelope, the sperm packet (essentially a

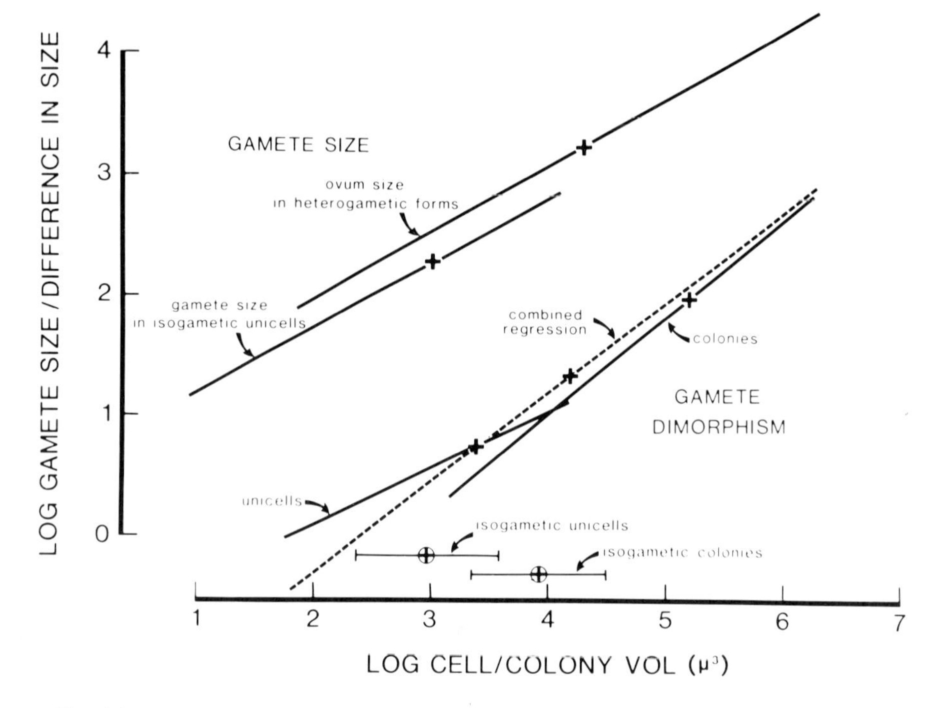

Fig. 15. Allometry of gamete size and gamete dimorphism in Volvocales. Horizontal lines at bottom of graph are mean size (± one standard deviation) for isogametic forms, showing them to be smaller than the corresponding heterogametic forms. The regression lines above these describe how the ratio of macrogamete to microgamete size increases with somatic volume both for unicells and for colonies, the broken line being the regression of the combined data. Upper regression lines are regressions for gamete size in isogametic unicells and macrogamete size in heterogametic unicells and colonies; ova are about twice as big as isogametes.

dwarf male, comparable to the dwarf males of rotifers) swims as a unit until it encounters a female colony; then it digests a passage through the somatic layer and dissociates into individual sperm which disperse into the colony lumen.

We have seen that the degree of gamete dimorphism increases with colony size; ovum volume also increases with colony size, though more slowly (Fig. 15); therefore, the larger colonies have absolutely as well as relatively smaller sperm. Because the sperm are smaller, more can be made; however, should this entail making the same number of packets and putting more sperm in each or keeping the number of sperm per packet constant and making more packets? Because under natural conditions it is most unlikely that any given female colony will be penetrated by more than one sperm packet at once, the number of sperm per packet should vary directly with the number of ova per colony; if more ova are available to any sperm packet which succeeds in encountering a female colony, each sperm packet should comprise more sperm

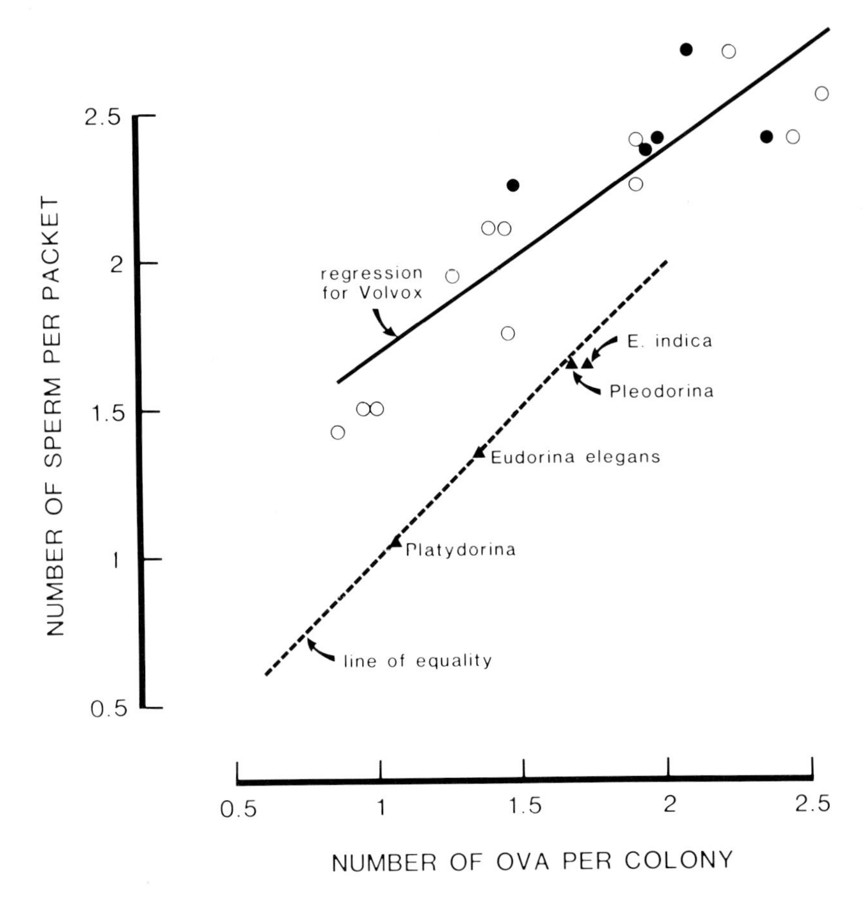

Fig. 16. The number of sperm per packet as a function of the number of ova per colony. Triangles are values for small coenobia, which seem to lie along the line of equality (broken line). Plotted points are for dioecious (hollow circles) and monoecious (solid circles) species of *Volvox;* the regression (solid line) is

$$y = 0.99 + 0.69x \ (r^2 = 0.81, P < 0.001)$$

both axes being logarithmic.

in order to fertilise them all. This hypothesis is readily tested and seems to be accurate (Fig. 16). The colonial forms seem to fall into two series, in both of which the number of sperms per packet is a steeply increasing function of the number of ova per colony. In small forms, such as *Eudorina* and *Platydorina,* the number of sperms in a packet is approximately equal to the number of ova per colony. In *Volvox* many more sperm are packaged together than there are ova for them to fertilize. This difference is presumably a reflection of the lower efficiency of fertilization in colonies with a very large lumen.

CONCLUSION

Within the Volvocales there exist three trends towards differentiation: between soma and germ, between sexual and asexual, and between male and female. In this essay, I have interpreted these trends as follows.

a) The distinction between soma and germ represents a division of labour between source and sink which has identifiable and measurable physiological benefits and makes possible a relatively large capacity for increase. Colonies can therefore realise the advantages of large size, perhaps primarily an escape from grazing by filter feeders, while evading much of the loss of fitness otherwise associated with size increase. Labour can be profitably divided only when the market is sufficiently extensive, which for these algae occurs in colonies of more than about 64 cells. Increased communication between cells by means of cytoplasmic connections further enhances the effect of dividing labour.

b) The interpretation of sex as a device for reducing competition between sibs in crowded environments leads to the prediction that sexuality will be more frequent among larger unicells, which is found to be the case.

c) Gamete dimorphism will arise through disruptive selection on gamete size if the zygote is small relative to the vegetative cell or colony and must therefore store reserves to fuel early growth. Because zygote volume increases less steeply than linearly with adult volume, larger cells or colonies produce relatively smaller zygotes. We therefore predict that isogametic forms should be smaller than heterogametic forms and that the degree of gamete dimorphism should be an increasing function of size within both unicells and colonies. These predictions are supported by the data.

In this way I have tried both to establish or confirm the functional basis of these major biological events and to interpret them synthetically as being, in large part, allometric consequences of an underlying tendency towards increase in size.

ACKNOWLEDGMENTS

My interest in the allometry of reproduction in the Volvocales was stimulated by a class project at McGill in spring, 1984; the data on *Chlamydomonas* habitats was collected by T. Mousseau, those on sex and size in *Chlamydomonas* were collated by C. Lacroix, and those on particle-size preference in filter feeders were in part collected by E.-M. Whitsun-Jones. I am grateful to M. Lechowicz for discussions on photosynthesis and to A. Vezina for showing me his data on phosphorus dynamics. This work was in part supported by an Operating Grant from the Natural Science and Engineering Research Council of Canada.

REFERENCES

Banse K (1982): Mass-scaled rates of respiration and intrinsic growth in very small invertebrates. Mar Ecol Prog Ser 9:281–297.

Bell G (1978): The evolution of anisogamy. J. Theor Biol 73:247–270.

Bell G (1982): "The Masterpiece of Nature." London: Croom-Helm.

Bisalputra T, Stein JR (1966): The development of cytoplasmic bridges in *Volvox aureus*. Can J Botany 44:1697–1702.

Blau PM (1974): "On the Nature of Organizations." New York: John Wiley.

Blau PM (1977): "Inequality and Heterogeneity." New York: The Free Press.

Blueweiss L, Fox H, Kudzma V, Nakashima D, Peters R, Sams S (1978): Relationship between body size and some life history parameters. Oecologia 37:257–272.

Bonner JT (1965): "Size and Cycle." Princeton: Princeton University Press.

Burns CW (1968): The relationship between body size of filter-feeding cladocerans and the maximum size of particle ingested. Limnol Oceanogr 13:675–678.

Charlesworth B (1978): The population genetics of anisogamy. J Theor Biol 73:347–358.

Coleman AW (1979): Sexuality in colonial green flagellates. In Levandowsky M, Hutner SH (eds): "Biochemistry and Physiology of Protozoa." New York: Academic Press, Vol. 1, pp. 307–340.

Darden WH (1966): Sexual differentiation in *Volvox aureus*. J. Protozool 13:239–255.

Ettl H (1976): Die Gattung *Chlamydomonas* Ehrenberg. Beihefte Nova Hedwiga 49:1–1122.

Fenchel T (1974): Intrinsic rate of natural increase: the relationship with body size. Oecologia 14:317–326.

Hino A, Hirano R (1980): Relationship between body size of the rotifer *Brachionus plicatilis* and the maximum size of particles ingested. Bull Jap Soc Sci Fish 46:1217–1222.

von Huber-Pestalozzi G (1961): Das Phytoplankton des Susswasser. Systematik und Biologie. Teil 5: Ordnung Volvocales. In Thienemann A, Die Binnengewasser, Band 16. Schweitzerbartsche verlangs-buchhandlungs, Stuttgart.

Huxley JS (1932): "Problems of Relative Growth." London: Methuen.

Iyengar MOP (1933): Contributions to our knowledge of the colonial Volvocales of South India. Bot J Linn Soc 49:323–373.

Karn RC, Starr RC, Hudock GA (1974): Sexual and asexual differentiation in *Volvox obversus* (Shaw) Printz, strains WD3 and WD7. Arch Protistenk 116:142–148.

Knowlton N (1974): A note on the evolution of gamete dimorphism. J Theor Biol 46:283–285.

Kochert G (1968): Differentiation of reproductive cells in *Volvox carteri*. J Protozool 15:438–452.

Kochert G (1982): Sexual processes in the Volvocales. In Round FE, Chapman DJ (eds): "Progress in Phycological Research." Amsterdam: Elsevier Biomedical Press, Vol. 1, pp 235–256.

Kryutchkova, NM (1974): The content and size of food particles consumed by filter-feeding planktonic animals. Hydrobiol J 10:89–94.

Madsen JD, Waller DM (1983): A note on the evolution of gamete dimorphism in algae. Am Natur 121:443–447.

Maynard Smith J (1978): "The Evolution of Sex." Cambridge: Cambridge University Press.

McQueen DJ (1970): Grazing rates and food selection in *Diaptomus oregonensis* (Copepoda) from Marion Lake, British Columbia. J Fish Res Bd Canada 27:13–20.

Meredith RF, Starr RC (1975): The genetic basis of male potency in *Volvox carteri* f. *nagariensis* (Chlorophyceae). J Phycol 11:265–272.

Moss B (1973): The influence of environmental factors on the distribution of freshwater algae: an experimental study. III. Effects of temperature, vitamin requirements and enorganic nitrogen compounds on growth. J Ecol 61:179–192.

Neales TF Incoll LD (1968): The control of leaf photosynthesis rate by the level of assimilate concentration in the leaf: a review of the hypothesis. Bot Rev 34:107–125.

Pall ML (1974): Evidence for the glycoprotein nature of the inducer of sexuality in *Volvox*. Biochem Biophys Res Comm 57:683–688.

Parker GA, Baker RR, Smith VGF (1972): The origin and evolution of gamete dimorphism and the male-female phenomenon. J Theor Biol 36:529–553.

Peters RH (1983): "The Ecological Significance of Body Size." Cambridge: Cambridge University Press.

Pilarska J (1977): Eco-physiological studies on *Brachionus rubens* Ehrbg (Rotatoria). I. Food selectivity and feeding rate. Pol Arch Hydrobiol 24:319–328.

Pinto CM (1980): Régulation de la photosynthèse par la demande d'assimilats: mécanismes possibles. Photosynthetica 14:611–637.

Pocock MA (1933a): *Volvox* and asociated algae from Kimberley. Ann S African Mus 16:473–521.

Pocock MA (1933b): *Volvox* in South Africa. Ann S African Mus 16:523–646.

Rich F, Pocock MA (1933): Observations on the genus *Volvox* in South Africa. Ann S African Mus 16:427–471.

Schindler DW (1975): Evolution of phosphorus limitation in lakes. Science 195:260–262.

Shaw WR (1916): *Besseyosphaera*, a new genus of the Volvocaceae. Bot Gaz 61:253–254.

Shaw WR (1919): *Campbellosphaera*, a new genus of the Volvocaceae. Phillippine J Sci 15:493–520.

Shaw WR (1922a): *Janetosphaera*, a new genus of the Volvocaceae. Phillippine J Sci 2a:477–508.

Shaw WR (1922b): *Copelandosphaera*, a new genus of the Volvocaceae. Phillippine J Sci

Shaw WR (1922c): *Merrillosphaera*, a new genus of the Volvocaceae. Phillippine J Sci 21:87–129.

Smith A (1776): The Wealth of Nations.

Smith GM (1944): A comparative study of the species of *Volvox*. Trans Amer Microscop Soc 63:263–310.

Starr RC (1969): Structure, reproduction and differentiation in *Volvox carteri* f. *nagariensis* Iyengar, strains HK9 and 10. Arch Protistenk 111:204–222.

Starr RC (1970a): Control of differentiation in *Volvox*. Symp Soc Devtl Biol 29:59–100.

Starr RC (1970b): *Volvox pocockiae*, a new species with dwarf males. J Phycol 6:234–239.

Starr RC (1971): Sexual reproduction in *Volvox africanus*. Contrib Physology 1971:59–66.

Starr RC, Jaenicke L (1974): Purification and characterization of the hormone initiating sexual morphogenesis in *Volvox carteri* f. *nagariensis* Iyengar. Proc Nat Acad Sci US 71:1050–1054.

Taylor WD, Shuter BJ (1981): Body size, genome size and intrinsic rate of increase in ciliated protozoa. Am Natur 118:160–172.

Vande Berg WJ, Starr RC (1971): Structure, reproduction and differentiation in *Volvox gigas* and *Volvox powersii*. Arch Protistenk 113:195–219.

Wiese L (1976): Genetic aspects of sexuality in Volvocales. In Lewin RA (ed): "The Genetics of Algae." Oxford: Blackwell Scientific Publications, pp 174–197.

Wilson EO (1971): "The Insect Societies." Cambridge: Harvard University Press.

The Origin and Evolution of Sex, pages 257–260

Evolutionary Patterns in Segregation of Germ Cells: Discussion

Leader: G. Bell

McGill University, Quebec, Canada

Dr. Bell opened the discussion by posing the question of what kinds of inquiries should be asked in regard to sex, mating patterns, gametes, and so forth. It is not clear why these should be so interesting; but, given that they are interesting, what sorts of questions can we ask about them? Bell suggested that basically, there are two sorts: the first is "why," and the second is "how." For the first, we do not wish to imply anything final or teleological; rather, we ask if we can ascribe any sort of motive force—presumably natural selection—to the systems that we observe. Why do we get some sorts of systems rather than others? Why do we normally observe male heterogamety, whereas female heterogamety is less common? Why do we get sexuality in some circumstances and asexuality in others? This is the subject matter of evolutionary biology.

"How," on the other hand, encompasses genetics. What is the structure of this locus? How is this character transmitted? So far, these are the only sorts of questions that we have looked at. We have heard in extremely erudite and sometimes brilliant detail the "how" of many disparate and interesting systems, but we should also consider the question "why." In short, questions of metagenetics—questions about how genetic systems themselves evolve—will be introduced, questions that deal with the reasons for the prevalence of certain systems and the rarity of others. Such questions are themselves of two sorts. Questions of the first sort are unanswerable, at least in a scientific sense; the second, we hope, can be answered.

Questions about origins fall into the first category. We have not heard any proposition about the origin of sex, or any other genetic system, for which an experiment or observation would falsify it. From a straightforward Popperian standpoint, therefore, we shall relegate questions about origins for the time being to the dustbin and instead insist that we must look at answerable questions, attempting to predict the circumstances in which we expect to find different sorts of genetic systems. Restricting ourselves to this sort of question, we can again make a useful twofold division, depending on whether we test our hypotheses by the comparative or by the experimental method. The choice that we make is dictated by the taxonomic level at which variation occurs. If we see differences only between different classes or phyla, we cannot conduct experiments. Either we cannot change things, or if we can the results are likely to be entirely inappropriate. On the other hand, we can easily do experiments if there are differences between individuals belonging to the same species. Although the experimental method is always more powerful than the comparative method, provided that it can be used at all, the comparative method is nevertheless entirely capable of giving clearcut answers to certain sorts of questions.

What sorts of questions? Bell remarked that sex seems to be what everyone is excited about, and one possible question might be, what is the function of sex? We might also include questions about gamete dimorphism and sexual bipolarity. At the other extreme are topics such as sex allocation and the sex ratio, where experiments are perfectly feasible. Somewhere betwixt and between fall things like hermaphroditism and sexual dimorphism; thus, we can move from questions that are relatively easy, indeed, largely solved, and which usually involve gender, to others which are extremely difficult, largely unsolved, and usually involve sex.

With this in mind, Bell asked all the speakers at the day's meeting: First, are "why" questions worth asking, or does every question of this sort turn out to be unanswerable? Secondly, if they are worth asking, what are the most important questions we should be pursuing, and do we have the answers to them?

Dr. Tardent pointed out that, in discussing recombination, we were talking about variability and the importance of variability in providing preadaptation, yet nobody has proved that variability is in fact valuable in this sense. We assume that it is; in the first lecture of the day, Dr. Zinder questioned the value of preadaptation, which is absolutely important to evolution.

Bell commented that we must have strong views about this, as about half the people present work on recombination, which would hardly be so unless they had strong views about the importance of its function.

Dr. Scofield pointed out that she had raised the problem of the function of sex, and argued that her ideas were open to comparative test. Dr. Mahowald explained how the inactivation of an X chromosome in male heterogametic animals might lead to a cyclical process through which the most advanced forms in a group would have the most highly specialized sex chromosomes.

Dr. Grosberg remarked that reference has often been made to "higher" and "lower" organisms and to what this implies for evolutionary tendencies; but the general consensus among invertebrate zoologists is that we have no idea what gave rise to what. All the major groups have histories going back to the Cambrian or Precambrian. For instance, it has been asserted that the tendency not to differentiate a special class of germ cells is a characteristic of "lower organisms"; yet it also occurs in ascidians, which, being protochordates, usually are not thought of as being "lower." This point goes back to what Weismann said in 1887, when he identified the crucial distinction as being between colonial and non-colonial organisms. Covariates such as coloniality may confound any interpretation based on simple evolutionary trees.

This leads to the general issue of homology. Grosberg was curious about the criteria for homology used by people working on different phenomena or organisms: do they base it on degree of similarity or on degree of analogy, functional or morphological analogy? He was curious about this because it seems that everyone uses different rules to infer homologies between structures and functions.

Scofield agreed on the first point. People use phylogenetic trees for simplicity. Grosberg replied that only if one follows a methodological approach that is clearly articulated can an analytical approach that is coincident with this methodological approach exist, as in systematics one might choose cladistics. The problem with phylogenetic trees is that we have no explicit means at our disposal for testing hypotheses about relationships; given this, and given the simplicity of the phylogenetic tree, we tend to think that the origins problem is answerable even though in a Popperian sense it is not. It may make us view putative similarities as though they were homologies when in fact they are not at all.

Scofield then reminded us that the issue being addressed at the end of the day's meeting was a gene and a sequence. Indeed, everyone who talks about homology makes this reservation. Grosberg, however, disagreed because one could have two genes that are entirely homologous and yet differ totally in their sequences. Cline remarked with regard to "why" there have been many speculations about the selective value of sex and recombination, and asked to hear more from those who have thoughts about the problem. One suggestion is that it is important to know how easy it is to take the first step towards parthenogenesis, because if we knew that, and how often it might happen, we might get some idea of whether a particular theory works or not. Cline then reminded us that the regression from a diploid sexual state to a parthenogenetic state might require as many new genes as to go in the reverse direction. A counter-example is the male-sterile mutation discovered recently in *Drosophila* in which the male mates and fertilizes the eggs, but the egg does not develop because the embryo remains haploid. However, approximately one of 500 embryos diploidizes itself and develops into a normal fly. This is the first step to parthenogenesis. In this way, it would be easy to select for more successful parthenogenetic strains, and systems such as those studied by Templeton could be used to test models of parthenogenesis.

Bell replied that the gradual acquisition of parthenogenesis by incorporating successive slight mutations applies to automixis only, which is basically a sort of internal self-fertilization. It does not apply to apomixis, where egg production is mitotic rather than meiotic, and which usually arises abruptly. Automixis and apomixis are polar opposites in that the one conserves heterozygosity and the other removes it and, in the case of the male-sterile mutant, abolishes it totally at all loci; to call both "parthenogenesis" can be misleading.

There is a theory of gender which will predict the sorts of organisms that should be hermaphroditic. There is certainly no unitary theory; rather, it looks as though we require specific hypotheses for specific cases. We need to test these hypotheses by using quantitative predictions which derive from the theory which predicted the pattern in the first place. This has been done in some cases, notably for sex ratio problems.

Eicher asked if hermaphroditism means that an organism can produce twice as many offspring. Bell responded that it does not. He pointed out that, in an outbreeding population at evolutionary equilibrium, each individual will spend equally on eggs and sperm, so that each will produce only half as many eggs as a pure female. Two hermaphrodites will therefore produce only as many eggs as a single female. If the population is inbreeding, then one can select for less sperm and more eggs, but this is just another example of the reproductive economy of parthenogenesis.

Cline responded that this is perhaps another way of asking the same question. Because all of these systems are extant, all of them are compatible with at least some type of advantage. Thus, it may be reasonable to ask whether there are any kinds of advantages which are incompatible with a particular system, perhaps leading to a theory which is other than history. Bell questioned whether this is just the technique of falsification? This if often not what happens, as people prefer to show instead how much evidence supports their case. Cline agreed, for a particular type of advantage for that system, rather than asking how many kinds of advantage might go along with a system.

In the discussion that followed it was pointed out that Tardent had found examples of bryozoans that had only an attached phase, others that had only a motile phase, and various combinations of these. Among hydrozoans, each of these situations is presumably advantageous in a certain ecological situation. There are many advantages in being attached, but there are compensations for being able to swim about. For some species, then, it is better to be a jellyfish, and for others a hydra. Bell remarked that it is impossible to disprove such a hypothesis.

SEX DETERMINATION AND DIFFERENTIATION OF VERTEBRATES

The Origin and Evolution of Sex, pages 263–270

Sexual Differentiation in Vertebrates

Sheldon J. Segal
Population Sciences, Rockefeller Foundation, New York, New York 10036

Sexual reproduction requires the differentiation of specialized germ cells that are distinguished from all other cells of the body. These are gametes—the egg of the female, the sperm of the male. The sex of an organism represents the aggregate of genetic, anatomical, physiological, and behavioral characteristics that are recognized as maleness or femaleness: the capacity of an individual to produce egg or sperm; the differentiation of the gonad itself from an undifferentiated, bipotential embryonic structure; the subsequent differentiation of the reproductive tract; and the characteristics of virtually all aspects of body form and function. One can speak of a male skeletomuscular configuration or a female pattern of body fat distribution. Even the chemistry of the blood is highly sex-specific. In part, these somatic features are influenced by the action of hormones produced by the male and female gonad, but the differences are more fundamental. An individual develops sexuality as a gradual process beginning with the establishment of genetic sex at fertilization, proceeding to the differentiation of the reproductive tract and other familiar secondary sex characteristics, and ultimately involving virtually all of the body, including the brain. Consequently, the process of developing sexuality influences attitudes, responses, and other aspects of behavior. In many lower animal forms, egg and sperm may be produced by a single, hermaphroditic individual. In these organisms, sexuality relates only to the genetic characteristics of the gametes, for there are no other discernable somatic variations among individuals.

SEX DETERMINATION

Genetic sex is established at fertilization; until differentiation of the gonads during embryonic development, chromosomal constitution may be considered the sole manifestation of sex. The term *sex determination* often is applied to the establishment of chromosomal constitution at time of fusion of egg and sperm. With each gamete contributing a haploid complement of chromosomes, the fertilization process establishes the diploid chromosome number characteristic of the species and found in the nucleus of all somatic cells.

The biochemical or cellular mechanism by which the genetic constitution for sex is expressed is not clearly understood. One theory proposes that sex is determined genetically on the basis of a quantitative balance between male-determining and female-determining factors. Although the sex chromosomes are the primary vehicles for sex-determining genes, in certain animal forms autosomal factors may contribute to the quantitative balance that decides sexuality [Goldschmidt, 1955]. The "balance" system works so that two doses of the female determiners on the X chromosomes (XX) overcome the male determiners outside of the X, but one dose (XY) is insufficient for the dominance of female factors. The quantitative balance theory of sex determination emphasizes that male and female sex determiners are present in both sexes, accounting for the fact that in each sex the potentiality for the other sex is present. A variation of this quantitative theory suggests that the male-determining gene on the Y chromosome is present in multiple copies, thus explaining the strong male-determining influence resulting from the presence of a Y chromosome. There are some interesting variants of the genetic mechanism involving separate sex chromosomes. In a few species, the Y chromosome is attached to one of the autosomes; this mechanism is found in the mongoose. In some marsupials and bats, it is the X chromosome that is joined to an autosome.

The X chromosome of humans is known to carry many genes that control traits other than sexuality. It is the locus, for example, of genes involved in blood-group types, color blindness, and certain inborn errors of metabolism. In contrast, it is a rarity for the Y chromosome to carry a gene controlling factor other than sex determination. Because the X chromosome carries many sex-linked genes, in contrast to the Y chromosome, it is evident that in XX females, such genes will occur in double doses, but in XY males, only single doses will be present. A dosage compensation mechanism appears to have evolved in female mammals. According to an hypothesis first proposed by Mary Lyons [Lyons, 1970], at a very early stage in embryonic development, one of the X chromosomes in each cell of the body becomes inactive. Consequently, all the progeny of that cell will reflect this inactivation of one X chromosome. This becomes the so-called heterochromatic X chromosome. In mice, the trophoblastic tissue selectively inactivates the paternal X chromosome, and the cells derived from the inner cell mass randomly inactivate one of the two parental X chromosomes [West et al., 1977]. In humans, the inactivation process appears to occur completely at random, although there is some evidence that if an X chromosome happens to be defective in some manner, it will be preferentially inactivated.

Primary Sex Differentiation

The sex glands serve the dual functions of gamete production and hormone secretion. They originate in both males and females as similar, undifferentiated primordia. Normally, the pathway of differentiation to form testes or ovaries is determined by the genetic sex established at fertilization. Nevertheless, this process can be influenced by nongenetic factors. It has been demonstrated in amphibia, birds, and mammals that testes very well may develop in genetic females, if the proper hormonal or physical influence is exerted while the embryonic gonad is still in the indifferent state. Conversely, ovaries may be induced to differentiate in genetic males. The factors that decide the alternatives of testicular or ovarian differentiation may be placed in three groups: genetic, environmental, and localized internal agents. In amphibia, environmental conditions, such as extreme temperatures or delayed fertilization, can reverse partly or completely the genetic determination. In many species of turtles, the sexual differentiation of the gonads is temperature-dependent [Bull and Vogt, 1979] In *Emys orbicularis,* for example, egg incubation below 27.5°C results in 100% males, whereas egg incubation above 29.5°C results in 100% females. At intermediate temperatures, both males and females, and also some intersexes (displaying ovotestes), differentiate. In lizard (*Agma agma*) and alligators (*Aligator mississippiensis*), in contrast, low temperatures produce females and high temperatures produce males [Charnier, 1966; Ferguson and Joanen, 1982]. Temperature-dependent sex determination also has been demonstrated in some species of fish. In nature, a striking example of the importance of environmental conditions during differentiation of the gonads is the case of the eel. Sex ratios of eel populations differ greatly in different geographic locales, in spite of the fact that the genetic mechanism for sex inheritance should result in a 50:50 sex ratio. Among certain European eels, females are more prevalent in the higher reaches of rivers, whereas in the estuaries and coastal seawaters, males are frequent. Also, in the farthest regions of the geographic area of distribution of a species, females tend to predominate. These examples serve to demonstrate that many of the vertebrates have genetic mechanisms so labile that natural or experimental alterations of the environment may suffice to produce sex reversals.

The role of localized internal factors in sex differentiation becomes evident through an analysis of experimental and naturally occurring sex reversal in vertebrates. Lillie's descriptions of different sexed cattle twins illustrate this concept [Lillie, 1916]. When the fetal membranes of male and female calf embryos become united in the uterus so that a common blood circulation develops, the female embryo is modified in the male direction, forming the so-called freemartin. So complete is this transformation that the female develops a sterile testis instead of an ovary. Gonadal sex reversal also has been observed when fetal rat gonads were transplanted into adult rats [Buyse, 1935]. However, sex reversal in mammals is restricted to somatic elements, and the resulting testes are sterile. Germ cell sex seems to be directed more strictly by genetic agents. In nonmammalian species, absolute sex reversal, to the extent of transforming genetic males or females into reproductively functional individuals of the opposite sex, has been achieved experimentally by the administration of hormonal substances before the embryonic gonad is differentiated. An example is the sex reversal of *Xenopus,* the African clawed toad. If young tadpoles are reared in aquarium water containing small quantities of estrogenic hormones, the expected 50:50 sex ratio does not prevail; instead, all larvae become ovary-bearing females. That genetic males are functioning as reproductive females without alteration of the original male genetic constitution can be proved by breeding experiments between sex-reversed males and normal males. Because in this species the male is homogametic (ZZ), 100% of the offspring of such a mating are male. Many species of amphibia and teleosts have been completely sex reversed in this fashion. Functional sex reversal may be produced also in birds. In mammals, modulations of sex differentiation by external hormones are limited to the marsupials (*Dipelphis virginiana, Macropus eugenii*); in eutherian mammals there has been little success in reversing the normal development by hormones.

A more specific interpretation of the biochemical expression of gene action that results in gonadal differentiation has been proposed recently [Ohno et al., 1979]. The so-called H-Y antigen is the histocompatibility related antigen determined by a locus on the Y chromosome. Antiserum to H-Y antigen exhibits cross-reactivity with cells derived from nonmammalian vertebrates, such as birds, amphibians, and fish. In these animals, the cross-reaction is strong in the heterogametic sex. Ohno has proposed that it may play a major role in testis (or gonad of heterogametic sex) differentiation. The postulate suggests that the H-Y antigen is the product of mammalian testis-inducing genes and that the mammalian primordial gonad organizes a testis in the presence of H-Y antigen and an ovary in its absence. The proposal is based on evidence that testicular architecture of the gonad is associated invariably with the presence of H-Y antigen despite primary sex genotype or secondary sex phenotype: XX males and XX true hermaphrodites are H-Y positive: XY human females exhibiting the syndrome of testicular feminization are also H-Y positive. These are all examples of testis-bearing individuals, irrespective of genotype or secondary sex phenotype. An example of a genotypic male without a testis is the fertile XY female of the wood lemming, *Myopus Schisticolor*. These animals are H-Y negative.

Most evidence ascribing a causal relationship between H-Y antigen and testis differentiation is circumstantial, based on correlations between gonadal morphology and H-Y antigen status in normal and intersexual individuals. Some experimental evidence in support of the theory has been forthcoming. It has been claimed that antibodies to H-Y antigen added to dissociated testicular cells of newborn rats or mice in culture will inhibit their reorganization as seminiferous tubules and cause them to aggregate, resembling ovarian follicles [Zenzes et al., 1978]. Furthermore, exogenous H-Y antigen induces testis formation when added in vitro to XX gonads of the fetal calf [Ohno et al., 1979]. These experiments, however, are based on morphological results that can be ambiguous. There are some experimental results that do not support the concept of the H-Y antigen as a systemically active gonad sex-reversing factor. In one experiment, fetal hamster ovaries were grown in organ culture in contact with fetal testes from animals of the same age; normal ovarian development was undisturbed.

It has been suggested that the structural gene for H-Y antigen is located on either X or autosomal chromosomes and regulated by gene(s) on Y chromosome. The evidence is as follows: In chicken and amphibia, the sex of gonads can be reversed by estradiol, and H-Y antigen is induced as a result of sex reversal even in the absence of W chromosome [Wachtel et al., 1980; Zaborski et al., 1981]. Furthermore, it has been shown that the expression of H-Y antigen in gonads is controlled by temperature in coincidence with ovarian development in turtles [Zaborski et al., 1982]. The same authors also showed that the H-Y type is different in the gonads and in the blood.

The precise role of H-Y antigen in gonadal differentiation, if any, is not yet established. Clarification will depend on additional experimental evidence, such as an analysis of H-Y status in experimental sex reversal and normal sex reversal, in species such as those mentioned above in which the bipotentiality of the embryonic gonad enables environmental factors to influence the direction of differentiation.

Secondary Sex Differentiation

Differentiation of the accessory sex structures follows the primary differentiation of the gonad and, almost without exception, throughout the vertebrate class, these organs are not under direct genetic control, but under the influence of secretions from the newly formed gonads. In the embryo during the indifferent stage, the oviducts (ducts of Müller) and mesonephric ducts (of Wolff) appear in both sexes. The oviduct is the primordium of the Fallopian tubes, uterus, and upper vagina. The mesonephric duct gives rise to the epididymis, vas deferens, and seminal vesicle. In the male, the oviduct degenerates during the process of secondary sex differentiation, whereas in the female it is the mesonephric duct that does not persist. External genitalia of each sex develop from bipotential primordia, the urogenital sinus, and the genital tubercle, which have the capacity to develop along masculine or feminine lines. Thus, each embryo possesses the potentiality to develop internal and external genital organs of either sex. Experiments employing the technique of fetal castration have elucidated the influence of the fetal gonad on the differentiation of the sex accessory organs [Jost, 1970]. In the absence of gonads, mammalian fetuses will develop female accessory structures regardless of whether the castrated individual is a genetic male or a genetic female. If genetic males are deprived of their fetal gonads after the masculinization of the upper portion of the genital duct system has begun, then only the lower regions and the external genitalia are of the feminine type. In this fashion, pseudo-hermaphrodites that possess both male and female genital structures can be produced experimentally. By unilateral castration, it can be demonstrated that the morphologic inductive capacity of the mammalian fetal testis acts locally to suppress the oviduct and activate the mesonephric duct. Removal of a single fetal testis results in the appearance of a female duct system on the operated side, whereas male development proceeds on the unmolested side. These experiments of Jost, with rabbit fetuses, have contributed greatly toward the understanding of human pseudo-hermaphroditism and intersexuality. Parallel experiments in birds reveal the interesting fact that in this group of vertebrates it is the ovary that must be present to suppress the tendency for masculine differentiation to occur [Wolff, 1953]. The principles regarding the inductive role of the gonad in the differentiation of the accessory sex organs are identical. Why the agonadal condition results in masculinization in one group of vertebrates and in another brings about feminization is problematical. In all likelihood, the explanation is linked to the fact that females are the homogametic sex in mammals, whereas male birds are homogametic. How this genetic factor acts in secondary sex differentiation is unknown.

Control of Adult Secondary Sex Characteristics

The establishment and maintenance of sexual patterns, both morphologic and behavioral, usually involve coordinated hormonal interactions. Some sexual dimorphisms, particularly among birds, are not controlled hormonally but are determined directly by genetic constitution. This is true of normal sex differences in plumage of a number of avian species. In the English sparrow, neither removal of the gonad nor hormone injection has any noticeable effect on plumage dimorphism. An intermediate type is represented by the pheasant, in which full development of sexually characteristic plumage is dependent upon simultaneous actions of both genetic and hormonal factors.

In contrast to those instances, mainly in birds, of direct genetic control is the vast majority of behavioral and morphologic expressions of sexuality that are established and controlled by hormonal mechanisms [Young, 1961]. The range of hormonally controlled sexual characteristics in vertebrates extends from exotic courtship rites of salamanders to such majestic ornaments as the antlers of the deer or the mane of the lion. Almost all these sex-specific features are under the influence of steroid hormones produced by the sex glands. However, exceptional situations, in which there is a direct influence of protein hormones from the pituitary on secondary sex characteristics, have been described. In several genera of African finches, the male bird assumes a bright nuptial plumage at the onset of the breeding season. It maintains this adornment for two or three months and then, after molting, dons the hen-type plumage, a constant characteristic of the female. This plumage change coincides with the cyclic change of the gonads from the quiescent to the breeding stage. Castrated males, however, continue to develop the colored plumage rhythmically. This indicates that feather pigmentation during the male phase of the plumage cycle is not controlled by the gonad. It has been established that a pituitary hormone, luteinizing hormone, directly controls this secondary sex characteristic. This represents an unusual case, perhaps an evolutionary transition, as pituitary hormones nearly always direct their action toward endocrine glands, and the secretions of the target glands influence the rest of the body soma.

As each vertebrate organism approaches the stage of gonadal maturation, greater and greater contrast between male and female becomes apparent. Male guppies develop a gonopod; thumbpads appear on the digits of male frogs. In the male turtle (*Pseudemys elegans*), the three middle foreclaws, which are used to stimulate the female during courtship, begin to elongate. Vocal changes, not unlike those of young boys approaching puberty, become apparent in such diversified vertebrates as the leopard frog, tree toad, prairie chicken, and domestic duck; in the male Virginia deer, antler growth begins. By the time these appendages are needed for fighting during courtship, they have shed the velvet and grown hard in response to increased production of testis hormones. The boar's tusks, the bull's horn and crest, the goat's odor gland, the ram's horns, and the rooster's comb and spurs are all well-known secondary sex characteristics that respond to the action of testicular hormones. Females are equally dependent upon hormonal stimuli from the gonads for the development of sex-contrasting characteristics. The thread-like oviducts of the female frog enlarge to fill most of the abdominal cavity as the first breeding approaches. The female opposum's vicious resentment of the male's advances is replaced by docile acceptance as ovarian function becomes established. The female xenopus, undistinguishable from males as young juveniles, responds to ovarian hormone production by a typically feminine growth pattern, just as the awakened ovary stimulates the developing of feminine contours in the human female at puberty.

CONCLUDING REMARKS

The foregoing pages have considered some selected issues and problems pertaining to the biology of sex. Without doubt, many other considerations could have been included. The selection, however, in itself presents a point of view. It suggests that to understand sexual reproduction is significant in an overall biological context but that we must also attempt to understand the regulation of the process.

REFERENCES

Bull JJ, Vogt RC (1979): Temperature-dependent sex determination in turtles. Science 206:1186–1188.

Buyse A (1935): The differentiation of transplanted mammalian gonad primordia. J Exp Zool 70:1–41.

Charnier M (1966): Action de la temperature sur la sex-ratio chez l'embryon d'Agama agama (*Agamidae, Lacertilien*). CR Soc Biol 160:620–622.

Ferguson MW, Joanen T (1982): Temperature of egg incubation determines sex in *Alligator mississippiensis*. Nature 296:850–853.

Goldschmidt RB (1955): "Theoretical Genetics." Berkeley, California: University of California Press.

Jost A (1970): Hormonal factors in the sex differentiation of the mammalian foetus. Philos Trans R Soc Lond 259:119–130.

Lillie FR (1916): The theory of the freemartin. Science 43:611–613.

Lyons MF (1970): Genetic activities of sex chromosomes in somatic cells of mammals. Philos Trans R Soc Lond 259:41–65.

Ohno S, Nagai Y, Ciccarese S, Iwata H (1979): Testis-organizing H-Y antigen and the primary sex determining mechanism of mammals. Recent Prog Horm Res 35:449–478.

Wachtel SS, Bresler PA, Koide SS (1980): Does H-Y antigen induce the heterogametic ovary?. Cell 20:859–864.

West JD, Frels WI, Chapman VM, Papaioannou VE (1977): Preferential expression of the maternally derived X chromosome in the mouse yolk sac. Cell 12:873–882.

Wolff E (1953): Le determinisme de l'atrophie d'un organe rudimentaire; le canal de muller des embryons males d'oiseaux. Experientia 9:121–133.

Young W (1961): "Sex and Internal Secretion." Baltimore: Williams and Wilkins.

Zaborski P, Guichard A, Scheib D (1981): Transient expression of H-Y antigen in quail ovotestis following early diethylstilbestrol (DES) treatment. Biol Cell 41:113–122.

Zaborski P, Dorizzi M, Pieau C (1982): H-Y antigen expression in temperature sex-reversed turtles (*Emys orbicularis*), Differentiation. 22:73–78.

Zenzes MT, Wolf V, Gunther E, Engel W (1978): Studies on the function of H-Y antigen: Dissociation and reorganization experiments on rat gonadal tissue. Cytogenet Cell Genet 20:365–375.

The Origin and Evolution of Sex, pages 271–287

Gonadal Sex Differentiation in Mammals

Teruko Taketo, S.S. Koide, and H. Merchant-Larios

Center for Biomedical Research, The Population Council, New York, New York 10021 (T.T., S.S.K.) and Instituto de Investigaciones Biomedicas, Universidad Nacional Autonoma de Mexico, Mexico 20, D.F., Mexico (H.M.-L.)

Genetic sex of mammals is determined at fertilization by the pairing of sex chromosomes, i.e., XX or XY. The genetic sex directs the differentiation of a fetal gonad into a testis or an ovary, which in turn secretes hormones that regulate phenotypic sex development, including internal and external genitalia and secondary sex characteristics. Thus, during normal development of a fetus, the gonadal and the phenotypic sexes are determined by the genetic sex. It has been suggested that genes on the Y chromosome regulate testicular differentiation by interacting with other genes on autosomal chromosomes [Simpson, 1982], whereas ovarian organization takes place in the absence of the Y chromosome. The mechanism involved in the regulation of gonadal organization remains to be elucidated. In this paper, gonadal sex reversal of mammals with normal chromosomes are reviewed and factors that influence gonadal development are discussed.

FREEMARTINS

Freemartinism occurs because of the formation of testicular components in female cattle, sheep, and goats that have shared the circulation with a male twin during fetal life [Lillie, 1916, 1917; Jost et al., 1972]. Other characteristics of the freemartin are the early regression of the Müllerian duct and the absence of germ cells in the seminiferous tubules. Several hypotheses have been proposed to explain the mechanism of freemartins: 1) Hormones secreted by the male fetus circulate into the female twin [Lillie, 1916, 1917]; 2) XY cells migrate from the male fetus to the female, resulting in XX/XY mosaicism [Fechheimer et al., 1963; Herschler and Fechheimer, 1967].

Humoral elements that are implicated in the masculinization of the freemartin ovary are androgenic steroids, Müllerian inhibiting substance (MIS), and histocompatibility-Y (H-Y) antigen. Although male fetuses produce androgenic steroids during early stages of testicular development [Wilson and Siiteri, 1973], they are unlikely participants of testis determination, because seminiferous tubules are formed before Leydig cell differentiation or testosterone production during normal testicular development [Jost and Magre, 1984]. Mammalian fetal and perinatal testes, probably the Sertoli cells, produce a hormone, MIS, which induces regression of Müllerian duct in the male fetus [Josso et al., 1977; Donahue et al., 1982; Swann et al., 1979]. MIS is probably not involved in gonadal sex determination, as it is produced after commencement of testicular differentiation [Josso et al., 1977]. In freemartin fetuses, regression of Müllerian duct begins before the development of testicular structures, suggesting that both MIS and a testis determining factor circulate from the male fetus to the female twin [Jost et al., 1972]. The H-Y antigen has been detected in the circulation from the male twin [Wachtel et al., 1977]. Much circumstantial evidence supports the hypothesis that H-Y antigen is the testis determining factor [Silvers and Wachtel, 1977; Wachtel, 1980; Polani and Adinolfi, 1983]. However, this hypothesis needs to be validated, as discussed below.

In the freemartin fetuses, XX/XY mosaicism is found in the blood and hemopoietic cells [Herschler and Fechheimer, 1967; Vigier, 1976]. The extent of the masculinization parallels the percentage of XY cells in the female twin [Herschler and Fechheimer, 1967]. However, freemartins did not develop when the circulation between twin fetuses was interrupted after male cells had migrated into the female fetus [Vigier, 1976]. It was concluded that freemartinism is not due to the transfer of male cells to the female fetus.

Freemartins show that ovarian primordia can develop testicular structures under the influence of fetal male factors. However, the mechanism is not clarified.

HISTOCOMPATIBILITY-Y (H-Y) ANTIGEN

H-Y antigen was first demonstrated by the rejection of male skin grafts placed on isogenic female recipients [Eichwald and Silmser, 1955]. Later, it was found that H-Y antigen can be assayed with complement-mediated cytotoxicity tests on male target cells [Goldberg et al., 1971]. This cytotoxicity test has been the most widely used to study the relation between the male antigen and testis development. However, it is now considered that the two principal methods of assay do not test the same male-specific antigen [Melvold et al., 1977], although these two antigens might be closely related [Polani and Adinolfi, 1983].

It has been proposed that H-Y antigen determines testicular differentiation, based on the finding that individuals with testes are H-Y antigen positive regardless of their karyotypic sex [Silvers and Wachtel, 1977; Ohno et al., 1979]. Although contradictory results have also been reported, several lines of evidence support this hypothesis as follows:

1) Dissociated testicular cells from newborn rats reaggregate to form follicle-like structures in culture with anti-H-Y antiserum (raised in female mice against male spleen cells) [Zenzes et al., 1978a]. However, this observation was not confirmed by others [Müller and Urban, 1981].

2) Dissociated ovarian cells from newborn rats organize tubular structures after incubation in the conditioned medium of neonatal testicular cells, which is positive in the H-Y assay [Zenzes et al., 1978b; Müller and Urban, 1981]. Moreover, like testicular cells, the reaggregates of ovarian cells thus obtained bind human chorionic gonadotropin (HCG), whose receptor is usually not present in ovaries until the eighth day after birth [Müller et al., 1978].

3) Indifferent gonads dissected from XX bovine embryos form tunica albuginea and seminiferous tubules in Daudi lymphoma cell culture medium that contains H-Y cross-reacting substance, whereas its accompanying gonad, cultured in serum alone, remained undifferentiated [Ohno et al., 1979]. These results were questioned by other investigators [Benhaim et al., 1982].

Because neither purified H-Y antigen nor monoclonal antibody against H-Y was used in these experiments, the factor(s) responsible for the morphological changes is not necessarily the only one. For example, the conditioned medium of testicular cells contain not only H-Y antigen but also many other substances. Furthermore, in the above studies, the modified gonadal structures were examined only with light microscopy. Therefore, it is not clear whether the observed changes involve the redifferentiation of sex specific cells. Thus, the role of H-Y antigen in gonadogenesis remains to be clarified.

FORMATION OF OVOTESTES FROM FETAL OVARIES FOLLOWING TRANSPLANTION

Ovotestis development can be induced on transplantation of ovarian primordia into various sites of adult hosts. This is the only experimental reversal of gonadal sex in mammals except for the series of experiments with H-Y antigen. Accordingly, it is important to analyze the factors involved in testicular development in fetal ovaries following transplantation. Previous studies in the rat and our recent studies in the mouse are reviewed below.

Morphology of Ovotestes

Ovotestis development after transplantation of fetal ovaries was reported in the rat [Buyse, 1935; Moore and Price, 1942; Holyoke, 1949; Torrey, 1950; MacIntyre, 1956; Turner and Asakawa, 1962; Mangoushi, 1975] and in the mouse [Turner and Asakawa, 1964]. The tubular structures in these ovotestes resemble seminiferous tubules when examined under a light microscope. However, they were not convincing as "testicular" structures for some investigators [MacIntyre, 1956; Ozdzenski et al., 1976]. The ovotestes induced by transplantation in the mouse were examined with transmission electron microscopy [Taketo et al., 1984a], and all types of testicular somatic cells, i.e., Serotoli, myoid, and Leydig cells were identified. Typical "intersertoli contact specializations" [Flickinger, 1967] were found between the cells inside of the seminiferous tubules. "Myoid cells" surrounding the outside of the basal lamina of tubular structure contain numerous microfilaments. In contrast, the theca cells enclosing the follicles do not contain microfilaments. Steroidogenic cells were found in the interstitial area of the testicular portion of ovotestes, whereas in ovaries or in the ovarian portion of ovotestes, steroidogenic cells were absent. To support the differentiation of Leydig cells, a positive correlation was found between testosterone level and the extent of testicular development [submitted for publication]. Occasional oocytes present inside the seminiferous tubules of ovotestes were often degenerated and spermatogenic cells were absent [Taketo et al., 1984a]. Turner and Asakawa [1964] reported the presence of spermatogenic cells in the tubular structures of ovotestes. However, our results, and those of others [Ozdzenski et al., 1976], disagree with such sex reversal of germ cells in transplanted ovaries. These findings show that testicular development in ovarian transplants involves the differentiation of "testicular" somatic cells, whereas the differentiation of germ cells appear to be controlled by different factors.

Grafting Sites

In general, ovotestis can develop at any grafting site, although there is a quantitative difference in the ovotestis-inducing capacity among different sites. In the rat, most fetal ovaries develop into ovotestes when transplanted into the anterior chamber of the eye [Torrey, 1950] or the scrotal position [Mangoushi, 1975]. On the other hand, the incidence of ovotestis development was lower when transplanted into the site beneath the kidney capsules [Buyse, 1935], in the subcutaneous position [Moore and Price, 1942], or in the omentum [Holyoke, 1949]. These results suggest that ovotestis development is induced by a factor present in the blood of host or by interaction with many types of host tissue.

Species and Strains

Gonadal sex reversal after transplantation has been studied only in rats and mice. When fetal ovaries are transplanted into a site beneath the kidney capsules, the frequency of ovotestis development appears to be higher in the mouse [Taketo et al., 1984a] than in the rat [Buyse, 1935; MacIntyre, 1956]. In the mouse, ovotestes developed more frequently in RU-NCS (Swiss, outbred) than in SJL/J or C57BL/6J strain; [Taketo et al., 1984a; unpublished data]. Ozdzenski et al [1976] found occasional development of testis cordlike structures in mouse A strain. Because of the similarities between ovotestis and freemartin development, gonadal sex reversal by transplantation should occur in other species and strains. However, we do not know yet why the capacity of ovotestis induction is variable among mouse strains.

Sex of Host

Transplantation experiments were performed to examine the influence of hormones of male or female hosts on the gonadal development. Buyse [1935] observed unusually large incidence of testicular development from gonadal primordia transplanted not only into male but also into female hosts. He concluded that some testes did develop from ovarian primordia on the assumption that the ratio of male to female embryos was 1 to 1. If only ovotestes are considered as evidence of sex reversal, the frequency was about 10% in male hosts but only 1% in female hosts. Moore and Price [1942] and Torrey [1950] reported that the frequency of ovotestis development in normal or castrated male or pregnant, nonpregnant, or ovariectomized female hosts was equivalent. On the other hand, Holyoke [1949] described that the tubules that developed in the ovarian grafts after transplantation into male hosts were longer and more numerous, more resembling testis cords, than in those transplanted into female hosts. Other investigators used only male hosts and did not carry out a comparative study on the influence of host sex.

The Mesonephros

The sex of host is probably the most controversial yet one of the most important factors to be considered for understanding the mechanism of ovotestis induction. If ovotestes can develop in a female environment, how could genes on the Y chromosome participate in testicular organization? In the mouse, it was found that when fetal ovaries on the 12th day of gestation (d.g.) were transplanted with mesonephroi attached, the frequency of ovotestis development was much higher in male hosts than in female hosts (Figs. 1a and b) [Taketo et al., 1984a]. In contrast, when fetal ovaries separated from mesonephroi were transplanted, ovotestes developed equally in male and

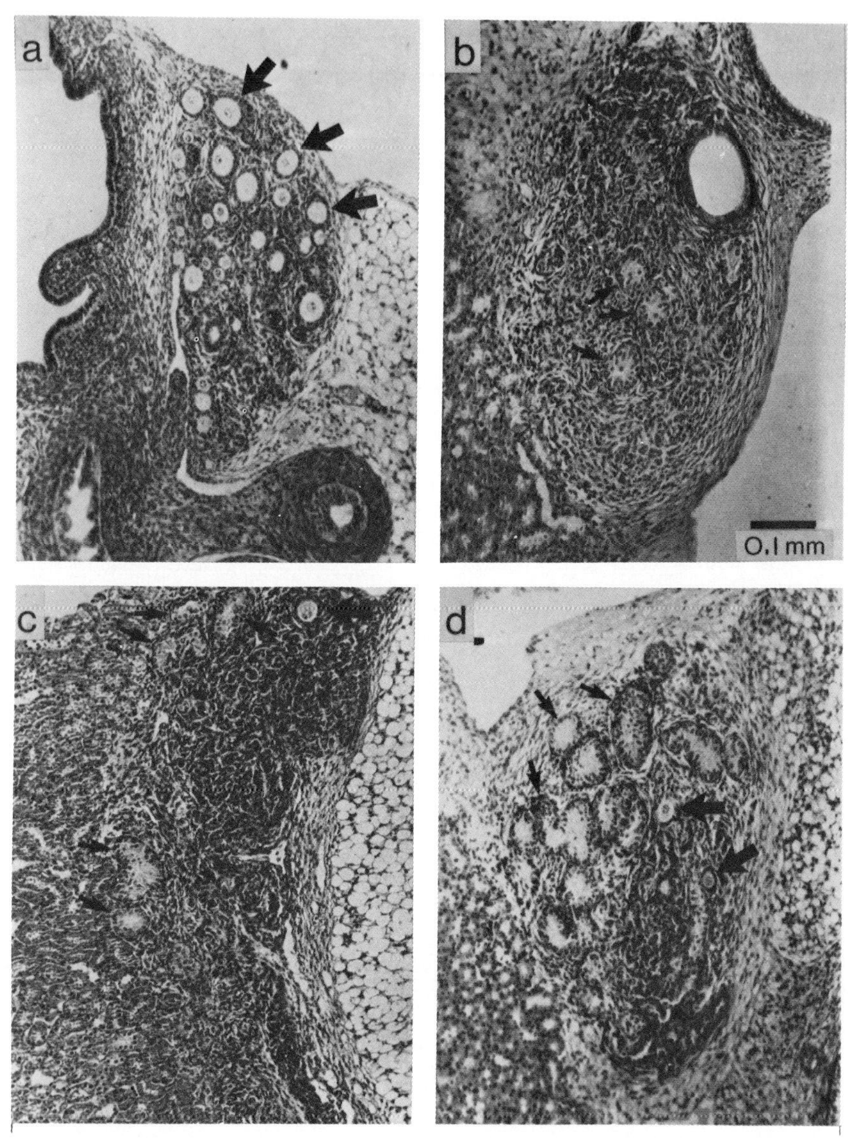

Fig. 1. Development of mouse fetal ovaries transplanted on the 12th d.g. into adult mice (SJL/J strain). Histologically examined on the 14th day after transplantation. Large arrows indicate typical follicular structures and small arrows testicular structures. Hematoxylin and eosin (H & E) stain. a) in a normal female host; b) in a normal male host; c) in a castrated male host; d) in a castrated male host, co-transplanted with a fetal testis on the 12th d.g. Magnification × 90.

female hosts [unpublished data]. Therefore, the mesonephros appears to play a key role in sex determination. Because fetal ovaries without mesonephroi developed normally in vitro [unpublished observation], mesonephros appear to influence ovarian development only *in vivo*. Based on the present findings, the authors proposed that ovarian primordia can develop testicular structures when placed in a host environment regardless of the host sex and that the mesonephros, when transplanted into female hosts, protects ovarian grafts from the masculinizing stimuli of the host. However, when transplanted into male hosts, the mesonephros does not prevent the development of testicular structures. The testis inducing factor in male hosts may be potent enough to overcome the feminizing effect of the mesonephros, or the mesonephros may function only in a female environment. The feminizing influence of the mesonephros on gonads has been suggested by other investigators [Byskov and Grinsted, 1981; Whitten et al., 1979].

In the rat, on the other hand, gonads transplanted with mesonephroi into female hosts develop into ovotestes [Buyse 1935; Holyoke 1949; Mangoushi 1975]. Torrey [1950] used gonadal ridges either with or without mesonephroi and did not find any significant difference. Therefore, the mesonephros appears not to be crucial for the development of ovarian grafts in the rat, although the mesonephros may not be functional under the conditions used by these investigators.

It has been suggested that both testis determining factors and ovary determining factors control the direction of gonadal development [Müller and Urban, 1981]. Such antagonizing factors have been considered in the sex determination of lower species [Witchi, 1934]; "cortexin" produced by gonadal cortex is postulated to induce ovarian development and "medullarin" produced by gonadal medulla may induce testicular development. A similar mechanism may remain in mammalian gonads, although the definition of cortex and medulla is not clear in mammals. Our observation is not at variance with the widely accepted theory of the Y chromosome-controlling sex determination, because the sex of hosts determines the direction of sexual differentiation when fetal ovaries develop in association with mesonephroi as in normal development. However, our results raise a question about the direct role of Y chromosome in gonadal sex determination. A gene(s) on Y chromosome probably regulates the expression of genes on autosomal or X chromosomes that directly influence gonadal organization [Polani and Adinolfi, 1983; Eicher and Washburn, 1983].

Male Factor(s) Promoting Ovotestis Development

To determine the male factor(s) that induces or promotes testicular development, the use of castrated hosts or co-transplantation with male tissues would be informative. Consistent with previous reports in the rat [Moore and Price, 1942; Torrey, 1950; Mangoushi, 1975], castration of male hosts did not change the frequency of ovotestis induction in the mouse (Fig. 1c) [Taketo et al., 1985]. This result indicates that adult testis is not the only source of the male factor promoting testicular development.

McIntyre [1956] did not observe any ovotestis development from rat ovarian primordia transplanted alone into castrated male hosts. However, when co-transplanted with fetal testes, many ovarian transplants developed tubular structures. Co-transplantation with other male organs (adrenals, spleen, or intestine), on the other hand, did not induce ovotestes formation. Similar observations were made in the mouse by Turner and Asakawa [1964] and in our laboratory (Figs. 1d, 2). Furthermore, McIntyre et al. [1956] showed that fetal testes affect ovaries only when placed in close proximity to each other. These results suggest that fetal testes produce a diffusible testis determining factor that is active either over a short distance, with a brief lifespan, or produced in trace amounts. It is noteworthy that the portion of ovarian grafts closest to the co-transplanted testes does not necessarily contain the most prominent testicular structures (Fig. 2). Since the same blood vessels may be shared by grafted ovary and testis, the male factor may be transported via local blood circulation rather than by diffusion between the two grafts.

Age of Ovarian Grafts

In both the rat and the mouse, the developmental age of ovarian transplants is critical for ovotestis formation. In the rat, fetal ovaries on the 16th d.g. or earlier develop into ovotestes. Within this period, age has little effect on the frequency of ovotestis development [Buyse, 1935]. In the mouse, the critical age is on the 13th d.g. (SJL/J strain) or 14th d.g. (NCS strain) [Taketo et al., 1985]. Testicular differentiation in male fetuses begins on the 13th d.g. in the rat and the 12th d.g. in the mouse. It should be pointed out that fetal ovaries are still bipotential when male sexual differentiation has already begun. After the 16th d.g. in the rat and the 14th d.g. in the mouse fetal ovaries appear to lose the capacity to undergo testicular development.

Time Course of Ovotestis Development

In most reports, ovotestis development was observed about the 14th d.g. or later after transplantation. McIntyre [1956] examined the ovarian grafts at different times and found that testicular structures became apparent by the third week after transplantation. Ozdzenski et al. [1976] followed the time course, but they focused on the differentiation of germ cells and did not give a detailed description of testicular structures. Because of the delayed formation of testicular structures in ovarian grafts in comparison to that in normal testes, these structures were considered secondary to the degeneration of oocytes

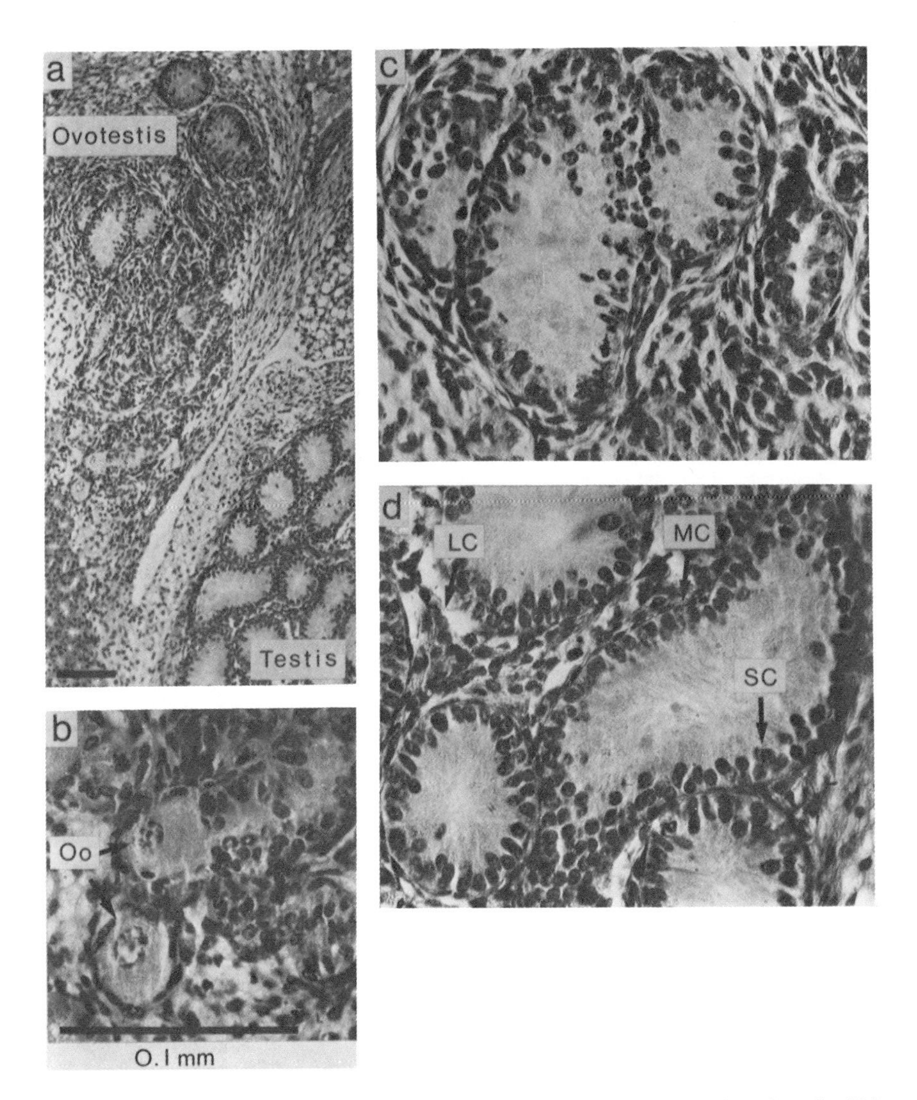

Fig. 2. Development of mouse ovarian and testicular primordia co-transplanted on the 12th d.g. into a castrated adult male mouse (SJL/J strain). H & E stain. a) An ovotestis and a testis developed in close association; b) an oocyte (Oo) is present in a testis cord of ovotestis; c) a part of ovotestis; d) a part of testis. LC, Leydig cells; MC, myoid cells; SC, Sertoli cells.

[MacIntyre, 1956; Ozdzenski et al., 1976]. However, as demonstrated, testicular cells do differentiate in the ovarian grafts. It is therefore important to examine in greater detail the development of testis cords at different stages during the growth of ovarian grafts. With this objective in mind, the time course of ovotestis development in fetal mouse ovaries transplanted on the 12th d.g. was examined [submitted for publication]. The results are summarized below.

Ovarian grafts appear to undergo normal ovarian development until the seventh day after transplantation. When these grafts are dissected at this stage and cultured, they develop as ovaries. Between the 7th and 11th day, oocytes are surrounded by sex cords, and some start to degenerate. The pregranulosa cells surrounding the healthy oocytes differentiate into granulosa cells and form follicles. When oocytes have degenerated, pregranulosa cells differentiate into Sertoli cells and form testis cords. The appearance of occasional epithelial cells with admixed characteristics of granulosa and Sertoli cell types supports this hypothesis. Morphological differentiation of myoid cells and steroidogenic (Leydig) cells follows Sertoli cell differentiation. When these ovarian grafts, on the 11th day after transplantation, are excised and cultured, they continue to develop as ovotestes and produce testosterone. Thus, the critical period for ovotestis formation is between the 7th and 11th day after transplantation, when sex cords develop into follicles in the course of normal ovarian development. Thus, the process of testis cord formation in ovotestes resembles in many aspects that of normal testicular development, except for the developmental stages of occurrence. In the future, it will be important to determine the factors that account for the time difference in sex cord formation between the male and female gonads, and direct the differentiation of gonadal epithelial cells into granulosa or Sertoli cells.

GONADAL DEVELOPMENT IN VITRO

To clarify the mechanism of gonadal differentiation, it would be advantageous to establish an in vitro culture system that promotes growth and development of gonadal primordia. Culture systems for gonadal tissues have been reported for the hamster [Baker, 1976], rat [Magre et al., 1981], and mouse [Byskov and Saxen, 1976; Byskov, 1978]. In most culture systems, serum is added to sustain development of the fetal gonads. The study will be facilitated by establishing a chemically defined medium so that the role of individual hormone or nutrient can be determined without the masking effects of serum [Barnes and Sato, 1980]. In the mouse, it was found that serum in the medium is essential for gonadal differentiation only during early stages of development [Bardin et al., 1982]. When gonadal tissues were explanted on the 12th d.g. or later, they continued to develop in Eagle's minimum essential medium (MEM) (without serum) directed by their genetic sex (Fig. 3e,f) [Taketo, 1984b]. On the other hand, when gonadal primordia were explanted on the 11th d.g., serum was essential (Fig 3a–d). Recently

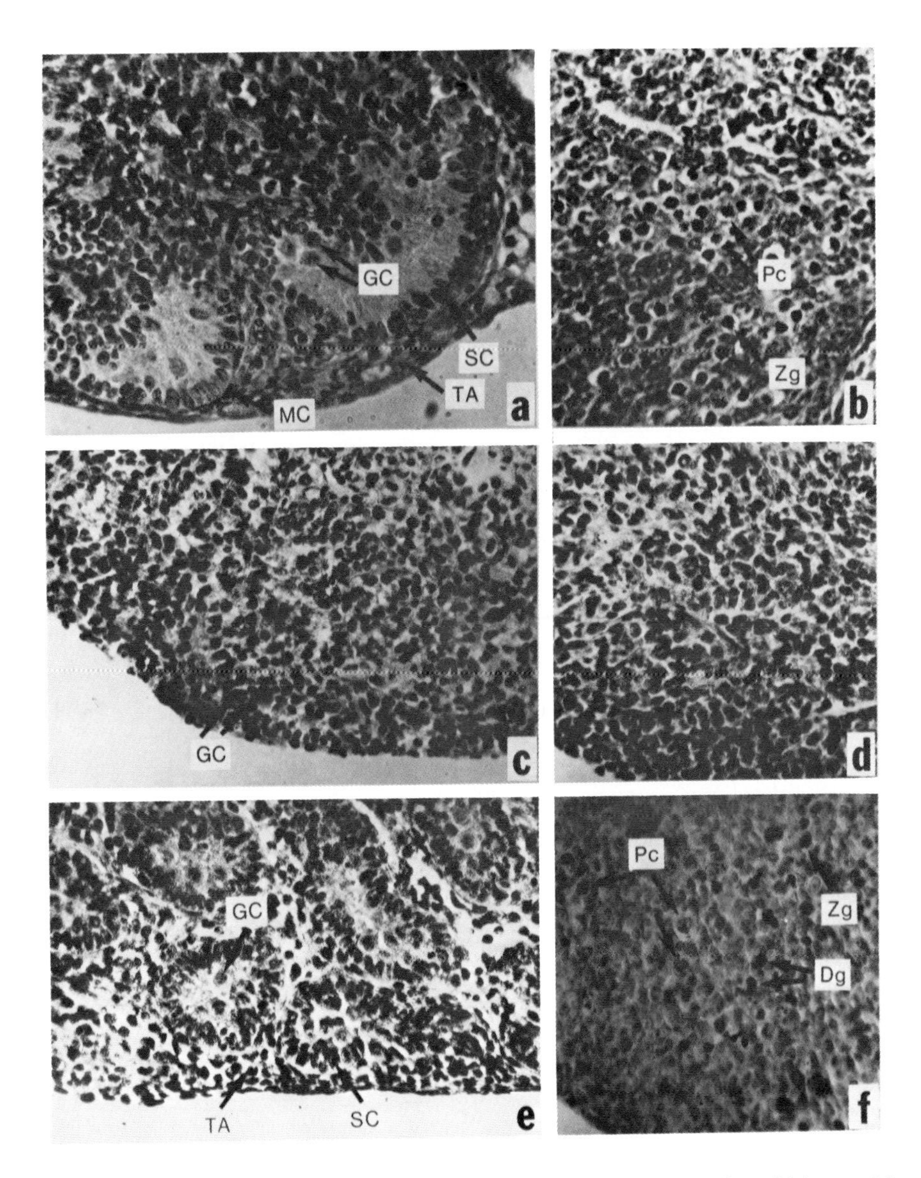

Fig. 3. Gonadal development after in vitro culture for seven days. H&E stain. a) Male gonadal primordium isolated on the 11th d.g. and cultured in MEM supplemented with 10% horse serum. Testis cords contain myoid cells (MC), fetal Sertoli cells (SC), and germ cells (GC). Leydig cells are present in the interstitial region. TA, tunica albugenia. b) Female gonadal primordium isolated on the 11th d.g. and cultured in serum-supplemented MEM. Many germ cells have reached zygotene (Zg) or pachytene (Pc) stage of meiotic prophase. c) Male gonadal primordium isolated on the 11th d.g. and cultured in serum-free MEM. No gonadal structure is seen other than occasional primordial germ cells. d) Female gonadal primordium isolated on the 11th d.g. and cultured in MEM alone. No germ cells are seen. e) Fetal testis isolated on the 12th d.g. and cultured in MEM alone. Fully developed testis cords are distributed in large area. f) Ovarian primordium on the 12th d.g. and cultured in MEM alone. Germ cells at zygotene or pachytene stage of meiotic prophase are seen among many degenerated germ cells (Dg). Magnification × 100.

the authors have studied how to replace serum with well-characterized serum components. Gonadal primordia explanted on the 10th d.g. could not differentiate even in the presence of serum [Taketo, 1981], whereas the isolated gonadal primordium on the 10th d.g. developed into an ovary or testis when transplanted into an adult host [unpublished data]. This finding suggests that the in vitro culture condition lacks an essential factor(s) for the differentiation of gonadal primordia on the 10th d.g.

Comparable Gonadal Development In Vivo and In Vitro

The sequence of events in the development of fetal mouse testis and ovary are as follows. Primordial germ cells appear in the yolk sac by the 8th d.g. and migrate via the gut and dorsal mesentry to congregate in the gonadal ridges situated on the coelomic side of the mesonephros by the 11th d.g. In the fetal testis, pre-Sertoli cells are identified on the 12th d.g. based on their position within the sex cords, their association with primordial germ cells, and the deposition of a basement membrane. Accumulation of myoid cells at the peripheral boundary of each cord is occasionally seen during the early stages of testicular development. Testosterone production is detectable with radioimmunoassay about the 12th d.g. [unpublished data], although fetal Leydig cells are recognized morphologically on the 13th d.g. or later. On the 15th d.g., the testis cords are well organized and contain germ cells in the prespermatogonia stage. These basic structures do not change until after birth, when germ cells start spermatogenesis.

Ovarian organization is delayed somewhat as compared to testicular differentiation. Ovigerous cordlike structures are seen clearly under the light microscope on about the 16th d.g. These structures are transitory and disappear when follicles are formed. In contrast, germ cells in fetal ovaries enter meiotic prophase by the 13th d.g. and become arrested at the dictyate stage when enveloped in follicles.

Gonadal development in vitro is comparable to that in the fetal mouse [Taketo and Koide, 1981], although the size of tissue, number of germ cells, testis cords, or follicles are always smaller after culture than those which develop in vivo. When gonadal primordia on the 11th d.g. are isolated and cultured in serum-containing medium, male explants develop sex cords by the third day of culture and well-organized testis cords by the fifth day (Fig. 3c). At this stage, Sertoli cells, myoid cells, and Leydig cells are recognized with light microscopy, and germ cells are arrested at the prespermatogonial stage. These explants start to produce testosterone by the second day of culture [unpublished data]. On the other hand, female explants usually do not begin follicle formation until the seventh day or later after explantation, whereas germ cells enter meiosis by the fifth day, and many are at the zygotene or pachycene stage around the seventh day (Fig. 3d). Thus, there is a delay in gonadal development in vitro, but morphologically and functionally, at least as to steroid production, there is no qualitative difference between in vitro and in vivo development. When fetal gonads on the 12th d.g. or later are explanted, their growth proceeds faster than those explanted on the 11th d.g., and their developmental stages approximate closely to the in vivo development [Taketo et al., 1984b].

Effect of Serum Components on Testicular Development

When gonadal primordia on the 11th d.g. were cultured in MEM alone for one or two days, and subsequently in serum containing medium, testis cords formed but contained a few or no germ cells [unpublished data]. This observation suggested that different serum components are required for germ cell survival and for somatic cells to organize into testis cords. It was found that human low density lipoprotein fraction increases the number of germ cells enclosed in testis cords, whereas lipoprotein-free serum fraction and progesterone at 20 ng/ml have weak activity. The ability to form testis cords decreases gradually during culture in serum-free medium. By culturing gonadal primordia on the 11th d.g. in MEM, various compounds could be screened for testis-cord organizing activity. It was found that the combination of human albumin and insulin can induce testis cord formation exceeding the capacity of whole serum [unpublished data] and testosterone promotes the testis cord formation. These serum components do not induce ovarian differentiation, i.e., onset of meiosis.

Inhibition of Gonadal Differentiation by cAMP Analogues

Testicular organization is prevented when cAMP analogues, forskolin, or prostaglandins are added to the culture medium [Taketo et al., 1984b]. This inhibitory effect by cAMP analogues is potentiated when combined with phosphodiesterase inhibitors. The most evident change induced by cAMP analogues is the disintegration of the basement membranes of the sex cords, whereas germ cells and Leydig cells appear to be morphologically normal. Germ cells reached the prespermatogonia (resting) stage in the presence or absence of cAMP analogue. The production of testosterone from testicular explants, probably by the Leydig cells, increases in the presence of cAMP analogues as expected for normal testes. These findings suggest that the formation of basement membranes plays a key role in testis organogenesis. It has been suggested that the interaction between Sertoli cells and myoid cells is important for the development of basement membrane [Tung, 1980]. Cyclic AMP analogues may affect either Sertoli cells or myoid cells, or both, and prevent organization of testis cords.

In the presence of dibutyryl cAMP, the germ cells of ovarian primordia explanted on the 12th d.g. showed high mitotic activity until the second or third day of culture. Thereafter, they degenerated and did not enter meiosis [unpublished data]. It has been suggested that the sex difference in the timing of meiosis is regulated by two competitive factors [Byskov, 1978; Byskov and Saxen, 1976]. One factor that induces meiosis is produced by the rete portion of the fetal urogenital complex, and the other that inhibits meiosis is produced by the fetal testis. The physico-chemical nature, or the role of these factors in gonadal differentiation, is not known. The effect of cAMP analogues on female germ cells may be related to that of meiosis inhibiting factors producted by testes.

SUMMARY

The work presented here indicates that fetal ovaries at early stages of development can form testicular structures even in the absence of any male-specific factors. Therefore, genes involved in testicular organization must be located on autosomal or X chromosomes. However, in some conditions, for example, when fetal ovaries are transplanted with mesonephroi attached or in the freemartin, the presence of male factors promotes testicular development. Thus, Y-chromosome-linked genes do have the capacity to induce (or promote) testicular development. The male factor appears to be circulating in male fetuses, male adults, and also to be produced by fetal testes.

The time course of ovotestis development and the study of the inhibitory effect of cAMP analogues on in vitro testicular development suggest that the differentiation of Sertoli cells, and thereby formation of basement membrane, play a key role in testis cord organization. However, in both systems, testosterone production also was found to be one of the earliest events during testicular development, even though morphologically typical Leydig cells may be recognized at somewhat later stages. In our in vitro system, low concentration of testosterone promoted testis cord formation if added to serum-free defined medium. It has been suggested that myoid cells differentiate late and are dependent on testosterone [Muller and Schindler, 1983]. Although exogenously applied steroids do not change the direction of gonadal differentiation, locally accumulated testosterone may be important for testicular organization.

Gonadal sex reversal induced by transplantation in the mouse is comparable to freemartinism. Therefore, a common mechanism may be involved in gonadal sex reversal in these two systems. On the other hand, the present findings are hardly explained on the basis of the sex-determining role of H-Y antigen, because ovotestes can develop in a female host environment in which H-Y antigen is presumably absent. The study of gonadal sex reversal following transplantation will continue to provide much valuable information on the mechanism of gonadal sex determination and differentiation in mammals.

ACKNOWLEDGMENTS

This study was supported by grants from the Rockefeller Foundation: GAPS 8418 and the National Institutes of Health: HD13184, HD18669.

REFERENCES

Baker BS (1976): The genetic control of meiosis. Annu Rev Genet 10:53–134.

Bardin CW, Taketo T, Gunsalus GL, Koide SS, Mather JP (1982): The detection of agents that have toxic effects on the testis and male reproductive tract. Banbury Report 11:337–351.

Barnes D, Sato G (1980): Methods for growth of cultured cells in serum-free medium. Anal Biochem 102:255–270.

Benhaim A, Gangnerau MN, Bettane-Casanova M, Fellous M, Picon R (1982): Effect of H-Y antigen on morphologic and endocrine differentiation of gonads in mammals. Differentiation 22:53–58.

Buyse A (1935): The differentiation of transplanted mammalian gonad primordia. J Exp Zool 70:1–41.

Byskov AG (1978): The meiosis inducing interaction between germ cells and rete cells in the fetal mouse gonad. Ann Biol Anim Biochim Biophys 18:327–334.

Byskov AG, Grinsted J (1981): Feminizing effect of mesonephros on cultured differentiating mouse gonads and ducts. Science 212:817–818.

Byskov AG, Saxen L (1976): Induction of meiosis in fetal mouse testis in vitro. Dev Biol 52:193–200.

Donohue PK, Budzik GP, Trelstad R, Mudgeth-Hunter M, Fuller Jr A, Hutson JM, Ikawa H, Hayashi A, MacLaughlin D (1982): Müllerian inhibiting substance: an update. Recent Prog Horm Res 38:279–326.

Eicher EM, Washburn LL (1983): Inherited sex reversal in mice: Identification of a new primary sex-detecting gene. J Exp Zool 228:297–304.

Eichwald EJ, Silmser WK (1955): Skin Transplant Bull 2:148–149.

Fechheimer NS, Herschler MS, Gilmore LD (1963): Sex chromosome mosaicism in unlike sexed cattle twins. In "Genetics Today." London: Pergamon Press, pp 265.

Flickinger JC (1967): The postnatal development of the Sertoli cells of the mouse. Z Zellforsch 78:92–113.

Goldberg EH, Boyse EA, Bennett D, Scheid M, Carswell EA (1971): Serological demonstration of H-Y (male) antigen on mouse sperm. Nature 232:478–480.

Herschler MS, Fechheimer NS (1967): The role of sex chromosome chimerism in altering sexual development of mammals. Cytogenetics 6:204–212.

Holyoke EA (1949): The differentiation of embryonic gonads transplanted to the adult omentum in the albino rat. Anat Rec 103:675–699.

Josso N, Picard J-Y, Tran D (1977): The antimüllerian hormone. Recent Prog Horm Res 33:117–163.

Jost A, Magre S (1984): Testicular developmental phases and dual hormonal control of sexual organogenesis. In Serio M (ed): "Sexual Differentiation: Basic and Clinical Aspects." New York: Raven Press pp. 1–15.

Jost A, Vigier B, Prepin J (1972): Freemartins in cattle: The first steps of sexual organogenesis. J Reprod Fertil 29:349–379.

Lillie FR (1916): The theory of the free-martin. Science 43:611–613.

Lillie FR (1917): The free-martin: A study of the action of sex hormones in the foetal life of cattle. J Exp Zool 23:371–452.

MacIntyre MN (1956): Effect of the testis on ovarian differentiation in heterosexual embryonic rat gonad transplants. Anat Rec 124:27–41.

Magre S, Agelopoulou R, Jost A (1981): Morphogenese Animale—Action du serum de foetus de veau sur la differentiation in vitro ou le maintien des cordons seminiferes du testicule de foetus de rat (F). C R Acad Sci (Paris) t 292 Serie III 85–89.

Mangoushi MA (1975): Scrotal allografts of fetal ovaries. J Anat 120:595–599.

Melvold RW, Kohn HI, Yerganian G, Fawcett DW (1977): Evidence suggesting the existence of two H-Y antigen in the mouse. Immunogenet 5:33–41.

Moore CR, Price D (1942): Differentiation of embryonic reproductive tissues of the rat after transplantation into post-natal hosts. J Expt Zool 90:229–265.
Müller U, Schindler H (1983): Testicular differentiation—a developmental Cascade. Morphogenetic effects of H-Y antigen and testosterone in the male mammalian gonad. Differentiation [Suppl] 23:S99–S103.
Müller U, Zenzes MT, Bauknecht T, Wolf U, Siebers JW, Engel W (1978): Appearance of hCG-receptor after conversion of newborn ovarian cells into testicular structures by H-Y antigen in vitro. Hum Genet 45:203–207.
Müller U, Urban E (1981): Reaggregation of rat gonadal cells in vitro: experiments on the function of H-Y antigen. Cytogenet Cell Genet 31:104–107.
Ohno S (1979): Major sex-determining genes. Monogr Endocrinol, vol. 2 pp. 19–87.
Ohno S, Nagai Y, Ciccarese S, Iwata H (1979): Testis-organizing H-Y antigen and the primary sex determining mechanism of mammals. Recent Prog Horm Res 35:449–478.
Ozdzenski W, Rogulska T, Batakier H, Brzozowska M, Rembiszewska A, Stepinska, HU (1976): Influence of embryonic and adult testis on the differentiation of embryonic ovary in the mouse. Arch D Anat Microsc 65:285–294.
Polani PE, Adinolfi M (1983): The H-Y Antigen and its functions: A review and a hypothesis. J Immunogenet 10:85–102.
Silvers WK, Wachtel, SS (1977): H-Y antigen: Behaviour and function. A cell surface component of vertebrates may be directly involved in primary sex determination. Science 195:956–960.
Simpson E (1982): Sex reversal and sex determination. Nature 300:404–406.
Swann DA, Donahoe PK, Ito V, Morikawa Y, Hendren WH (1979): Extraction of müllerian inhibiting substance from newborn calf testis. Dev Biol 69:73–84.
Taketo T, Koide SS (1981): In vitro development of testis and ovary from indifferent fetal mouse gonads. Dev Biol 84:61–66.
Taketo T, Merchant-Larios H, Koide SS (1984a): Induction of testicular differentiation in the fetal mouse ovary by transplantation into adult male mice. Proc Soc Exp Biol Med 176:148–153.
Taketo T, Thau RB, Adeyemo O, Koide SS (1984b): Influence of Adenosine 3′:5′-cyclic monophosphate analogues on testicular organization of fetal mouse gonads in vitro. Biol Reprod 30:189–198.
Taketo T, Koide SS, Merchant-Larios H (1985): Induction of testicular development in the fetal mouse ovary. Ann New York Acad Sci 438:671–674.
Torrey TW (1950): Intraocular grafts of embryonic gonads of the rat. J Exp Zool 115:37–58.
Tung PS, Fritz JB (1980): Interactions of Sertoli cells with myoid cells in vitro. Biol Reprod 23:207–217.
Turner CD, Asakawa H (1962): Differentiation of fetal rat ovaries following transplantation to kidneys and testes of adult male hosts. Am Zoolog 2:270.
Turner CD, Asakawa H (1964): Experimental reversal of germ cells in ovaries of fetal mice. Science 143:1344–1345.
Vigier B, Locatelli A, Prepin J, du Mesnilde Bussion F, Jost A (1976): Les premieres manifestation du "Freemartinisme" chez le foetus de veau ne dependent pas du chimerisme chromosomique XX/XY. C R Acad Sci (Paris) 282:1355–1358.
Wachtel SS (1980): The dysgenetic gonad: Aberrant testicular differentiation. Biol Reprod 22:1–8.
Wachtel SS, Koo GC, Ohno S (1977): H-Y Antigen and male development. In Troen P, Nankin HR (eds): "The Testis in Normal and Infertile Men." p. 35–43.
Whitten WK, Beamer WG, Byskov AG (1979): The morphology of fetal gonads of spontaneous mouse hermaphrodities. J Embryol Exp Morphol 52:63–78.

Wilson JD, Siiteri PK (1973): Developmental pattern of testosterone synthesis in the fetal gonad of the rabbit. Endocrinology 92:1182–1191.

Witschi E (1934): Genes and inductors of sex differentiation in amphibians. Biol Rev 9:460–488.

Zenzes MT, Wolf U, Engel W (1978a): Organization in vitro of ovarian cells into testicular structures. Hum Genet 44:333–338.

Zenzes MT, Wolf U, Gunther E, Engel W (1978b): Studies on the function of H-Y antigen: Dissociation and reorganization experiments on rat gonadal tissue. Cytogenet Cell Genet 20:365–372.

The Origin and Evolution of Sex, pages 289–300

Relation of Germ Cell Sex to Gonadal Differentiation

Anne McLaren

Medical Research Council Mammalian Development Unit, Wolfson House, London NW1 2HE

The sexual characteristics which impinge most directly on our attention, and which have been celebrated by artists and writers over the centuries, are what are termed secondary sexual characteristics. Some aspects of the sexual phenotype develop autonomously, even in the absence of the gonads; this is especially true of the female phenotype. However, most mammalian sexual characteristics depend on sex hormones produced by the gonads, estrogens from the ovaries, and androgens from the testes. It is for this reason that gonadal differentiation is of such paramount importance in any consideration of sexual phenotype in mammals.

GENESIS OF THE MAMMALIAN GONAD

The first appearance of the future gonad consists in the formation of a blastema immediately beneath the coelomic epithelium of each urogenital ridge and adjacent to the mesonephric kidney. The blastema, together with the overlying epithelium, constitutes the genital ridge. Much discussion has centered on the question of the origin of this blastema. Some authorities have claimed that it originates by proliferation of the cells of the coelomic epithelium or from the adjacent mesenchyme [Gruenwald, 1942; Franchi et al., 1962), others that the blastemal cells all come from the mesonephros [Zamboni et al., 1979, 1981]. Others again have supported a dual origin [Merchant, 1975; Pelliniemi, 1975]. In species in which a cortical and a medullary region can be distinguished, some authorities derive the cortex from the coelomic epithelium and the medullary region from the mesonephric system [Witschi,

1951]. Others believe that within each tissue (e.g., Sertoli cells in the testis, granulosa cells in the ovary) the cells have a dual origin [Wartenberg, 1981; Byskov, 1975].

The diverse viewpoints quite inadequately summarized above are based entirely on histological and ultrastructural studies. Except perhaps with the aid of a very precisely timed series of pictures, this type of static investigation is unlikely to provide conclusive evidence as to the direction in which cells are moving. However, recent findings suggest that much of the controversy stems from differences between species. Zamboni was one of the strongest proponents of the mesonephric origin of the gonadal blastema, on the basis of the detailed study that he and his colleagues carried out on the sheep [Zamboni et al., 1979, 1981]. In this species the most anterior element of the mesonephric kidney is a giant nephron; this ceases to be functional shortly before the genital ridge develops, and cells of the giant nephron appear to dedifferentiate and to migrate out of the mesonephric region towards the coelomic epithelium so as to constitute the gonadal blastema. An equally careful study by the same group on the early development of the rhesus monkey gonad suggests that here the gonadal blastema is formed in part by proliferation of the nephrogonadoblastic cells lining the surface of the genital ridge, and in part by the nephrogonadoblastic cells that in previous stages had differentiated into mesonephric elements (L. Zamboni, personal communication).

Analogous observations were made by the same authors on the mouse, in which the somatic cells of the gonadal blastema are contributed by the tubules situated in the most cranial region of the mesonephros [Upadhyay et al., 1979, 1981]. It may be no coincidence that in the two species where mesonephric cells are thought to be involved, the mesonephric kidney is either never functional (mouse) or has ceased to be functional at the level of the genital ridge by the time that the blastema forms (sheep), whereas in the species where gonadal cells appear to be recruited from the coelomic epithelium (rhesus monkey), the mesonephros continues to function actively as an excretory organ beyond the time of gonadal blastema formation.

The gonadal blastema in the mouse begins to form at about 10 days post coitum. Within 2–3 days the testis is morphologically distinguishable from the ovary: testis cords have formed, consisting of densely packed germ cells surrounded by a layer of pre-Sertoli cells and an outer sheath of very elongated peritubular cells, and between the cords are situated the precursors of the Leydig cells. Thus, at least three distinct somatic cell types already have differentiated from the gonadal blastema. Within another couple of days, it becomes possible to detect the first hormone to be produced by the testis, namely anti-Müllerian hormone [Josso et al., 1977]. This is a glycoprotein responsible for the regression of the Müllerian duct in males. The developing gonads also have a rich blood supply, with a particularly striking blood vessel around the periphery of the testis.

The somatic cells of the gonadal blastema appear initially to be homogeneous, but it is not known whether they are homogeneous with respect to developmental potential. Perhaps before leaving the mesonephros (if this is indeed their origin) some are already determined in the male to become Sertoli cells and others to become Leydig cells. Alternatively, they may be equipotent at the time of leaving the mesonephros and may undergo a process of serial induction within the blastema. The first recognizable cell type to differentiate in the rat testis are the pre-Sertoli cells; these are first seen close to the mesonephros [Jost, 1972; Magre and Jost, 1980], and this cellular differentiation step may take place even though the formation of testis cords is inhibited [Agelopoulou et al., 1984]. Perhaps, therefore, the first cells to leave the mesonephros encounter some chemical stimulus from the coelomic epithelium and are induced to develop as pre-Sertoli cells. These in turn, as they undergo aggregation to form cords, may induce the neighbouring cells to differentiate as peritubular cells, whereas those cells not in contact with a pre-Sertoli cell may develop subsequently into Leydig cell precursors. It is easy thus to speculate; but it is hard to see how any real understanding of the causal basis of early gonadal differentiation can be obtained in the absence of the type of in vitro system that has proved so informative in the case of kidney development [Ekblom, 1981].

DETERMINATION OF GONADAL SEX

At what point are the somatic cells that go to form the gonadal blastema determined to form either a male or a female gonad? In the normal male these cells have an XY chromosome constitution, and in the female they are XX, but we know from studies on XX ↔ XY chimeras that neither XY nor XX somatic cells develop autonomously. When an XX and an XY cell population coexist in an embryo, only rarely is an ovotestis with separate ovarian and testicular regions seen. More usually, the gonads develop as phenotypically normal testes. A review of the literature on XX ↔ XY chimeric mice shows that testicular development tends to be found in individuals that in the adult have 25–30% or more XY cells, ovarian development in those that have 15–20% or fewer XY cells, whereas ovotestes are associated with the intermediate range [McLaren, 1984].

The results on chimeras suggest that XY cells produce some inducing factor that ensures that all cells (XY or XX) in the gonadal blastema differentiate as testicular rather than ovarian cell types (e.g., pre-Sertoli and Leydig cells rather than follicle and interstitial cells). The inducing factor could act within the gonadal blastema itself or at an earlier stage, e.g., in the mesonephros, or perhaps still earlier, during gastrulation. The cells that produce the inducing factor need not be located within the gonad, but could be anywhere within the embryo. It is not known whether XX cells produce a comparable factor that induces ovarian development or whether the gonadal blastema develops into an ovary whenever the level of testis-inducing substance is below some threshold.

Because XXY embryos develop testes and XO embryos develop ovaries, the testis-inducing substance must be controlled by the presence of a Y rather than by the number of X chromosomes. From the in situ hybridization studies of Singh and Jones [1982] using a DNA probe that hybridizes to the testis-determining sequences of the mouse, it is now known that these sequences normally are located near the centromere of the Y chromosome. In the condition known as sex reversal (*Sxr*), an extra copy of these sequences is located at the distal end of the Y chromosome beyond the pairing segment, and is therefore transferred to the X chromosome by crossing over between the X and Y during male meiosis [for review, see McLaren, 1983c]. The resulting XX *Sxr* individuals develop as phenotypic males. A full account of this and other sex-reversed conditions has been given by Eicher [1982].

The hypothetical testis-inducing substance is presumably either absent from females or at least present in a lower concentration than in males. The first male-specific substance to be reported in mammals was H-Y antigen, a histocompatibility antigen identified by grafting male skin onto female mice of the same inbred strain. It appeared to be controlled by the Y chromosome and proved to be H-2 restricted, as are many antigens eliciting T-cell-mediated responses [Simpson, 1983]. Females that have rejected male grafts may show anti-male cytotoxic antibody in their serum; the antigen (or antigens) that elicit this B-cell-mediated response is often referred to also as H-Y but more properly should be termed serologically detectable male antigen (SDM antigen) [Silvers et al., 1982].

Whether H-Y and SDM represent the same or different antigens, and whether either can be equated with the testis-determining substance, has been hotly debated [for discussion see Silvers et al., 1982; Wachtel and Koo, 1981; Stewart, 1983]. Certainly neither antigen constitutes a sufficient condition for male development. Female XO mice are reported to be positive for SDM [Engel et al., 1981] though negative for H-Y [Simpson et al., 1982]. Other female mice are positive for H-Y, including some that are XY in sex chromosome constitution, with a Y chromosome on an unaccustomed genetic background that does not allow it to exercise its testis-determining function [Eicher et al., 1982], and others that carry *Sxr*. The latter females are thought to be mosaics, with one population of cells in which *Sxr* is expressed and another in which the *Sxr* region is inactivated, together with the X chromosome to which it is attached. Such mosaic mice may develop as females or males or intersexes [McLaren and Monk, 1982; Cattanach et al., 1982], but whatever their phenotypic sex they are positive for H-Y antigen [Simpson et al., 1984]. Recently a variant *Sxr* region (*Sxr′*) has been identified that has lost its H-Y antigenicity but retains its testis-determining capacity [McLaren et al., 1984]. XX *Sxr′* individuals develop as phenotypically normal males but are negative for H-Y histocompatibility antigen. Their SDM antigen status has not yet been determined [for a discussion of the methodological problems involved in testing for SDM antigen, see Zenzes and Reed, 1984]. This result suggests that H-Y antigenicity is not the inducing stimulus for testis development; however, the role of SDM, if it is indeed a distinct antigen, remains to be clarified.

ROLE OF GERM CELLS

Against this background of gonadal differentiation, what can we say about the role of the germ cells?

Primordial germ cells in the mouse can be identified first at about eight days post coitum in the mesoderm at the base of the allantois. From there they migrate or are carried along the wall of the invaginating hind gut, finally travelling up the dorsal mesentery and into the genital ridges at just about the time that the gonadal blastema is forming. By 11 days post coitum, germ cells and somatic cells are closely intermingled within the blastema.

The question arises, does the presence of the germ cells affect the formation of the gonad? There is no evidence that it does. The number of germ cells entering the genital ridges can be reduced to a very low level, either by genetic means (mouse embryos homozygous for the mutants Steel or White-spotting) or by treatment of the mother with the drug Busulfan during the period of germ cell migration [Merchant, 1975], without affecting the differentiation of the gonadal blastema in either sex. In particular, the pre-Sertoli cells that normally enclose the germ cells in the testis cords still aggregate to form a cord, even though no germ cells are present within that cord. However, neither in mutant nor in Busulfan-treated embryos is it possible to ensure that no germ cells at all reach the genital ridges [McCoshen, 1983], so some signalling or inducing role for the germ cells, though unlikely, cannot be excluded.

When aggregation chimeras are made between normal embryos and those homozygous for White-spotting, so that the germ cell population in the genital ridges is almost entirely drawn from one component, the phenotypic sex of the gonad, and hence of the embryo, does not necessarily coincide with the chromosomal sex of the predominating germ cell population [McLaren and Buehr, unpublished observations]. This suggests that any inducing role that the germ cells might play is not concerned with sexual differentiation and in particular that the hypothetical testis-inducing substance does not emanate from the germ cells.

Postnatal differentiation of the gonads is much more dependent on the presence of germ cells. In XX sex-reversed male mice, the germ cells degenerate within a few days of birth. The adult testis is small and appears to contain a disproportionate amount of interstitial tissue. This is partly because the seminiferous tubules, although present and apparently normal in structure, are devoid of germ cells and hence greatly reduced in size; however, in older *Sxr* males, the interstitial tissue becomes hyperplastic, and ultrastructural alterations have been detected in the Leydig cells [Chung et al., 1972]. Closer investigation reveals that the Sertoli cells are also abnormal, and some aberrant cells free in the lumen of the tubules have been identified as immature Sertoli cells [Chung, 1974]. The ovary is much more obviously affected by absence of germ cells. Without oocytes around which to aggregate, no follicles are formed; hence no corpora lutea develop, and all that remains is the "streak" ovary characteristic of the female mule or the adult Turner's patient.

GERM CELL SEX

Although germ cells do not influence the phenotypic sex of the gonad, it seems that the gonad may influence the phenotypic sex of the germ cells.

By definition, germ cells that undergo oogenesis are phenotypically female, and those that undergo spermatogenesis are phenotypically male. We know that XY germ cells undergo spermatogenesis in the normal testis, and XX and XO germ cells undergo oogenesis in the ovary. However, is it the germ cell's chromosome constitution or its environment that determines the direction of differentiation? It is increasingly clear that environmental factors are largely or even entirely responsible for whether a germ cell is phenotypically male or female. XY germ cells can undergo oogenesis in a variety of situations: in the ovaries of female XX $\leftrightarrow$ XY chimeras [Ford et al., 1975; Evans et al., 1977], in the disrupted testes of XY male embryos grafted under the kidney capsule [Ozdzenski, 1972; Ozdzenski and Presz, 1984], and in the adrenal glands of male mice [Upadhyay and Zamboni, 1982; McLaren, 1985]. Spermatogenesis, in contrast, occurs as far as we know only in the testis, but is not confined to XY germ cells; in an XO *Sxr* testis, the XO germ cells pass through the early stages of spermatogenesis in an apparently normal manner, though few if any normal spermatozoa are formed [Cattanach et al., 1971]. Neither XX nor XXY germ cells in a testis undergo spermatogenesis, as the second X chromosome leads to degeneration of the spermatogonia soon after birth, at a premeiotic stage, but up to this point their development has been phenotypically male rather than female.

In all circumstances the germ cells that pursue a female pathway of development, and embark on oogenesis, are those that in the mouse enter meiosis before birth, at the time that germ cells in meiotic prophase are first seen in the normal fetal ovary. Those that fail to enter meiosis before birth either degenerate or undergo spermatogenesis, with the first meiotic stages seen during the second week of postnatal life. It may therefore be concluded that the phenotypic sex of a germ cell depends on how early it enters meiosis, and this appears to be independent of the cell's chromosomal sex.

ENTRY INTO MEIOSIS

What then determines the stage at which a germ cell enters meiosis? Relevant observations have been made on germ cells in extragonadal sites and in gonads maintained in vitro.

Not all primordial germ cells succeed in colonizing the genital ridges. Some end up instead in the nearby adrenal primordium, and others remain in the mesonephric region of the urogenital ridge. Upadhyay and Zamboni [1982], in an ultrastructural study of fetal and postnatal mouse adrenals, identified germ cells in the adrenals of most individuals up to about two weeks of age. The germ cells in the adrenals entered meiotic prophase at the same stage of fetal development as did those in the ovaries, and this occurred not only in female but also in male adrenals. In both sexes the meiotic germ cells went on after birth to develop into growing oocytes, surrounded by zonae pellucidae as in the ovary [Zamboni and Upadhyay, 1983]. The findings of Upadhyay and Zamboni [1982] were confirmed by identifying the germ cells histochemically by their high alkaline phosphatase activity with a new fixation technique that allows the meiotic status of the nucleus also to be ascertained [McLaren and Hogg, unpublished observations]. We have examined also the germ cells within the fetal testis but outside the testis cords, and those outside the testis in the mesonephric and mesenchymal tissue of the urogenital ridge. These form a mixed population, some in meiotic prophase (as in the ovary) and some in mitotic arrest (as in the testis) [McLaren, 1985]. Similar observations have been made recently by Francavilla and Zamboni [1985].

The simplest interpretation of these findings is that all germ cells enter meiosis autonomously at the same stage of fetal life as in the ovary, unless they are prevented from doing so by some inhibitory substance emanating from the testis. Within the testes, the inhibition is usually total, though a few meiotic germ cells have been reported in the fetal testes of XX $\leftrightarrow$ XY chimeras and XX *Sxr* males [see McLaren, 1980; 1983b], and of normal males of certain inbred strains of mice [Byskov, 1978]. Outside the testes, the inhibitory influence evidently does not extend as far as the adrenals and is only partially effective in the mesonephric region of the urogenital ridge.

A contrary view has been put forward by Byskov [1978], according to which germ cells do not enter meiosis autonomously, but only if stimulated to do so by a meiosis-inducing substance (MIS) produced by the mesonephros or by cells derived from the mesonephros. Again, the testis must be postulated to exert some inhibitory influence. The original finding that suggested the presence of a meiosis-inducing substance was that fetal germ cells appeared to enter meiosis more readily in grafted or cultured ovaries when the mesonephric region was retained than when it was removed. More recently, conditioned media and fluids supposedly containing MIS have been claimed to induce meiosis in male gonads cultured from the sexually indifferent stage [Byskov, 1985]. Such treatments tend, however, to prevent the development of normal testis cords and could at the same time also interfere with the inhibitory influence normally exerted by the testis.

The evidence for and against the two hypotheses has been reviewed in greater detail by McLaren [1985]. At the present time it is not obvious that either fits the available data better than the other. The postulated inhibitor or inducer will need to be better characterized, and its origin demonstrated, before a conclusion can be reached.

EFFECT OF GONADAL DIFFERENTIATION ON GERM CELL SEX

Some further information can be gleaned from studies in which genital ridges from embryos of different ages have been maintained in culture.

Taketo and Koide [1981] used an organ culture system to study the development in vitro of mouse genital ridges taken with the attached mesonephric region at a stage prior to sexual differentiation. When the ridges were taken from embryos $11\frac{1}{2}$ days post coitum (incorrectly termed by them "the 11th day of gestation"), development in vitro was qualitatively similar to in vivo in that ridges from female embryos formed ovaries in which all the germ cells entered meiosis, and those from male embryos formed testes with germ cells in mitotic arrest (T-prospermatogonia), enclosed within testis cords. In contrast, ridges taken one day earlier ($10\frac{1}{2}$ days post coitum) showed no gonadal differentiation and little if any germ cell development in culture. Neither T-prospermatogonia nor germ cells in meiotic prophase were seen.

Using a slightly different organ culture system, the same results were obtained as Taketo and Koide at $11\frac{1}{2}$ days post coitum, whether or not the mesonephric region was left attached to the genital ridge [McLaren, 1983a]. In the male ridges, testis cords were formed, but they tended to be few in number and abnormally wide, and they failed to lengthen, so that in some explants only a single wide short cord developed. None of the germ cells in the cultured testes entered meiosis; they appeared as T-prospermatogonia in mitotic arrest, just as in vivo. Within some cords, large groups of prospermatogonia persisted, though never as many as in a normal testis; but in others the somatic cells of the cord seemed to overgrow the germ cells, so that after a few days in culture only a few germ cells could be identified. Recently [McLaren and Hogg, unpublished observations], the same culture system was used to grow ridges from embryos $10\frac{1}{2}$ days post coitum, with better results than those reported by Taketo and Koide [1981]. Each explant included not only both entire urogenital ridges, but also the intervening tissue, including the dorsal aorta. In the female explants, large numbers of germ cells were seen after six days of culture, all in meiotic prophase. In male explants, no testis cords were formed, but pre-Sertoli cells were identified, aligned in small groups with a prominent basement membrane. Many fewer germ cells persisted than in the female explants, but again all were in meiotic prophase, and no T-prospermatogonia were seen.

These results support the observation of Agelopoulou et al. [1984] that pre-Sertoli cells can be found in cultured rat gonads in which testis cord formation has been prevented. Thus, cytodifferentiation, at least with respect to pre-Sertoli cells, is not dependent on the morphogenetic event of testis cord formation. It seems that the presence of pre-Sertoli cells is not in itself sufficient to inhibit the entry of germ cells into meiosis. The inhibitory influence exerted in the fetal testis in vivo must therefore stem from some other aspect of gonadal differentiation, some event that is unable to take place in vitro under our present culture conditions and that perhaps occurs in vivo between $10\frac{1}{2}$ and $11\frac{1}{2}$ days post coitum. Our results shed no light on whether entry of the germ cells into meiosis occurred autonomously or was induced.

SUMMARY

The phenotypic sex of a germ cell, i.e., whether it undergoes oogenesis or spermatogenesis, depends crucially on the stage at which it enters meiosis. If a germ cell that has entered meiotic prophase before birth survives at all, it develops into an oocyte and undergoes oogenesis. This is so of XY as well as XX germ cells, in the testis as well as in the ovary. So it seems that germ cells develop in the female direction because they have entered meiosis before birth, rather than entering meiosis before birth because they are female germ cells. The time of entry into meiosis appears to be a function of the tissue environment rather than the chromosome constitution of the germ cell, but it is not yet clear whether meiosis has to be induced by the environment or whether, on the contrary, entry into meiosis before birth is an autonomous property of germ cells. In either case, some inhibitory influence must be exerted by the testis. The differentiation of pre-Sertoli cells seems not in itself to be sufficient to block entry into meiosis. The origin of the gonadal blastema has been studied extensively, but nothing is known of the mechanisms of cell differentiation and tissue organization within the blastema. The formation of a testis rather than an ovary is probably dependent on the presence at some stage of development of a testis-inducing substance. This must be controlled by the Y chromosome, but it appears not to be associated with H-Y histocompatability antigen.

REFERENCES

Agelopoulou R, Magre S, Patsavoudi E, Jost E (1984): Initial phases of the rat testis differentiation *in vitro*. J Embryol Exp Morphol 83:15–31.

Byskov AGS (1975): The role of the rete ovarii in meiosis and follicle formation in different mammalian species. J Reprod Fertil 45:201–209.

Byskov AG (1978): Regulation of initiation of meiosis in fetal gonads. Int J Androl Suppl 2:29–38.

Byskov AG, Westgaard L (1985): Effect of meiosis inducing substances in vitro and in vivo. In Dickinson HG (ed): "Controlling events in meiosis." Society of Experimental Biology Symposium 38. Cambridge: Cambridge University Press, (in press).

Cattanach BM, Pollard CE, Hawkes SG (1971): Sex reversed mice: XX and XO males. Cytogenetics 10:318–337.

Cattanach BM, Evans EP, Burtenshaw MD, Barlow J (1982): Male, female and intersex development in mice of identical chromosome constitution. Nature 300:445–446.

Chung KW (1974): A morphological and histochemical study of Sertoli cells in normal and XX sex-reversed mice. Am J Anat 139:369–388.

Chung KW, Blackburn W, Bullock L, Ohno S, Bardin CW (1972): Testicular structure-function relationships in the mouse with XX-sex reversal. Anat Rec 172:290 (abstr).

Eicher EM (1982): Primary sex determining genes in mice. In Amann RP, Seidel GE (eds): "Prospects for Sexing Mammalian Sperm." Boulder: Colorado Assoc Univ Press, pp 121–135.

Eicher EM, Washburn LL, Whitney TB, Morrow KW (1982): Mus poschiavinus Y chromosome in the C57BL/6J murine genome causes sex reversal. Science 217:535–537.

Ekblom P (1981): Determination and differentiation of the nephron. Med Biol 59:139–160.

Engel W, Klemme B, Ebrecht A (1981): Serological evidence for H-Y antigen in XO female mice. Hum Genet 57:67–70.

Evans EP, Ford CE, Lyon MF (1977): Direct evidence of the capacity of the XY germ cell in the mouse to become an oocyte. Nature 267:430–431.

Ford CE, Evans EP, Burtenshaw MD, Clegg HM, Tuffrey M, Barnes RD (1975): A functional 'sex-reversed' oocyte in the mouse. Proc R Soc Lond B 190:187–197.

Francavilla S, Zamboni L (1985): Differentiation of mouse ectopic germinal cells in intr- and perigonadal locations. J Exp Zool 233:101–109.

Franchi LL, Mandl AM, Zuckerman S (1962): The development of the ovary and the process of oogenesis. In Zuckerman S, Mandl AM, Eckstein P (eds): "The Ovary." London: Academic Press, Vol. 1, pp 1–88.

Gruenwald P (1942): The development of the sex cords in the gonads of man and mammals. Am J Anat 70:359–397.

Josso N, Picard JY, Tran D (1977): The antimüllerian hormone. Recent Prog Horm Res 33:117–167.

Jost A (1972): Données préliminaires sur les stades initiaux de la differenciation du testicule chez le rat. Arch Anat Microsc Morphol Exp 61:415–438.

Magre S, Jost A (1980): The initial phases of testicular organogenesis in the rat. An electron microscopy study. Arch Anat Microsc Morphol Exp 69:297–318.

McCoshen JA (1983): Quantitation of sex chromosomal influence(s) on the somatic growth of fetal gonads in vivo. Am J Obstet Gynecol 145:469–473.

McLaren A (1980): Oocytes in the testis. Nature 283:688–689.

McLaren A (1983a): Studies on mouse germ cells inside and outside the gonad. J Exp Zool 228:167–171.

McLaren A (1983b): Does the chromosomal sex of a mouse germ cell affect its development? In McLaren A, Wylie CC (eds): "Current Problems in Germ Cell Differentiation." Cambridge: Cambridge University Press.

McLaren A (1983c): Sex reversal in the mouse. Differentiation [Suppl] 23:S93–S98.

McLaren A (1984): Chimeras and sexual differentiation. In Le Douarin N, McLaren A (eds): "Chimeras in Developmental Biology." New York and London: Academic Press, pp 381–399.

McLaren A (1985): Meiosis and differentiation of mouse germ cells. In Dickinson HG (ed): "Controlling Events in Meiosis". Society of Experimental Biology Symposium 38. Cambridge: Cambridge University Press (in press).

McLaren A, Monk M (1982): Fertile females produced by inactivation of an X chromosome of "sex-reversed" mice. Nature 300:446–448.

McLaren A, Simpson E, Tomonari K, Chandler P, Hogg H (1984): Male sexual differentiation in mice lacking H-Y antigen. Nature 312:552–555.

Merchant H (1975): Rat gonadal and ovarian organogenesis with and without germ cells. An ultrastructural study. Dev Biol 44:1–21.

Ozdzenski W (1972): Differentiation of the genital ridges of mouse embryos in the kidneys of adult mice. Arch Anat Microsc Morphol Exp 61:267–278.

Ozdzenski W, Presz M (1984): Precocious initiation of meiosis by male germ cells of the mouse. Arch Anat Microsc Morphol Exp 73:1–7.

Pelliniemi L (1975): Ultrastructure of the early ovary and testis in pig embryos. Am J Anat 144:89–112.

Silvers WK, Gasser DL, Eicher EM (1982): H-Y antigen, serologically detectable male antigen and sex determination. Cell 28:439–440.

Simpson E (1983): Immunology of H-Y antigen and its role in sex determination. Proc R Soc Lond B 220:31–46.

Simpson E, McLaren A, Chandler P (1982): Evidence for two male antigens in mice. Immunogenet 15:609–614.

Simpson E, McLaren A, Chandler P, Tomonari K (1984): Expression of H-Y antigen by female mice carrying *Sxr*. Transplantation 37:17–21.

Singh L, Jones KW (1982): Sex reversal in the mouse (Mus musculus) is caused by a recurrent non-reciprocal crossover involving the X and an aberrant Y chromosome. Cell 28:205–216.

Stewart A (1983): The role of the Y chromosome in mammalian sexual differentiation. In Johnson MH (ed): "Development in Mammals," Vol 5. Amsterdam: Elsevier, pp 321–367.

Taketo T, Koide SS (1981): In vitro development of testis and ovary from indifferent fetal mouse gonads. Dev Biol 84:61–66.

Upadhyay S, Luciani JM, Zamboni L (1979): The role of the mesonephros in the development of indifferent gonads and ovaries of the mouse. Ann Biol Anim Biochim Biophys 19(4B):1179–1196.

Upadhyay S, Luciani JM, Zamboni L (1981): The role of the mesonephros in the development of the mouse testis and its excurrent pathways. In Byskov AG, Peters H (eds): "Development and function of reproductive organs." Amsterdam: Excerpta Medica: pp 18–30.

Upadhyay S, Zamboni L (1982): Ectopic germ cells. A natural model for the study of germ cell sexual differentiation. Proc Natl Acad Sci USA 79:6584–6588.

Wachtel SS, Koo GC (1981): H-Y antigen in gonadal differentiation. In Austin CR, Edwards RG (eds): "Mechanisms of sexual differentiation in animals and men." London: Academic Press, pp 255–300.

Wartenberg H (1981): Differentiation and development of the testes. In Burger H, de Kretser D (eds): "The Testis." New York: Raven Press, pp 39–80.

Witschi E (1951): Embryogenesis of the adrenal and the reproductive gland. Recent Prog Horm Res 6:1–27.

Zamboni L, Bézard J, Mauléon P (1979): The role of the mesonephros in the development of the sheep fetal ovary. Ann Biol Anim Biochim Biophys 19(4B):1153–1178.

Zamboni L, Upadhyay S, Bézard J, Mauléon P (1981): The role of the mesonephros in the development of the sheep testis and its excurrent pathways. In Byskov AG, Peters H (eds): "Development and function of reproductive organs." Amsterdam: Excerpta Medica, pp 31–40.

Zamboni L, Upadhyay S (1983): Germ cell differentiation in mouse adrenal glands. J Exp Zool 228:173–193.

Zenzes MT, Reed TE (1984): Variability in serologically detected male antigen titer and some resulting problems: A critical review. Hum Genet 66:103–109.

The Origin and Evolution of Sex, pages 301–327

Primary Events in the Determination of Sex in *Drosophila melanogaster*

Thomas W. Cline

Biology Department, Princeton University, Princeton, New Jersey 08544

INTRODUCTION

An argument can be made that every sexual species carries in its DNA a record of the origin and evolution of its sex-determination mechanism. In order to extract that information, we must first discover how sexual development is genetically programmed, and perhaps do so in considerable detail. *Drosophila melanogaster* is an organism whose sex-determination mechanism we can expect to understand to such an extent.

Ironically, J.J. Bull [1983] begins his recent book, *Evolution of Sex Determining Mechanisms,* with a chapter that includes the following statement: "The chief reason for excluding a serious consideration of the details of sex development and differentiation is that evolution of the inherited basis of sex may be described without understanding these various details." In the absence of such understanding, however, one can discuss the evolution of sex determination only in a rather narrow and predominantly theoretical context. Moreover, such discussion would be limited generally to only one aspect of sex-determining mechanisms and their evolution: the initial signals that trigger the developmental choice between male and female. Even understanding sex-determination signals, and particularly understanding their susceptibility to change, may require understanding many genetic details of sexual development. Hodgkin's recent report [1983] on *C. elegans* sex determination well illustrates this point. He showed how an XX/XO hermaphrodite/male sex determination system could be changed to a ZW/ZZ female/male system with just two mutations in one gene: *tra-1*. The *tra-1* gene itself seems unlikely to be part of the primary sex-determination signal in the wild-type animal. Instead, it appears to function in the cells' response to that signal; nevertheless, it could become part of the primary sex signaling system through simple mutation. How one might distinguish operationally between regulatory genes that serve as primary sex-determination signals and regulatory genes that act in the response of cells to those signals often is not included in reviews on sex determination. Such a distinction is important in considering not only how sex-determination mechanisms work, but also how they might change.

By working out many of the "details of sex development and differentiation" in a few model organisms that lend themselves to genetic and molecular analysis, one may obtain genes and gene products that can serve as tools for comparative studies with organisms that are less amenable to direct genetic and/or molecular analysis. From such comparative studies of DNA and proteins, one might be able to infer what actually has happened during the course of evolution to particular sex-determination mechanisms. It seems likely that such information cannot help but be relevant to the question of *why* those changes might have taken place.

The scope of our discussion of *Drosophila* sex determination is limited and is concerned primarily with the general nature of developmental steps that appear to control fruit fly sex determination and with the criteria that one might use to distinguish among the different categories of regulatory genes involved. Making such distinctions is an important step not only in understanding the genetic programming of sexual dimorphism in this organism, but also for relating regulatory mechanisms involved in sexual development to mechanisms that operate in other fundamental developmental processes, such as embryonic segmentation [see Lawrence, 1981]. The special nature of the master regulatory gene, Sex-lethal, the object of our attention for many years, will be discussed. The central position of this gene in *Drosophila* sex determination makes it of particular interest with respect to the evolution of sex determination mechanisms.

SEXUAL PATHWAY CHOICE AND SEXUAL PATHWAY EXPRESSION

Two Categories of *Drosophila* Intersexes

Early in this century, Bridges [1925] showed that the ratio of the number of X chromosomes to the number of sets of autosomes determines sex in *Drosophila*. A value $\geqq 1$ (XX, $\pm$Y, AA) elicits female development, whereas a value $\leqq 0.5$ (X, $\pm$Y,AA) elicits male development. The cell-autonomous nature of sexual development in hypodermal cells (cells that form the exterior cuticle of the fly) was demonstrated in gynandromorphs. Gynandromorphs are mixtures of cells, some with the male ratio of 0.5 and some with the female ratio of 1.0. Such mosaics are generated by loss of an X chromosome in some embryonic nuclei in the first few divisions of a diplo-X (female) zygote's nucleus. For such genetic mosaics, it was found that the sexual phenotype of terminally differentiated cuticular cells reflected their individual genotype with respect to the X/A ratio [Morgan and Bridges, 1919]; thus, gynandromorphs develop as adults with a mixture of normal male and normal female cells. The lack of phenotypic interactions between cells of different X/A ratios in gynandromorphs was illustrated vividly in the development of the sexually dimorphic, first tarsal segment of the foreleg [Tokunaga, 1962]. Males carry a sexcomb in the distal part of this segment (Fig. 1B). The teeth of this comb consist of a row of very distinctive bristles running longitudinally. Each bristle shaft is the product of a single cell which differentiates within the foreleg imaginal disc during the pupal stage. The corresponding region of the female foreleg lacks these distinctive bristles and has instead additional transverse rows of bristles indistinguishable from those on more proximal regions of the segment in both sexes (Fig. 1A). It was found that only genetically male (X/A = 0.5) cells produce sexcomb teeth, but they can do so even when surrounded by genetically female (X/A – 1) tissue.

The mosaic sexual development of gynandromorphs results from the autonomous differentiation of genetically dissimilar cells; however, similar mosaic intersexual phenotypes can be generated within individuals whose cells are genetically identical. Hannah-Alava and Stern [1957] first recognized the mosaic nature of the development of triploid intersexes (XX ± Y; AAA), individuals whose X/A ratio has a sexually ambiguous value of 0.67. Subsequent studies have shown a remarkable similarity in foreleg phenotypes between gynandromorphs and triploid intersexes under conditions where the intersexes and gynandromorphs have nearly equal proportions of female foreleg tissue [Cline, 1984]. Combinations of mutant genes have been discovered recently that also produce such mosaic-type intersexual development in individuals whose cells are all genetically identical [Skripsky and Lucchesi, 1982; Uenoyama et al., 1982; Cline, 1984]. The foreleg first tarsal segment of one such phenotypically mosaic intersex is shown in Figure 1D, with a patch of two female bristles separating two patches of male comb teeth. All the cells in this mosaic are "genetically" female (XX AA).

Such intersexes that differentiate as phenotypic mosaics of normal male and normal female tissue stand in striking contrast to a different class of intersexes whose phenotype is truly intermediate between male and female at the level of individual cells. An example of the development exhibited by such true intersexes is shown in Figure 1C. Note that a range of sexcomb tooth phenotypes is observed in various gradations between normal male and normal female bristle morphology. Such individuals can be produced by

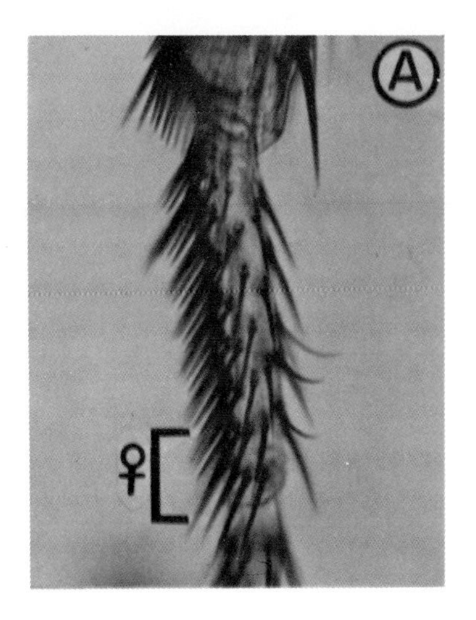

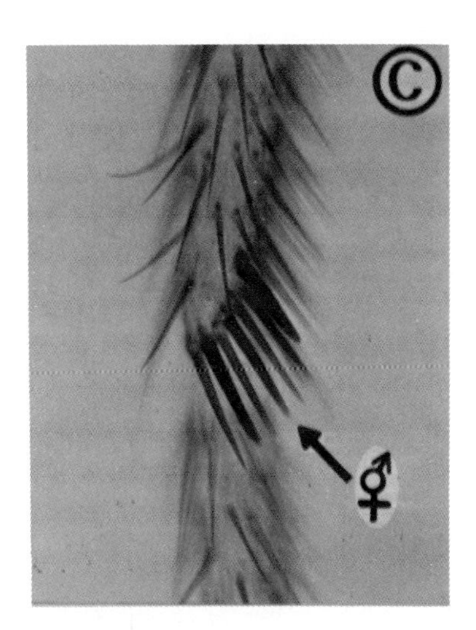

Fig. 1. Normal and abnormal first tarsal leg segments (basitarsi) of *Drosophila melanogaster*. This sexually dimorphic region of the adult [bracketed area in panels A and B) lends itself particularly well to analysis, as sexual differentiation can be observed at the level of individual bristle-forming cells. The *true intersex* phenotype in panel C resulted from a partial loss-of-function mutation in the gene Sex-lethal. The arrow points to a bristle that is morphologically intermediate between the male and female phenotypes. The *mosaic intersex* phenotype shown in panel D resulted from mutational perturbation of both the maternal genotype with respect to the daughterless gene and the zygotic genotype with respect to Sex-lethal and other genes involved in dosage compensation (see Fig. 3). The arrow identifies a patch of phenotypically female tissue interrupting two patches of phenotypically male tissue. All cells in this intersexual leg are identical and are "genetically female" (i.e., XX AA).

mutations in a variety of genes that control sexual development [Morgan et al., 1943; Hildreth, 1965; Fung and Gowen, 1957; Belote and Baker, 1982; Cline, 1984].

A mosaic intersex phenotype for an individual whose cells are genetically identical indicates perturbation of the processes by which cells become sexually committed to one developmental pathway or the other; however, the processes by which either pathway is actually expressed when cells differentiate seems not be be perturbed in such intersexes. Thus, although the individual is intersexual overall, sexual differentiation at the level of individual cells appears normal. The mosaic intersex phenotype is found in situations where the signals that trigger sexual pathway choice, or the response of the gene that seems to be most immediately affected by those signals, are ambiguous [see Cline, 1983a, 1984]. The true intersex phenotype, on the other hand, represents abnormal sexual pathway expression; differentiation, even at the level of individual cells, is neither male nor female. This phenotype seems to be generated in situations where the sexual differentiation instructions within cells are abnormal, i.e., where cells apparently are being instructed to be both male and female [see Baker and Belote, 1983]. The fact that cells can differentiate at all under such circumstances, coupled with the fact that their morphological parameters sometimes appear to be nearly the arithmetic mean of normal male and normal female values, must have important implications for the mechanisms by which sexually divergent morphologies are generated.

Sex "Determination" as Related to Developmental "Determination"

Descriptions of sexual development generally include the terms sex determination and sex differentiation, respectively, in order to separate discussion of the signals (primary or secondary) that dictate cellular choices between male and female developmental pathways from discussion of the responses of cells to those signals to produce differentiated cell structures and functions. The term "determination" can be used, however, in a more strict developmental sense to denote heritable restrictions in developmental potential that can arise among mitotically active cells. Such restrictions arise in dividing cells in response to signals, but they are maintained subsequently by those cells and by their mitotic progeny independent of the initiating signals and independent of the cells' environment. Such a response is a special type of differentiation: It is regulatory rather than structural in nature, and it is differentiation in mitotically active cells that is passed on to their cellular descendents. Such differentiation can be considered separately from the overt kind of differentiation that is accompanied by gross alterations in cell structure and function. Indeed, the former may precede the latter by a considerable time and number of cell divisions. The classical example of determination in this sense is that exhibited by *Drosophila* imaginal discs, the anlagen that give rise to parts of the adult integument. Embryonic cells that contribute to these discs acquire their segmental identity very early in response to their position in the egg. They can then maintain that identity in ectopic locations over decades of culture as "undifferentiated" tissue [Gehring, 1972].

There is no reason a priori to expect sex determination to be a determinative process in this sense of a pathway commitment maintained independently of the signal (X/A balance) that initiated it. All cells would seem to carry throughout development an X/A signal to which they could refer whenever they were required to chose between developmental alternatives that depended on their perceived sex. There would seem to be no need for dividing cells to "remember" their X/A balance, no need for them to make a heritable (irreversible) sexual pathway commitment. Nevertheless, increasingly compelling evidence is accumulating to suggest that such commitments are made by cells at early stages.

Evidence for Stable Sexual Pathway Commitments in *Drosophila*

That cells make sexual commitments early in development was suggested by the degree of interspersion of male and female tissue in the forelegs of mosaic intersexes whose cells are genetically identical and by the similarity of these phenotypes to the phenotypes exhibited by gynandromorphs [Cline, 1984]. Recall that gynandromorphs are genetic as well as phenotypic mosaics whose intersexuality reflects mosaicism at the cellular blastoderm stage of embryonic development. A similar conclusion was suggested by experiments in which the X/A balance was changed at progressively later stages in development [Sanchez and Nöthiger, 1983]; after early embryonic stages, the X/A balance could not be changed without killing cells, presumably due to upsets in dosage compensation. Dosage compensation is the process by which male cells hyperactivate X-linked genes at the transcriptional level [reviewed by Stewart and Merriam, 1980]. In this way, male cells compensate for the fact that they have only a single dose of each X-linked gene, yet they have requirements for the products of these genes that are similar to those of females who have two doses of each locus. Dosage compensation is intimately related to sex determination; indeed, it is required only as a consequence of the sex-determination signal—the X/A balance. Both dosage compensation and sex determination are controlled by the same gene, Sex-lethal [Lucchesi and Skripsky, 1981; Cline, 1983a]. A cell that failed to adjust quickly its level of X-linked gene expression following a change in the X/A balance should become functionally aneuploid and be at a considerable developmental disadvantage. Baker and Belote [1983] report having found small, very unhealthy clones of tissue which, though they seem to have lost one of their two X chromosomes late in development, appear nevertheless to have differentiated as if they were still genetically female; unfortunately, the only criterion for sexing such clones was pigmentation, not a very reliable criterion for small patches of poorly growing abdominal tissue.

Although the results of both kinds of experiments suggested that cells make a sexual pathway commitment early and subsequently maintain that decision in a heritable fashion, alternative explanations for the results were not excluded. More recently, a different approach has been used regarding the question of sexual pathway commitments. If, as suggested, mosaic intersexual phenotypes can result from genetically identical cells becoming stably committed to different developmental pathways long before they actually differentiate, one should find that cells that are genetically tagged *after* the sexual pathway choice is made should not be able to give rise to marked clones of progeny cells that include structures of different sexes within the same clone. *Drosophila* cells can be tagged by the technique of mitotic recombination: Exposure to ionizing radiation induces an abnormal mitotic division in which a mother cell heterozygous for alleles of a marker gene gives rise to daughters who are homozygous for (and thus tagged by) one or the other of the maternal cell's marker alleles. These homozygous cells in turn produce homozygous progeny, which eventually differentiate into a clone of tissue in the adult that is distinguished from its heterozygous surroundings by its recessive marker-allele phenotype (bristle color, shape, etc.) [see Postlethwait, 1978]. The experiment can be designed so that the event that tags cells also gives them a considerable growth advantage over their neighbors. As a consequence, they can be made to divide many more times than they would normally, extending any test for possible restrictions in developmental capacity.

Figure 2 presents the logic of our test for heritable restrictions in the sexual potential of cells that might arise during the development of XX AA mosaic intersexes. Rather complicated genetic conditions were used to generate these intersexes, conditions that also minimized upsets in dosage compensation that normally might be expected to arise if an XXAA cell chose a male developmental pathway (see Fig. 3 legend for genotypes). The design of the experiment also gave marked cells a potential growth advantage over their unmarked neighbors. Such a clonal test of sexual potential must be carried out in a region where sex can be unambiguously ascertained at the level of individual cells. Moreover, for the test to be valid, cell lineage relationships must be such as would allow clones induced even late in development to include cell structures of both sexes if the cells' sexual fates were not restricted. The sexcomb region of the foreleg appears to satisfy both criteria well.

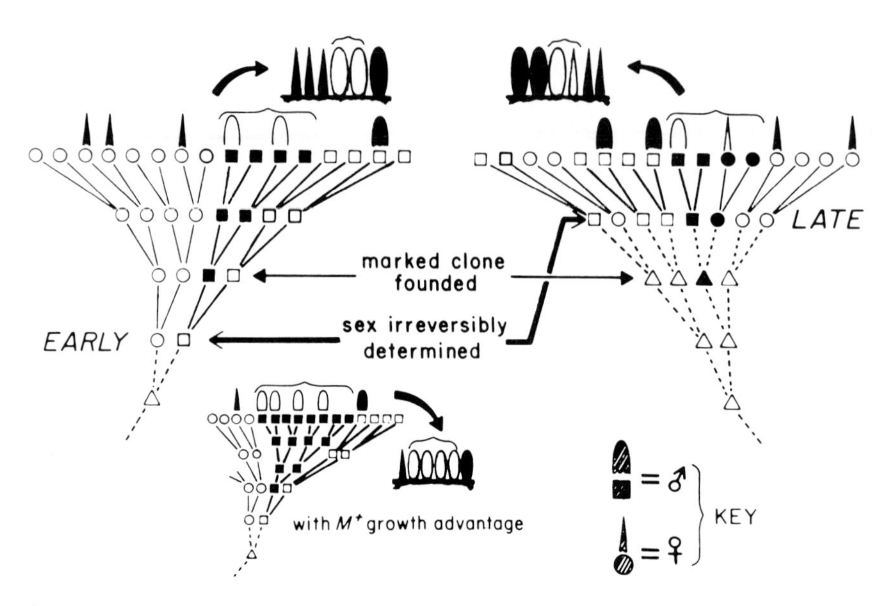

Fig. 2. Experimental strategy used to determine whether mosaic intersexual foreleg phenotypes, such as the one shown Figure 1D, arise from stable sexual pathway commitments made by cells early in development, long before overt sexual differentiation occurs. The sexually dimorphic sexcomb region of the foreleg was chosen for the analysis. If the phenotypic sex of a cell is determined irreversibly prior in development to an event that marks a cell and all its progeny (somatic recombination), the clone produced from the marked cell cannot include adult differentiated structures (in this case, bristles) of both sexes. On the other hand, if irreversible determination of sex occurs after the marking event, or perhaps not at all, one expects to recover adult forelegs with marked clones of tissue that include bristles of both sexes. The experiment can set up ("With M^{+} . . .) so that the progeny of the marked cell have a considerable growth advantage over those of its unmarked siblings and consequently divide more times than they would otherwise before terminally differentiating. This increases the rigor of such a test for the sexual potential of a cell and its progeny.

In a study still in progress, marked clones were founded by irradiating young mosaic intersex larvae between the middle of the first instar to the early part of the second instar (48–72 h.a.e.l. in *M*/+ larvae). This is long before cells stop dividing and overt sexual differentiation takes place. To be candidates for our analysis, the marked clones had to run through a sexcomb region that was clearly intersexual; moreover, clones were considered only when it was possible to ascertain both the sex of the tissue in the clone and the sex of the tissue adjacent to the clone. Fifty-one clones (39 female and 12 male) satisfied these stringent criteria. Not a single one of these clones appeared to include both male and female tissue, despite the fact that in 36 of the 51 cases clones of one sex were immediately *adjacent* to tissue of the opposite sex. Examples of such marked clones in intersexual legs are shown in Figure 3. It should be noted that the growth advantage of M^+/M^+ clones induced at this stage is such that the marked cells can include a region of an imaginal disc at least as large as that occupied by the progeny of a single blastoderm stage cell (the first stage at which somatic cells exist in the developing embryo) in a normal (M^+) individual. Currently, several different rigorous tests are being run to determine the validity of the criteria used for assigning the sex of bristles in the foreleg. (Sexcomb bristles, of course, are unambiguous, but the sex of some other types of leg bristles must be determined by their position.) If, as expected, these criteria hold up, this experiment will have demonstrated unequivocally that cells make a stable sexual pathway commitment long before genes involved in sexual differentiation are known to direct sex-specific bristle morphogenesis (during the pupal stage). The ideal control for this experiment may be provided by a genetic disruption in the normal mechanism for sex determination which breaks down the clonal restriction in cell sexuality seemingly observed here. Recently, we found one mutant *Sxl* allele that, unlike all (13) others examined, appears to generate a mosaic intersex phenotype for homozygous mutant clones founded at the same time as the Sxl^+ clones generated in this experiment. With this mutant, clones do include foreleg tissue of both sexual phenotypes as defined by our positional criteria.

In light of these experiments, we can consider that the regulation of imaginal disc sexual development includes the three categories of steps illustrated in Figure 4: 1) sexual pathway initiation with genes that provide the signal for the sexual pathway choice and genes that most immediately respond to that signal; 2) sexual pathway maintenance with genes that allow the cell to remember the pathway choice decision made earlier and pass that memory on to progeny cells; and 3) sexual pathway expression, with genes that elicit overt sexual differentiation. The distinction between sexual pathway choice and maintenance on the one hand and pathway expression on the other can be reflected in the two general classes of sexually ambiguous phenotypes described earlier. Indeed, understanding how sexual development is programmed in *Drosophila* requires attention to the type of intersexuality that various genetic perturbations may induce. The importance of intersexuality in distinguishing the specific role of particular genes in the control of sex determination was anticipated by Stern [1966] who wrote, "There seem to be different levels of sex determination, one in which a basic decision between maleness and femaleness is made, and a variety of others in which a developmental modification is secondarily superimposed on the basic decision."

REGULATORY GENES THAT CONTROL *DROSOPHILA* SEXUAL DEVELOPMENT

The Hierarchical Nature of Sex-Regulating Genes

A large number of mutants that affect *Drosophila* sexual development have been recovered fortuitously. Application of more systematic methods for identifying such genes is certain to reveal many more [see Baker and Belote, 1983; Steinman-Zwicky, 1984]; nevertheless, even with the mutants currently

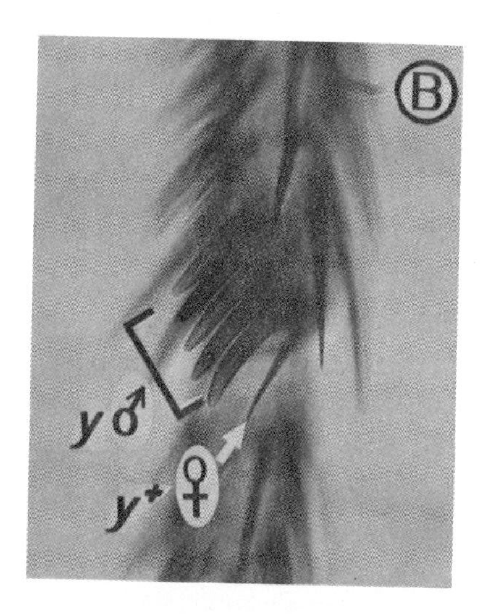

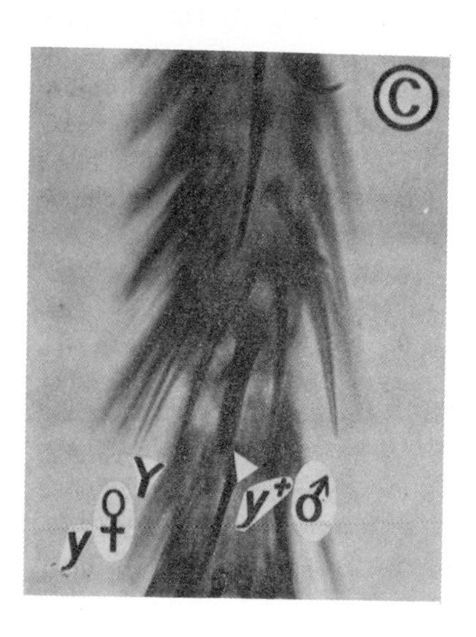

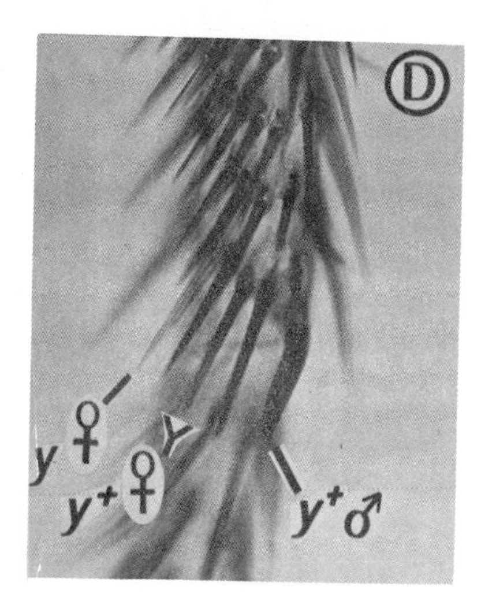

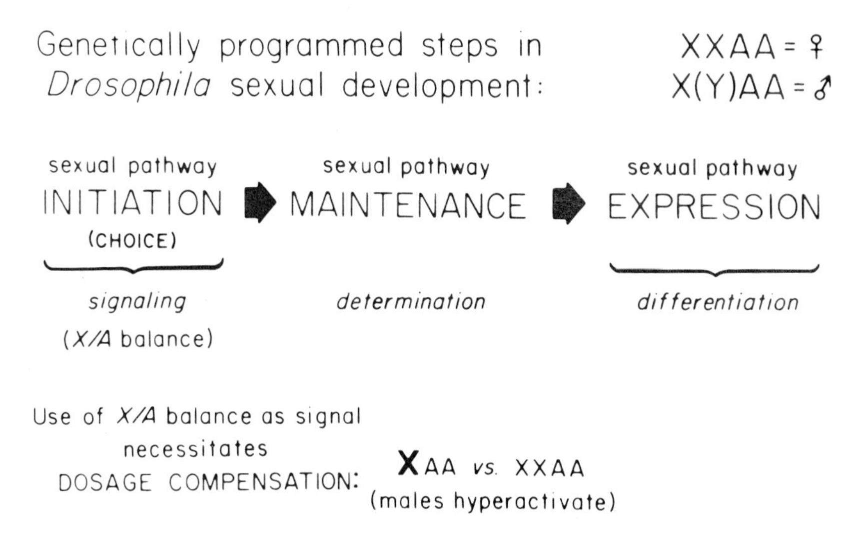

Fig. 4. A logical context in which to consider the different functions that regulatory genes controling *Drosophila* sexual development might have. Sexual development is divided into three broad categories of steps that take place within cells. Sex determination and dosage compensation in cells of the adult integument are cell autonomous processes: the sexual phenotype of one cell generally is not influenced by that of its neighbors.

Fig. 3. Examples of the types of genetically marked clones observed in mosaic intersex forelegs of adults. No clones included tissue of both sexes; moreover, the majority of clones satisfying the criteria for the experiment (see text) ran adjacent to tissue of the opposite sex. A) A marked male clone including two of the five (male) sexcomb teeth and running adjacent to a female bristle outside of the clone. B) A male clone including all five (male) sexcomb teeth, running adjacent to a female bristle outside of the clone. C) A marked female clone running adjacent to a male bristle outside of the clone. D) A marked female clone *not* adjacent to the male tissue outside of the clone.
Clones were induced by exposing developing larvae of assorted ages between middle-first and early-second instar stage to a short burst (four minutes; ca 1400 rad) of gamma radiation. Irradiated progeny ($cm\ f^{+}$) were from the following cross which generates mosaic intersexes at a high rate: $cm\ Sxl^{fm\#7.M\#1}$ *M*(1)*o*/+ + *f*; *msl-2 da b msl-1 mle*/+ + + + +; *e*/*e* ♀♀ X ♂♂ *y cm v f car*/Y; *msl-2 msl-1 mle bw*/+; *e*/*e*. The clones induced (*y*/*y*; *e*/*e*) are phenotypically yellow against a black background (y^{+}/y; *e*/*e*) of unmarked tissue. Cells within the clones were generally M^{+}/M^{+} and thus had a potential growth advantage over their M^{+}/M unmarked neighbors. It should be noted also that clones consisted of Sxl^{+}/Sxl^{+} XXAA (genetically female) cells.

in hand, we are already beginning to see the outlines of the strategy that *Drosophila melanogaster* uses to control the development of sexual dimorphism. This outline is illustrated in Figure 5. With Figure 4, it provides a context within which functional distinctions between regulatory genes may be discussed.

A primary distinction among the genes presented in Figure 5 is whether they participate in both sex determination and dosage compensation or in only one or the other of these processes. Loss-of-function mutations in genes required only for sexual differentiation cause gross alterations in sexual phenotype (e.g., transforming genotypic females into phenotypic males) without effects on dosage compensation or viability [data reviewed in Stewart and Merriam, 1980]. Conversely, sex-specific lethal loss-of-function mutations can upset dosage compensation without effects on sex determination [Belote, 1983]. At one level, then, these two processes are clearly under separate control; only fairly recently has it been appreciated that both processes are controlled by the same gene, Sex-lethal, located higher up in a regulatory gene hierarchy, [Lucchesi and Skripski, 1981; Cline, 1983a; 1984].

Two factors may have contributed to the delay in understanding the genetic link between these two processes. First, dosage compensation is a cell-vital process whose disruption can kill mutant animals and even kill mutant clones within an otherwise nonmutant individual long before the differentiation of unambiguous sexually dimorphic characters takes place. As a consequence, unless special measures are taken, a mutant's lethal dosage compensation phenotype may mask its sex determination phenotype. Second, only quite recently has the relevance of sex-specific lethal mutations to the study of dosage compensation been appreciated. The two classical hypotheses for dosage compensation mechanisms ("negative control by compensators" and "positive control by an autosomal substance") [see Stewart and Merriam, 1980] viewed hyperactivation of the male X as part of a process that was essential for proper X-linked gene transcription in both sexes. According to these models, one generally would not have expected to recover sex-specific lethal mutations affecting dosage compensation, except perhaps for an occassional partial loss-of-function mutation; certainly, null mutations affecting the mechanism should have been lethal to both sexes. A variety of studies indicated that the transcription rate per X-linked gene was a rather smooth function of the total amount of euchromatic X-chromosome material present [Maroni and Lucchesi, 1980]. These data fit the classical models well; however, characterization of *Sxl* and the autosomal male-specific lethals (the "late-vital" genes of Fig. 5) indicated that the X-linked gene hyperactivation observed in males seems instead to be a truly male-specific process, which when disrupted consequently kills only males. Reconciliation of these different

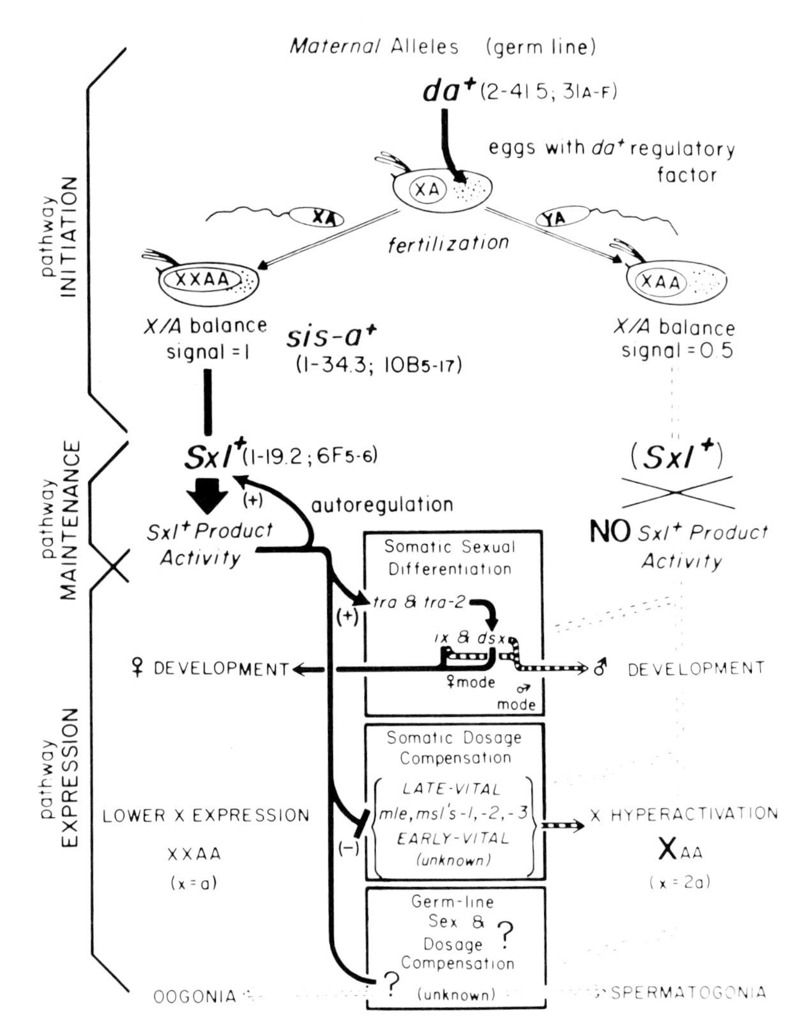

Fig. 5. A model for the control of sex determination and X-chromosome dosage compensation in *Drosophila melanogaster*. The female-specific Sex-lethal (*Sxl*) gene occupies a key position among regulatory genetic elements known to participate in these processes. Its products impose female development on cells that will follow the male alternative in their absence. Sexual pathway initiation appears to involve the establishment of this gene's activity state in response to the X/A balance signal. In somatic cells, this step requires the positive regulatory activity of a maternally acting gene, daughterless (*da*), and a zygotically acting locus, sisterless-a (*sis-a*). The latter may be a dose-sensitive element in the X/A ratio itself. Maintenance of the sexually determined state may involve the propagation of *Sxl*'s activity state independent of the initiating X/A signal, a process that could involve the positive, autoregulatory activity recently suggested for *Sxl* product. Expression of the sexually determined state appears to result from the presence in females, or absence in males, of several apparently distinct *Sxl* product activities: one that interacts with genes that control only sexual differentiation, another that interacts with genes that control only dosage compensation, and perhaps a third activity that interacts with unknown genes involved in some aspect of germ cell development. It should be mentioned that there is a possibility that *da* and *sis-a* may not function to control *Sxl*'s germline function(s).

results is bound to provide important new insights into the mechanism(s) of dosage compensation.

The role in sex determination of the three sex-specific vital genes in Figure 5, (daughterless, sisterless-a, and Sex-lethal) was ascertained by four ploys designed to overcome the complications of their involvement in dosage compensation: 1) analysis of triploid intersex individuals (XXAAA) for whom dosage compensation upsets are necessarily less severe than for diploids; 2) analysis of genetic mosaics in which only a small fraction of an individual's tissue was mutant, and/or in which the wild-type vital gene was allowed to function for some time before it was removed experimentally; 3) analysis of partial-loss-of-function alleles which affected dosage compensation and sex determination functions differentially; and 4) analysis of suppressing and/or enhancing interactions between male-specific and female-specific lethals. Study of these and the other genes shown in Figure 5 often focuses, for practical reasons, on their effects in external cuticular tissue of adults; the extent to which certain of the observations can can be generalized to include sexually dimorphic internal tissues, such as those of the nervous system, musculature, and gonads, remains, in many cases, to be determined.

Sxl seems to be rather special among the genes shown in Figure 5. Not only does it participate in both somatic sex determination and dosage compensation, but it also is the only gene of all those shown in Figure 5 whose product function in the germline appears to be consistent in some respects with its product function in somatic tissue [Cline, 1983b; Schüpbach, 1985]. In contrast, the genes *tra, dsx,* and *ix* seem to have no vital germline function [see Schüpbach, 1982], and *tra-2*'s germline function is paradoxical [Belote and Baker, 1983]. Based on genetic criteria, *Sxl*'s functions seem to be truly sex-specific. Though a very large variety of female-specific lethal loss-of-function *Sxl* alleles exist, no loss-of-function, male-specific lethal alleles have been found, though such were sought [Nicklas and Cline, 1983]. Nor do alleles exist that are lethal to both sexes. Recently, we were able to show that males who carry a deletion of *Sxl* are both viable and fertile (unpublished). *Sxl* appears to act like a sexual switch, imposing the female developmental sequence on cells when in its functionally active state, and the male sequence, by default, when it is not. Male-specific lethal *Sxl* alleles do exist, but they are cis-dominant, gain-of-function mutations that impose female-specific Sxl^+ activities on genetic males; this kind of male-specific lethality is entirely consistent with a female-specific role for *Sxl* [Cline, 1981; 1984]. Curiously, although *Sxl*'s functioning appears to be female-specific, the heart of the sex differentiation system seems to be a bifunctional gene, doublesex, that is active in both sexes [see Baker and Belote, 1983]. The *ix* gene as well seems to have functions in males and females [McRobert and Tompkins, 1985].

Control of Sexual Pathway Expression

Figure 5 shows the five genes (*Sxl, tra, tra-2, dsx,* and *ix*) known to be involved in the expression of the sexual pathway choice—the sexual differentiation steps which can be disrupted to generate true intersexes. Of these, only *Sxl* appears also to be involved in sexual pathway initiation and maintenance. The specific hierarchical relationships between the functioning of these five genes is important in the discussion of the control of sex determination. The relationships shown are deduced from an extensive amount of work, only the broad outlines of which will be reviewed here [see Baker and Belote, 1983, for more details]. For such analysis, one wants first to establish the true null phenotype for a gene, then compare it to the phenotypes of gain-of-function and partial loss-of-function alleles where possible. The null condition for all five genes except *ix* has been established with a fair degree of confidence. A particularly important aspect of the analysis involves determination of the phenotypes of various double mutant combinations to ascertain how they fit such a hierarchical regulatory model that appears to have the bifunctional negative regulatory element *dsx* at its heart. A third approach employs somatic recombination to remove wild-type alleles from growing cells at progressively later times in development. This approach, in combination with temperature shifts of temperature-conditional alleles, can reveal much about the time of action of individual genes, their interaction with other regulatory elements, and their spectrum of phenotypic effects.

Particular attention has been devoted to *tra-2*, as an excellent temperature-conditional allele exists for this gene [Belote and Baker, 1982]. Genetic females (XXAA) homozygous for *tra-2* null alleles develop as sterile, phenotypic males. This gene has been found not to act early to establish a sexually determined cellular state, but rather it is required at different times throughout development, apparently at any point when a wild-type cell must normally face a growth or differentiation alternative that depends on its sex. Analysis of the effects of *tra-2* on genital disc development supported the belief that all somatic sexual differentiation decisions that depend on the X/A signal also depend on *tra-2* function. Put another way, the male terminal somatic phenotype exhibited by sexually transformed XX; *tra-2/tra-2* "genetic females" appears to arise from the same sequence of terminal differentiation as it does in a "true" (XY;AA) male.

Wild-type alleles of all five genes, including *Sxl*, have been shown to be required quite late in development for proper sexual differentiation. Indeed, for temperature-sensitive alleles of both *tra-2* and *Sxl*, temperature shifts made even as late as the adult stage can alter sex-specific functions [Baker and Belote, 1983; Tompkins, 1984]. On the other hand, early functions in the process of sex determination are suggested by one heteroallelic combination of *Sxl* alleles, Sxl^{f9}/Sxl^{2593}, which has a temperature-sensitive lethal period that ends within the first half of embryonic development [Cline, unpublished]. One of the processes disrupted even during this early period may involve sexual development. In this particular case, the early temperature effects for this late-acting gene may reflect an aspect of *Sxl* regulation in which the gene's own product plays a role (see below), a sexual pathway initiation and/or maintenance step rather than an expression step. This contrasts with temperature effects on other mutant *Sxl* alleles that clearly involve later aspects of this gene's functioning, aspects involved in sexual pathway expression.

Sxl as a Candidate for "Carrier" of the Sexually Determined State

According to the model shown in Figure 5, the most immediate event in the response of *Drosophila* somatic cells to the X/A balance signal is the determination of *Sxl*'s activity state. The X/A balance appears to control the probability of individual cells stably activating Sxl^+ early in development, with higher values of this signal being associated with an increased probability of activation [see Cline, 1984]. The term "activation" is used rather loosely here to refer to expression of the female-specific functions of Sxl^+; we do not mean to imply that regulation of *Sxl*'s activity state is necessarily at the level of transcription initiation. Indeed, we have evidence (unpublished) from molecular studies of this female-specific gene that some species of *Sxl* transcripts may be present in both sexes, and others may be in males alone!

Sxl and the four downstream loci known to control somatic sexual differentiation (*tra, tra-2, ix,* and *dsx*) have been referred to as "sex determination" genes [Baker and Belote, 1983]. If, however, the term "determination" is used in the sense of an epigenetic process in which a regulatory gene maintains itself metastably in a particular functional state independent of the processes that initiated that state, only *Sxl* would seem to be a good candidate for a "sex determination" gene, an individual genetic element acting as the "carrier" of the sexually committed state. *Sxl* is the only one of these regulatory genes in which different mutations can cause reciprocal transformations: the loss-of-function allele transforming females into males and the gain-of-function allele transforming males into females. A particularly important test for any potential carrier of the sexually determined state is whether its functioning remains unaffected by changes in regulatory elements upstream of it when those changes are made after the point in development when cells normally become sexually committed. Sex transformations are produced by changes in *Sxl* made late in development (either by somatic recombination in heterozygotes or by temperature shifts of conditional alleles) long after the time at which cells appear to have made their sexual pathway commitment. Such *Sxl*-induced sex transformations imply changes in the functioning of all four wild-type regulatory genes downstream of *Sxl*, suggesting that their activity states or their products' activities are not set irreversibly at the time when the pathway commitment is normally made.

The phenotypically reversible behavior of these four genes contrasts with the behavior of *Sxl* itself in response to regulatory genes upstream of it in Figure 5: *da* and *sis-a*. Though the maternal genotype with respect to *da* is critical, the zygotic *da* genotype seems not to affect *Sxl* functioning (see below). Likewise, changes in *sis-a* made after the embryonic period do not alter sexual differentiation. On the other hand, until and unless these results are confirmed with amorphic rather than hypomorphic mutant *da* and *sis-a* alleles, the observations must be interpreted with caution. This discussion is complicated also by the fact that we do not yet know the level at which these changes in *Sxl* affect the functioning of downstream genes. They could reflect anything from pretranscriptional to posttranslational effects. Thus, although the present data seem to suggest that only the activity state of *Sxl would* have to be established metastably to maintain a sexual pathway commitment, we cannot yet know whether it *is* the only gene of the group to be regulated in this way.

Because sex determination appears to occur early, one would anticipate that *Sxl* must also act early as well as late if it is directly involved in that process. An early period of activity was suggested by the fact that the lethal phase for null *Sxl* alleles is embryonic. Moreover, a very early temperature-sensitive period for certain *Sxl* heteroallelic combinations was mentioned above. A new line of experimental evidence suggests that *Sxl* may be among the very earliest genes expressed in the embryo. The runt gene controls embryonic segmentation and appears to act only during the blastoderm stage [Nüsslein-Volhard and Wieschaus, 1980]. P. Gergen (unpublished) has discovered recently that this early acting, X-linked gene is dosage compensated and that its level of expression seems to depend on the zygotic functioning of *Sxl* just as expected from the model in Figure 5.

Among the group *Sxl, tra, tra-2, dsx,* and *ix,* only mutations in *Sxl* are known to generate mosaic intersexual phenotypes within a mutant clone or in individuals who are not genetic mosaics. The fact that mutations in *Sxl* can produce both kinds of intersexes in whole animals is consistent with a role of *Sxl* both in sexual pathway maintenance and pathway expression. Indeed, the type of intersex that is generated when the functioning of *Sxl* is perturbed can indicate the nature of that perturbation, i.e., whether it affects *Sxl* regulation or *Sxl* product functions with respect to genes downstream.

In light of *Sxl*'s possible role as carrier of the sexually determined state, it is interesting that a positive autoregulatory function of Sxl^{+} product has recently been suggested [Cline, 1984]. The role that such an autoregulatory function would be likely to have is not yet clear. It could serve to maintain the active expression state of *Sxl;* alternatively, it could act only early to facilitate a transition from an initially unstable expression state to a stable one.

Sexual Pathway Initiation

The role of daughterless. In order for Sxl^+ to respond positively to the X/A balance signal, the product of a maternally acting gene called daughterless (*da;* 2-41.5) must be present (see Fig. 5). In its absence, progeny Sxl^+ alleles appear to be locked into the male ("nonfunctional") state. Our studies on *Sxl* and the nature of sexual pathway commitments were based on studies of the female-lethal maternal effect of a heat-sensitive, partial loss-of-function *da* allele. We determined the consequences of this mutant's female-lethal maternal effect for the development of progeny who were genetic mosaics of XX and XO tissue. Though most such gynandromorphic progeny of *da/da* mothers died, the phenotype of the rare survivors was striking: The defect in functioning of the *maternally* acting gene caused abnormal bristle shaft differentiation by XX cells in the progeny at the pupal stage, at least nine cell divisions after the blastoderm stage when the temperature-sensitive period for the maternal effect ended [see Fig. 4e in Cline, 1976]. The cells in these mosaics developed abnormally as a consequence of their X/A balance, even though they themselves carried a da^+ allele and could be growing alongside phenotypically normal XO cells. This indication of a long-lasting, cell-autonomous, epigenetic change on progeny cell development, one that depended specifically on the cells' X/A balance, suggested the importance of discovering the basis for the *da* female-lethal effect. It has since been discovered that this long-lasting, maternally imposed, developmental handicap seems to be a consequence of XX cells having chosen the wrong sexual pathway and suffering upsets in dosage compensation as a consequence [Cline, 1983a, 1984]. In the absence of maternally produced da^+ activity, progeny cells seem unable to activate Sxl^+ and are therefore blind to their X/A balance.

The functional relationship between the material *da* locus and the zygotic *Sxl* gene is indicated most strongly by the behavior of the dominant, male-specific lethal allele, $Sxl^{M\#1}$ [Cline, 1978, 1979, 1980, 1981, 1983a, 1984]. This gain-of-function allele rescues daughters from the otherwise lethal *da* maternal effect. Like the *da* maternal effect, it appears to make zygotes blind to their X/A signal, but in just the opposite way: It expresses female-specific Sxl^+ functions regardless of the X/A signal—and regardless of the maternal *da* genotype. Dosage compensation functions appropriate only for female (diplo-X) individuals are lethal when expressed in male (haplo-X) flies. The fact that a lesion that makes *Sxl* insensitive to the X/A signal also eliminates the zygotic requirement for maternally produced da^+ activity suggests an intimate relationship between maternal da^+ function and the X/A signal.

Daughterless gene activity is required both maternally and zygotically by individuals of both sexes; in fact, it is only because the one extant *da* mutant allele is leaky that one can observe the striking maternal effect after which the gene is named. How are the maternal and zygotic functions of this gene related? Baker and Belote [1983] discussed the possibility that the functions carried out by maternally and zygotically specified da^+ products might be the same. The observation that the maternal genotype with respect to *da* does have some influence on zygotic da^+ functions is consistent with this view. A variety of other data, particularly the lack of effects by $Sxl^{M\#1}$ on the recessive zygotic lethal aspect of the *da* phenotype, favored the alternative hypothesis of multiple functions for da^+ product(s) with a specific requirement for maternally synthesized da^+ product in the regulation of zygotic Sxl^+ alleles [Cline, 1976, 1980]. With the finding that males with *Sxl* deletions are viable and fertile, the multiple function hypothesis would seem inescapable: da^+ clearly has functions vital to both sexes, yet clearly one of its vital functions involves control of the sex-specific gene *Sxl*. The modest maternal effect observed on zygotic da^+ functions is consistent with the recent findings of Robbins [1984] and need not detract from the hypothesis of a maternal-specific function for da^+ in the regulation of progeny Sxl^+ alleles. The nature of the zygotic da^+ function required by both sexes is unclear, though it may involve control of genes in heterochromatic regions [Sandler, 1972; Mange and Sandler, 1973]. Isolation and characterization of true null *da* alleles is in progress and should soon allow us to explore further the developmental role of zygotic da^+ functions and their relation to maternal functions.

X/A signal elements: what and how do flies count to two? A fundamental question regarding the primary determination of sex and dosage compensation in *Drosophila* is whether there are genetic elements that are part of an X/A signaling system that are distinct from the regulatory genes that appear to respond to the X/A signal. By what operational criteria might these two classes of genes be distinguished? Dobzhansky and Schultz [1934] approached the question by asking if there were specific regions on the X chromosome whose dose strongly influenced the sexual phenotype of triploid intersexes. They concluded (on relatively limited data) that the signaling system was polygenic, with many individual elements in the euchromatic regions of the X that contributed additively to the overall X/A balance. Baker and Belote [1983] pointed out difficulties with that classical approach that arise from the link between sex determination and dosage compensation. An important problem that was not discussed, however, is the point that the classical triploid intersex approach used by Dobzhansky and Schultz does not necessarily distinguish between X/A signal elements and the gene(s) that respond most immediately to such elements. Neither does it distinguish between elements whose specific dose establishes the numerator or denominator of the X/A balance and elements like *da* whose product(s) seem more likely to be involved in making the target gene(s) responsive to the X/A signal.

Based on our current understanding of the central role *Sxl* plays in *Drosophila* sex determination and dosage compensation, there would seem to be two general criteria which together should identify what could most reasonably be termed true X/A balance signal elements, if such exist: 1) dose-dependent, sex-specific lethality stemming from effects on Sxl^+ expression; and 2) an early time of action consistent with evidence on the timing of the sex determination decision.

Figure 6 illustrates the first category of criteria for X-linked signal elements. If there are relatively few numerator signal elements in proportion to the overall number of X-linked genes, by manipulating those elements, one should be able to alter substantially the X/A signal perceived by *Sxl* without substantially altering the overall ratio of X-linked to autosomal genes. Thus, a sufficiently large reduction of the number of such functional elements (Df(X)s.e.'s in the figure) should block activation of Sxl^+ and thereby allow dosage-compensated gene hyperactivation that would be inappropriate for a cell with a gross X/A balance of nearly one. Such loss-of-function mutations in numerator signal elements should behave, therefore, as female-specific

Genotype	Gross X/A	Perceived X/A Signal	Developmental Consequences	
XXAA, *Df(X)s.e.*	1.0	<1.0	Sxl^+ OFF when it should be ON	♀-Lethal XXAA
XAA, *Dp(X)s.e.*	0.5	>0.5	Sxl^+ ON when it should be OFF	♂-Lethal xAA

Expect Sxl^{M^c} and *Dp(X)s.e.* to suppress ♀-lethality of *Df(X)s.e.*

Expect da^- *(maternal)* and Sxl^{f-} to suppress ♂-lethality of *Dp(X)s.e.*

Fig. 6. Sex-specific lethal effects predicted for duplications and deficiencies of X/A balance numerator signal elements [(X)s.e.], provided that there is a limited number of discrete signal elements. In this case, one can expect to be able to change the signal element dose, thereby changing the X/A signal perceived by *Sxl*, without making a significant change in the overall ratio of X-linked to autosomal genes (gross X/A balance). The sex-specific lethal upsets that are expected to arise from levels of dosage compensation that are inappropriate for the gross X/A balance should be suppressed by particular mutations in *Sxl* or *da* or by compensating changes in other signal elements.

(XXAA) lethals. In order to distinguish bona fide signal elements from other positive regulators of *Sxl*, the converse result must also be observed, namely, increasing the dose of putative numerator signal elements (Dp(X)s.e.'s in Fig. 6) should cause Sxl^{+} to be activated when the gross X/A balance is still nearly 0.5. In this case, activation of Sxl^{+} should suppress dosage-compensated gene hyperactivation, an effect that would cause a lethal deficit of X-linked gene products for cells with a gross X/A balance of nearly 0.5. Duplications of such signal elements should behave, therefore, as male-specific (XAA) lethals. The magnitude of the changes in element dose that would be required to elicit sex-specific lethality cannot be predicted a priori but would depend on the number of putative signal elements, their relative strength, and the nature of the interactions between them.

A critical part of the dose-dependent, sex-specific lethality criterion is that lethality should arise from effects on *Sxl* and thus should be affected by mutations in this gene. Female-specific lethal effects should be suppressed by $Sxl^{M\#1}$, the mutation that expresses female-specific functions regardless of the X/A balance; conversely, male-specific lethality should be suppressed by $Sxl^{f\#1}$, a null allele that eliminates all female-specific functions regardless of the X/A balance. The *da* maternal effect can be expected also to suppress male lethality arising from inappropriate activation of Sxl^{+} by signal element changes, as *da* also renders progeny blind to the X/A signal. Suppression of male-specific lethality by a loss-of-function mutation in *Sxl* is a particularly stringent criterion for any putative numerator signal element, as males normally do not need a functional allele of this gene to survive or reproduce. As a corollary to the criterion of sex-specific lethality in diploid animals, duplications and deletions of these elements should have the predicted effects on triploid intersex (XXAAA) sexual phenotype (feminizing and masculinizing, respectively); moreover, these influences on triploid intersex phenotype should be suppressed by mutations in *da* and *Sxl*. The sex-specific effects of changes in denominator signal elements should be the inverse of those of numerator elements.

The second major criterion for signal elements is based on the increasingly compelling evidence that sex is determined by a response to signal elements early in development and is maintained thereafter without reference to the signal. In this case, changes in signal element dose or function made after the point in development when sex is determined should have no effect on sexual differentiation. The clonal analysis of mosaic intersexes discussed above establishes the mid-second larval instar period as an outer time limit for the sex determination decision; the decision may be made much earlier.

The first set of criteria involving dose-sensitive, sex-specific lethality can establish a gene's identity as a signal element if the predicted results are obtained; however, as the signaling system may be polygenic, and, as the range over which gene dose may be changed generally is rather limited, failure to satisfy the criteria of both female-specific *and* male-specific lethality do not necessarily exclude a signal element role for a given gene. On the other hand, failure to satisfy the second signal element criterion, that of timing, should be useful in eliminating the possibility of a signal element role for a given genetic entity. *Sxl*$^{+}$ itself clearly is excluded as a signal element by this second criterion, as loss of functional *Sxl*$^{+}$ alleles even as late as the pupal stage can masculinize genetically female cuticular cells.

Recently, a genetic element has been identified at 1–34.3 (between chromomeres 10B5 and 10B17) that satisfies these criteria for a numerator signal element and appears to be part of a highly polymorphic, polygenic system (in preparation). It has been named sisterless-a, anticipating that additional elements will be discovered. A loss-of-function mutation in this element is recessive, female-specific lethal; it exhibits strong female-specific lethal dominant synergism with loss-of-function mutations in *Sxl* and/or (maternal) *da*. Indeed, an X chromosome carrying both *Sxl*$^{f\#1}$ and *sis-a* is fully viable in males and yet can behave as a dominant lethal chromosome in females. The female-specific recessive lethality of *sis-a* is suppressed nearly completely by *Sxl*$^{M\#1}$, and less so by *Sxl*$^{+}$ duplications. The *sis-a* mutation, like the *da* maternal effect and *Sxl*$^{f\#1}$, masculinizes triploid intersexes (XXAAA).

A key element in the analysis of *sis-a* was the discovery of its interaction with the 2-proximal half of the translocation, *T*(1;2)*Hw*^{bap}, a "nearly" reciprocal exchange between the tip of the X chromosome and the tip of 2L [referred to in Cline, 1984]. Combination of the 2-proximal half of this translocation with any duplication of *sis-a*$^{+}$ acts as a dominant, male-specific lethal; either element alone has only minor effects on male and female viability. Most importantly, the dominant, male-specific lethality of the "double duplication" combination is suppressed by loss-of-function mutations at either the zygotic *Sxl* locus or at the maternal *da* locus. Both the dominant male-lethal effect of a double duplication chromosome and the dominant female-lethal effect of the double-mutant X chromosome mentioned above depend very much on genetic background. This suggests that the X/A signaling system may be both polygenic and highly polymorphic. The 2-proximal half of *T*(1;2)*Hw*^{bap} that is male-lethal with duplications of *sis-a*$^{+}$ suppresses the female-specific lethality of *sis-a*, suggesting additive effects.

The basis for these effects of this chromosome rearrangement is unclear. They do not appear to reflect simply the X-chromosome genes carried on the rearrangement or the second chromosome genes that are missing. They may instead reflect something unusual about the translocation breakpoint, perhaps involving a second numerator signal element candidate besides *sis-a*. Selection for suppressors of these sex-specific lethal effects should lead to other candidates for numerator and denominator signal elements.

The mutant allele *sis-a* has no maternal effect; its interaction with *Sxl* is strictly zygotic. An interesting question is whether *sis-a* is really a gene or instead is only a regulatory site. Experiments discussed above suggested that *Sxl* must be transcribed and perhaps translated prior to the expression of a blastoderm-stage specific gene, runt. If the earliest zygotic gene expression in *Drosophila* starts at the syncytial blastoderm stage [see Hafen et al., 1984], there would hardly seem to be enough time to transcribe and translate *sis-a* prior to the time that *Sxl* must first function.

SEX-LETHAL, SEX, AND EVOLUTION

There appears to be considerable genetic complexity both upstream of *Sxl* in the X/A signaling system for sexual pathway choice and downstream of *Sxl* in the various genes involved in the expression of all aspects of the sexually determined state: somatic sexual differentiation, dosage compensation, size dimorphism, and germ line differentiation. *Sxl* itself seems to represent a regulatory "constriction" in this whole genetic scheme—one gene that connects the complex sex signaling system with the complex sex differentiation system. Although we clearly have only just begun to identify the many regulatory and structural genetic elements that control sexual dimorphism in *Drosophila,* the central (perhaps unique) role that *Sxl* appears to play in this genetic program seems unlikely to be an artifact of our incomplete knowledge. The phenotype of the dominant, male-specific lethal $Sxl^{M\#1}$ is particularly significant in this connection. The fact that this allele will suppress a variety of genetic perturbations in the sex signaling system upstream of it argues against there being other genes besides *Sxl* that must act in concert at the same point in the regulatory gene hierarchy. If Sxl^{+} were responsible for only one of a number of mutally essential subunits in a multimeric regulatory protein or if the X/A signaling system acted simultaneously on several genes parallel to *Sxl*, it is difficult to imagine how a mutation in *Sxl* alone could exhibit such a phenotype.

With respect to the operation of *Sxl*, the control of *Drosophila* sex determination appears to be very efficiently designed (highly evolved?). The link between sex determination and dosage compensation is not too surprising in light of the fact that the latter process is required only as a consequence of the particular signal that *Drosophila* has evolved to control the former; however, this link does have an important consequence for the fly: Any sloppiness in the determination of sex by the X/A balance signal, i.e., any mistakes in the regulation of Sxl^{+} at the cellular level, should be eliminated by cell competition prior to sexual differentiation. Thus, the fidelity of the sex-signaling system in *Drosophila* need not be nearly as great as one would infer from the remarkably uniform sexual phenotype of wild-type individuals.

Even the chromosomal location of *Sxl* on the X makes sense from the standpoint of minimizing the possibility of mistakes in this gene's regulation: There are necessarily fewer doses of the gene in the sex which must keep the gene nonfunctional.

On the other hand, some aspects of the *Drosophila* sex determination system would appear to limit its evolutionary plasticity. In particular, the coupling of sex determination and dosage compensation and the fundamental differences between the control of sex in the germ line and its control in somatic tissue make changes in either the sex signaling system or the primary response to the sexual signaling system difficult. So far, every gene change that has a substantial effect on either system seems to be either lethal or sterile.

From an evolutionary standpoint, the question of how genetic programming for sexual dimorphism is related for different organisms is particularly intriguing. With respect to the genetic and environmental signals for sexual pathway choices, there appears to be a bewildering variety of mechanisms used [see Bull, 1983]; nevertheless, one can wonder whether the genes that respond to those signals might not have common evolutionary roots. There would appear to be little similarity betwen sex-determining mechanisms in humans and fruit flies. Humans have an active Y system of sex signaling, and much of sexual differentiation depends on hormones; the fruit fly uses an X/A balance signaling system, and much of its sexual differentiation is cell autonomous. With respect to dosage compensation, fruit flies hyperactivate the male X, whereas humans inactivate all but one X. In the face of such differences, however, it does seem curious that fruit flies appear to possess a specific mechanism for X-chromosome inactivation that functions in primary spermatocytes [see Lindsley and Tokuyasu, 1980]. It is also curious that if genes on the human X chromosome were hyperactive relative to autosomal genes, it would account for the fact that we tolerate both actual (in normal males) and functional (in normal females with X-chromosome inactivation) monosomy for the X chromosome but seem unable to tolerate any autosomal monosomy. Could it be that these differences in dosage compensation mechanisms simply reflect a different emphasis on two genetic systems that were present in the common ancestor of flies and man? With respect to the mechanisms of sex determination, it is curious that both mammals [see McLaren et al., 1984] and fruit flies have master switch genes that globally regulate sexual differentiation; furthermore, both species have hormonal as well as cell autonomous aspects to their sexual development.

The isolation and molecular characterization of master regulatory genes that fruit flies use to control segmental determination has revealed what seems to be a remarkable evolutionary relationship between developmental regulation in humans and fruit flies [Levine et al., 1984]. Molecular characterization of genes involved in the regulation of fruit fly sex determination is presently underway in a number of labs. It seems reasonable to anticipate that these studies may be equally enlightening from a developmental and evolutionary standpoint.

ACKNOWLEDGMENTS

I thank C. Cronmiller, E. Maine, H. Salz, and P. Schedl for allowing me to discuss their unpublished results, and I thank N. Lyons for her excellent technical assistance with various aspects of this work. This study was supported by the (U.S.) National Institute of General Medical Sciences (GM23468).

REFERENCES

Baker BS, Belote JM (1983): Sex determination and dosage compensation in *Drosophila melanogaster*. Annu Rev Genet 17:345–393.

Belote JM (1983): Male-specific lethal mutations of *Drosophila melanogaster*. II. Parameters of gene action during male development. Genetics 105:881–896.

Belote JM, Baker BS (1982): Sex determination in *Drosophila melanogaster:* Analysis of transformer-2, a sex-transforming locus. Proc Natl Acad Sci USA 79:1568–1572.

Belote JM, Baker BS (1983): The dual functions of a sex determination gene in *Drosophila melanogaster*. Dev Biol 95:512–517.

Bridges CB (1925): Sex in relation to chromosomes and genes. Am Natur 59:127–137.

Bull JJ (1983): "Evolution of Sex Determining Mechanisms." Menlo Park: Benjamin/Cummings.

Cline TW (1976): A sex-specific temperature-sensitive maternal effect of the daughterless mutation of *Drosophila melanogaster*. Genetics 84:723–742.

Cline TW (1978): Two closely linked mutations in *Drosophila melanogaster* that are lethal to opposite sexes and interact with daughterless. Genetics 90:683–698.

Cline TW (1979): A male-specific lethal mutation in *Drosophila* that transforms sex. Dev Biol 72:266–275.

Cline TW (1980): Maternal and zygotic sex-specific gene interactions in *Drosophila melanogaster*. Genetics 96:903–926.

Cline TW (1981): Positive selection methods for the isolation and finestructure mapping of cis-acting, homeotic mutations at the Sex-lethal locus of *D. melanogaster*. Genetics 97:s23.

Cline TW (1983a): The interaction between daughterless and Sex-lethal in triploids: a lethal sex-transforming maternal effect linking sex determination and dosage compensation in *Drosophila melanogaster*. Dev Biol 95:260–274.

Cline TW (1983b): Functioning of the genes daughterless and Sex-lethal in *Drosophila* germ cells. Genetics 104:s16–s17.

Cline TW (1984): Autoregulatory functioning of a *Drosophila* gene product that establishes and maintains the sexually determined state. Genetics 107:231–277.

Dobzhansky T, Schultz J (1934): The distribution of sex factors in the X-chromosome of *Drosophila melanogaster*. J Genet 28:233–255.

Fung SC, Gowen JW (1957): The developmental effect of a sex-limited gene in *Drosophila melanogaster*. J Exp Zool 134:515–532.

Gehring W (1972): The stability of the determined state in cultures of imaginal disks in

Drosophila. In Ursprung H, Nöthiger R (eds): "The biology of imaginal disks." New York: Springer-Verlag, pp 35–58.

Hafen H, Kuroiwa A, Gehring WJ (1984): Spatial distribution of transcripts from the segmentation gene fushi tarazu during *Drosophila* embryonic development. Cell 37:833–841.

Hannah-Alava A, Stern C (1957): The sexcombs in males and intersexes of *Drosophila melanogaster*. J Exp Zool 134:533–556.

Hildreth PE (1965): Doublesex, a recessive gene that transforms both males and females into intersexes. Genetics 51:659–678.

Hodgkin J (1983): Two types of sex determination in a nematode. Nature 304:267–268.

Lawrence PA (1981): The cellular basis of segmentation in insects. Cell 26:3–10.

Levine M, Rubin GM, Tijan R (1984): Human DNA sequences homologous to a protein coding region conserved between homeotic genes of *Drosophila*. Cell 38:667–673.

Lindsley DL, Tokuyasu KT (1980): Spermatogenesis. In Ashburner M, Wright TRF (eds): "The Genetics and Biology of *Drosophila*." London: Academic Press, Vol 2d, pp 225–294.

Lucchesi JC, Skripsky T (1981): The link between dosage compensation and sex differentiation in *Drosophila melanogaster*. Chromosoma 82:217–227.

Mange AP, Sandler L (1973): A note on the maternal effect mutants daughterless and abnormal oocyte in *Drosophila melanogaster*. Genetics 73:73–86.

Maroni G, Lucchesi JC (1980): X-chromosome transcription in *Drosophila*. Chromosoma 77:253–261.

McLaren A, Simpson E, Tomonari K, Chandler P, Hogg H (1984): Male sexual differentiation in mice lacking H-Y antigen. Nature 312:552–555.

McRobert S, Tompkins L (1985): The effect of transformer, doublesex, and intersex mutations on the sexual behavior of *Drosophila melanogaster*. Genetics (in press).

Morgan TH, Bridges CB (1919): Contribution to the genetics of *Drosophila melanogaster*. I. The origin of gynandromorphs. Carnegie Inst Publ Wash 278:1–122.

Morgan TH, Redfield H, Morgan LV (1943): Maintenance of a *Drosophila* stock center, in connection with investigations on the constitution of the germinal material in relation to heredity. Carnegie Inst Wash Yearbook 42:171–174.

Nicklas JA, Cline TW (1983): Vital genes that flank Sex-lethal, an X-linked sex-determining gene of *Drosophila melanogaster*. Genetics 103:617–631.

Nüsslein-Volhard C, Wieschaus E (1980): Mutations affecting segment number and polarity in *Drosophila*. Nature 287:795–801.

Postlethwait JH (1978): Clonal analysis of *Drosophila* cuticular patterns. In Ashburner M, Wright TRF (eds): "The Genetics and Biology of *Drosophila*." London: Academic Press, Vol 2c, pp 359–441.

Robbins LC (1984): Developmental use of gene products in *Drosophila:* the maternal-zygotic transition. Genetics 108:361–375.

Sanchez L, Nöthiger R (1983): Sex determination and dosage compensation in *Drosophila melanogaster:* production of male clones in XX females. The EMBO Journal 2:485–491.

Sandler L (1972): On the genetic control of genes located in the sex-chromosome heterochromatin of *Drosophila melanogaster*. Genetics 70:261–274.

Schüpbach T (1982): Autosomal mutations that interfere with sex determination in somatic cells of *Drosophila* have no direct effect on the germline. Dev Biol 89:117–127.

Schüpbach T (1985): Normal female germ cell differentiation requires the female X-chromosome-autosome ratio and expression of Sex-lethal in *Drosophila melanogaster*. Genetics 109:529–548.

Skripsky T, Lucchesi JC (1982): Intersexuality resulting from the interaction of sex-specific lethal mutations in *Drosophila melanogaster*. Dev Biol 94:153–162.

Steinmann-Zwicky M (1984): The role of the X chromosome in sex determination and dosage compensation in *Drosophila melanogaster*. Ph.D. thesis, University of Zürich.

Stern C (1966): Pigmentation mosaicism in intersex of *Drosophila*. Rev Suisse Zool 73:339–355.

Stewart B, Merriam J (1980): Dosage compensation. In Ashburner M, Wright TRF (eds): "The Genetics and Biology of *Drosophila*." London: Academic Press, Vol 2d, pp 107–140.

Tokunaga C (1962): Cell lineage and differentiation on the male foreleg of *Drosophila melanogaster*. Dev Biol 4:489–516.

Tompkins L (1984): The effect of Sex-lethal mutations on the sexual behavior of *D. melanogaster*. Genetics 107:s107.

Uenoyama T, Uchida S, Fukunaga A, Oishi K (1982): Studies on the sex-specific lethals of *Drosophila melanogaster*. V. Sex transformation caused by interactions between a female-specific lethal, $Sxl^{f\#1}$, and the male-specific lethals *mle*(3)*132*, $msl\text{-}2^{27}$, and *mle*. Genetics 102:233–243.

The Origin and Evolution of Sex, pages 329–330

Sex Determination and Differentiation of Vertebrates: Discussion

Leader: Horacio Merchant-Larios

Instituto de Investigacions Biomédicas, U.N.A.M., Apartado Postal 70228 D.F. 04510, México

This discussion focused on X chromosome inactivation and its relation to meiosis, embryo implantation, and development. If primordial germ cells (PGC) enter meiosis in fetal testes as in fetal ovaries, should these gonads be considered to be ovotestis? Recently it has been found that PGC outside of the testis cords can undergo meiosis. This observation suggests that whereas X and XY PGCs can undergo meiosis, the recurrence of meiosis in PGC is prevented by the preSertoli cells in testis cords.

In the mouse, one of the X chromosomes of PGC is inactivated during migration and proliferation in the gonadal anlagen and reactivated at the onset of meiosis. The question of whether X chromosome reaction is associated with meiosis was asked. In XX sex-reversed male mice, the inactive X chromosome of the germ cells is reactivated in the seminiferous cords at approximately the same developmental stage when the process of reactivation occurs in normal ovaries, although these germ cells are arrested in a resting stage as XY PGC in normal testes. This observation suggests that PGC X chromosome reactivation is not necessarily linked to the initiation of meiosis. Furthermore, it has been demonstrated that XO and XY germ cells can differentiate into oocytes. However, these germ cells cannot develop fully, suggesting that both X chromosomes must be active in order for the development of normal fertilizable eggs.

The relation between X chromosome inactivation and totipotency of PGC was discussed. The inner cell mass and the epiblast from which the PGCs originate have a restricted potential of differentiation because they are unable to develop into trophoblastic cells. At this stage, one of the X chromosomes of PGC is inactive in a manner similar to that of the somatic cells. Nevertheless, when the PGCs undergo meiosis after migrating into the gonadal anlage, the X chromosome becomes reactivated and recovers totipotency to assure development of the next generation.

Sex chromosome inactivation occurs only in mammals. X chromosome of some species of kangaroos shows incomplete inactivation, whereas in other species only the paternal X chromosome is inactivated. However, the inactivation might have been secondary to the culture conditions of the cells.

Another question that has been raised is whether a segment of the inactive X chromosome might be active. The two genes located at the end of the short arm of the inactive X chromosome of human cells, ie, loci for HGA blood group and steroid sulfatase, are expressed. However, these genes are probably partially inactivated because their measured activities are below normal values. It is likely that these two genes were translocated recently to the human X chromosome, because both genes are present in autosomes of lower species of mammals.

The relationship between preferential inactivation of paternal or maternal X chromosome and the failure of embryo implantation and of normal development was discussed also. Either maternal or paternal X chromosome was found to be inactivated in human XO fetuses that were aborted or carried to term (approximately 3%). Accordingly, failure of human embryos to implant is not dependent on which X chromosome is inactivated. In the mouse, however, paternal X chromosome is preferentially inactivated in extraembryonic membranes. Moreover, there is an inbred strain of mice in which 50% of the females develop ovarian teratomas from parthenogenetic oocytes. Some of these oocytes are ovulated and develop into embryos that can implant, but perish by the 10th day of gestation. Because these parthenogenetic embryos have two X chromosomes of maternal origin, the presence of paternal X chromosome is not required for implantation but probably essential for the normal development of fetuses.

Index